Advances in Intelligen

Volume 973

The series "Advances in Intelligent Systems and Computing" contains publications on theory, applications, and design methods of Intelligent Systems and Intelligent Computing. Virtually all disciplines such as engineering, natural sciences, computer and information science, ICT, economics, business, e-commerce, environment, healthcare, life science are covered. The list of topics spans all the areas of modern intelligent systems and computing such as: computational intelligence, soft computing including neural networks, fuzzy systems, evolutionary computing and the fusion of these paradigms, social intelligence, ambient intelligence, computational neuroscience, artificial life, virtual worlds and society, cognitive science and systems, Perception and Vision, DNA and immune based systems, self-organizing and adaptive systems, e-Learning and teaching, human-centered and human-centric computing, recommender systems, intelligent control, robotics and mechatronics including human-machine teaming, knowledge-based paradigms, learning paradigms, machine ethics, intelligent data analysis, knowledge management, intelligent agents, intelligent decision making and support, intelligent network security, trust management, interactive entertainment, Web intelligence and multimedia.

The publications within "Advances in Intelligent Systems and Computing" are primarily proceedings of important conferences, symposia and congresses. They cover significant recent developments in the field, both of a foundational and applicable character. An important characteristic feature of the series is the short publication time and world-wide distribution. This permits a rapid and broad dissemination of research results.

** Indexing: The books of this series are submitted to ISI Proceedings, EI-Compendex, DBLP, SCOPUS, Google Scholar and Springerlink **

More information about this series at http://www.springer.com/series/11156

Tareq Ahram
Editor

Advances in Human Factors in Wearable Technologies and Game Design

Proceedings of the AHFE 2019 International Conference on Human Factors and Wearable Technologies, and the AHFE International Conference on Game Design and Virtual Environments, July 24–28, 2019, Washington D.C., USA

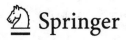 Springer

Editor
Tareq Ahram
Institute for Advanced Systems Engineering
Orlando, FL, USA

ISSN 2194-5357 ISSN 2194-5365 (electronic)
Advances in Intelligent Systems and Computing
ISBN 978-3-030-20475-4 ISBN 978-3-030-20476-1 (eBook)
https://doi.org/10.1007/978-3-030-20476-1

This Springer imprint is published by the registered company Springer Nature Switzerland AG
The registered company address is: Gewerbestrasse 11, 6330 Cham, Switzerland

Advances in Human Factors and Ergonomics 2019

AHFE 2019 Series Editors

Tareq Ahram, Florida, USA
Waldemar Karwowski, Florida, USA

10th International Conference on Applied Human Factors and Ergonomics and the Affiliated Conferences

Proceedings of the AHFE 2019 International Conference on Human Factors and Wearable Technologies, and the AHFE International Conference on Game Design and Virtual Environments, held on July 24–28, 2019, in Washington D.C., USA

Advances in Affective and Pleasurable Design	Shuichi Fukuda
Advances in Neuroergonomics and Cognitive Engineering	Hasan Ayaz
Advances in Design for Inclusion	Giuseppe Di Bucchianico
Advances in Ergonomics in Design	Francisco Rebelo and Marcelo M. Soares
Advances in Human Error, Reliability, Resilience, and Performance	Ronald L. Boring
Advances in Human Factors and Ergonomics in Healthcare and Medical Devices	Nancy J. Lightner and Jay Kalra
Advances in Human Factors and Simulation	Daniel N. Cassenti
Advances in Human Factors and Systems Interaction	Isabel L. Nunes
Advances in Human Factors in Cybersecurity	Tareq Ahram and Waldemar Karwowski
Advances in Human Factors, Business Management and Leadership	Jussi Ilari Kantola and Salman Nazir
Advances in Human Factors in Robots and Unmanned Systems	Jessie Chen
Advances in Human Factors in Training, Education, and Learning Sciences	Waldemar Karwowski, Tareq Ahram and Salman Nazir
Advances in Human Factors of Transportation	Neville Stanton

(continued)

(continued)

Advances in Artificial Intelligence, Software and Systems Engineering	Tareq Ahram
Advances in Human Factors in Architecture, Sustainable Urban Planning and Infrastructure	Jerzy Charytonowicz and Christianne Falcão
Advances in Physical Ergonomics and Human Factors	Ravindra S. Goonetilleke and Waldemar Karwowski
Advances in Interdisciplinary Practice in Industrial Design	Cliff Sungsoo Shin
Advances in Safety Management and Human Factors	Pedro M. Arezes
Advances in Social and Occupational Ergonomics	Richard H. M. Goossens and Atsuo Murata
Advances in Manufacturing, Production Management and Process Control	Waldemar Karwowski, Stefan Trzcielinski and Beata Mrugalska
Advances in Usability and User Experience	Tareq Ahram and Christianne Falcão
Advances in Human Factors in Wearable Technologies and Game Design	Tareq Ahram
Advances in Human Factors in Communication of Design	Amic G. Ho
Advances in Additive Manufacturing, Modeling Systems and 3D Prototyping	Massimo Di Nicolantonio, Emilio Rossi and Thomas Alexander

Preface

Successful interaction with products, tools, and wearable technologies is facilitated by the application of usable design principles. The needs of potential users should be met, and costly training avoided. This book is concerned with emerging technology for wearable devices and covers concepts, theories, and applications of human factor knowledge to understand and improve human interaction with products and systems.

The first part of the book focuses on the human aspects of wearable technologies. It reports on the proceedings of the *AHFE International Conference on Human Factors and Wearable Technologies*. Contributions in this part cover both research and best practices, showing how user-centered design and practices can optimize wearable experience, thus improving user acceptance, satisfaction, and engagement toward novel wearable gadgets.

The second part and the third part of the book gather the proceedings of the *AHFE International Conference on Human Factors in Game Design and Virtual Environments*. The game industry has been rapidly expanding in the past decades, and games became more appealing to a wider audience. The level of complexity in games control interfaces and graphics has increased exponentially, in addition to the growing interest in integrating augmented reality in gaming experience. As a result, there is a growing demand for human factors and ergonomics practitioners to ensure users' engagement.

Overall, the book shows how human factors and ergonomics can be applied to the development of sensors, wearable technologies and in game design, highlighting results obtained upon the applications of different wearability principles such as aesthetics, affordance, comfort, contextual-awareness, customization, ease of use, intuitiveness, privacy, reliability, responsiveness, satisfaction, subtlety, and user friendliness.

This book provides a timely reference to professionals, researchers, and students in the broad field of game design, human modeling, human–computer interaction and human systems integration. It offers extensive information on feedback generated by devices' interfaces (visual and haptic), user-centered design, and design for special populations, particularly the elderly.

Chapters included in the book have been reviewed by members of the International Editorial Board. Our sincere thanks and appreciation goes to those board members, which are listed below:

Wearable Technologies

Akram Alomainy, UK
Waseem Asghar, USA
Wolfgang Friesdorf, Germany
S. Fukuzumi, Japan
Sue Hignett, UK
Wonil Hwang, S. Korea
Muhammad Ismail, Qatar
Yong Gu Ji, Korea
Bernard C. Jiang, Taiwan
Ger Joyce, UK
Chee Weng Khong, Malaysia
Zhizhong Li, PR China
Nelson Matias, Brazil
Valerie Rice, USA
Emilio Rossi, Italy
Masood ur Rehman, UK
Alvin Yeo, Malaysia
Wei Zhang, PR China

Game Design and Virtual Environments

Wonil Hwang, S. Korea
Yong Gu Ji, Korea
Bernard C. Jiang, Taiwan
Ger Joyce, UK
Chee Weng Khong, Malaysia
Zhizhong Li, PR China
Nelson Matias, Brazil
Delfina Gabriela Ramos, Portugal
Valerie Rice, USA
Nuno F. Rodrigues, Portugal
Emilio Rossi, Italy
João Vilaça, Portugal

This book is not only intended to be informative, but also to be thought provoking. We hope it inspires, leading the reader to contemplate other questions, applications, and potential solutions fostering good design for all.

July 2019   Tareq Ahram

Contents

Human Factors in Game Design

Human Factors and Wearable Technologies

Comparative User Experience Analysis of Pervasive Wearable Technology

Nicholas Caporusso[(✉)], Angela Walters, Meng Ding, Devon Patchin,
Noah Vaughn, Daniel Jachetta, and Spencer Romeiser

Fort Hays State University, 600 Park Street, Hays 67601, USA
{n_caporusso, awalters, m_ding6. se}@fhsu. edu,
{djpatchin, nqvaughn, ddjachetta,
s_romeiser}@mail. fhsu. edu

Abstract. Although the growing market of wearable devices primarily consists of smartwatches, fitness bands, and connected gadgets, its long tail includes a variety of diverse technologies based on novel types of input and interaction paradigms, such as gaze, brain signals, and gestures. As the offer of innovative wearable devices will increase, users will be presented with more sophisticated alternatives: among the several factors that influence product adoption, perceived user experience has a significant role. In this paper, we focus on human factors dynamics involved in the pre- and post-adoption phase, that is, before and after customers buy or use products. Specifically, objective of our research is to evaluate aspects that influence the perceived value of particularly innovative products and lead to purchasing them. To this end, we present the results of a pilot study that compared performance expectancy, effort expectancy, social influence, and facilitating conditions, in the pre- and post-adoption stages of three types of wearable technology, i.e., brain-computer interface, gesture controllers, and eye-tracking systems.

Keywords: Technology acceptance model ·
Unified Theory of Acceptance and Use of Technology ·
Performance expectancy · Effort expectancy · Social influence · Eye tracking ·
Gaze tracking · Gesture controllers · Brain-computer interface

1 Introduction

In the last decade, the evolution of mobile devices and their market penetration resulted in the capillary adoption of the smartphone as a wide-spread platform for human-computer interaction; also, the introduction of smartwatches and fitness bands demonstrated the potential of wearable consumer electronics and simultaneously rendered them mainstream. As a result, the market of wearable technology has experienced a constant growth, in the last years. Moreover, the success of smartphones and smartwatches accelerated the development of miniaturized components (e.g., cameras and inertial measurement units) that can be embedded into eyewear, head-mounted displays, clothing, wristbands, gloves, and several types of body attachments. Nowadays, more sophisticated components, such as, electroencephalograph, myoelectric

© Springer Nature Switzerland AG 2020
T. Ahram (Ed.): AHFE 2019, AISC 973, pp. 3–13, 2020.
https://doi.org/10.1007/978-3-030-20476-1_1

sensors, vibration and ultrasound actuators are incorporated in consumer wearable devices that enable novel interaction modalities. Consequently, the segment of smartwatches and fitness bands has reduced significantly, as demand has recently shifted to head mounted displays (HMD) and novel types of devices, such as, gestures controllers, body cameras, glasses that support augmented reality, and smart clothing. Recent statistics about the next 5 years project that 25% of the wearable device market will consist of a fragmented ecosystem of different types of products, each representing a niche consisting of innovative products offering non-conventional interfaces and interaction paradigms [1]. Moreover, as new technology will be developed, the segment of non-mainstream wearable devices will grow and, simultaneously, its internal fragmentation will increase, producing a broader and richer galaxy of choices.

The market of wearables is unique in terms of breadth and variety: the presence of recent and non-conventional technology, in addition to the many available alternatives results in more difficult buying decisions. This, in turn, is particularly interesting from a user experience standpoint, in specific regard to the human factors that influence the decision-making process for adopting a non-mainstream technology and buying a specific product. Indeed, marketing and sales strategies are crucial for creating awareness and for incentivizing prospect customers. Also, users might finalize a purchase considering aspects, such as, technical specifications (e.g., product longevity, battery life), opinions and reviews, and price. These factors are especially significant for intra-category decisions, that is, choices between two alternatives of the same product (e.g., different brands, versions, or technical characteristics). Nevertheless, several other factors, more strictly related to the intrinsic aspects of user-product interaction, might play a significant role in influencing buying patterns and have a broader impact in inter-category decisions.

Several studies investigated human factors dynamics that occur during or after product adoption. Conversely, in our work, we are especially interested in analyzing user experience in the context of innovative and non-conventional pervasive wearable devices, with particular regard to user-product interaction in the pre-purchase (or pre-adoption) phase. Specifically, in this paper, we investigate the human factors involved in the awareness and evaluation stages of the sales funnel. To this end, we use the Unified Theory of Acceptance and Use of Technology (UTAUT) [2] to analyze measures that help assess the behavioral intention to use a technology, such as, performance expectancy, effort expectancy, social influence, and facilitating conditions, which might be utilized as a predictor of acquisition and discontinuation dynamics in early adopters as well as other types of customers. Moreover, we present the results of a study that analyzed three types of non-mainstream consumer technology in the context of pervasive wearable devices, that is, brain-computer interfaces (BCIs), eye-tracking systems, and gesture controllers. Incorporating human factors in the earliest stages of the sales funnel might help vendors design marketing strategies that align better with user expectations before the purchase, deliver consistent user experiences without solution of continuity, and, consequently, decrease post-adoption dissatisfaction, frustration, discontinuation, and product abandonment.

2 Related Work

Nowadays, the design of novel, user-centered products is supported by a growing repository of methodologies, guidelines, and patterns. Furthermore, research on usability and acceptance led to inspection methods (e.g., heuristic analyses, interviews, and observational studies) that enable the evaluation of devices and software before they are marketed. In contrast, several communities are still contributing to define the fuzzy boundaries of user experience: as UX covers a broader spectrum of aspects, the design and evaluation of technology and processes often rely on tools and practices from different domains rather than on a coherent set of practices [3]. Moreover, typically user experience is regarded in the post-adoption phase, and earlier stages of the sales funnel deserve more research. Studies that observed the pre-sale stage of products focused on extrinsic factors of user-product interaction, such as, marketing. Indeed, ancillary aspects of the product, such as, package design and other factors related to brand identity, influence user experience and buying patterns [4]. In the last decades, holistic models and frameworks have been developed to extend the analysis of user experience to aspects beyond product interaction that occurs after purchase or adoption. Specifically, customer journey [5] takes into consideration prospect users as well as loyal customers, and helps investigate factors that incentivize purchase, motivate adoption, and reduce abandonment. Conversely, the Unified Theory of Acceptance and Use of Technology (UTAUT) [2] was developed with the aim of synthesizing several models (e.g., the technology acceptance model) that describe the behavioral intention to use a technology and help compare it with dimensions perceived in the actual adoption. The meaningful role of temporality in product use is analyzed in [6], which showed that early experiences and prolonged use are characterized by very different user experience dynamics. The authors of [7] investigated sensory modalities in user-product interaction with three categories of articles (i.e., high-tech devices, shoes, and coffee maker) before purchase and at different stages of adoption: sensory dominance (i.e., the relative importance of vision, audition, touch, smell, and taste) change over time depending on product type, relevance of the sensory channel to product interactions, and familiarity with the product. Also, it significantly varies between pre- and post-purchase and, thus, demonstrate the need of investigating user experience of prospect customers from earlier stages of the sales funnel.

Nevertheless, most of research focused on direct user-product interaction, whereas indirect user experience, which typically occurs in the pre-adoption phase, received less attention [8]. The relationship between extrinsic elements, such as, online customer reviews, on user experience and their relationship with buying patterns was investigated by [9], which showed correlation between customer decisions and features, such as, level of experience of the author and social metrics of the of the review. The role of simulated user-product experience has been investigated in [10]: virtual reality (VR) is utilized for enabling prospect customers to experience products, with specific regard to items that introduce novelty in the market and devices that incorporate highly innovative features. Also, simulated products were utilized in [11] to create hedonic user experiences that could influence user satisfaction and increase willingness to adopt product. Unfortunately, there is a lack of studies that compare indirect and direct dynamics that specifically regard user-product relationships at the individual level in the pre-purchase phase.

3 Experimental Study

The objective of our study was to evaluate the human factors that influence the adoption of non-mainstream innovative devices, with specific regard to pervasive wearable technology. To this end, we analyzed intrinsic and extrinsic aspects that characterize user-product experience in the pre- and post-adoption phases and compared indirect and direct experiences. In our study, we modeled human factors utilizing the UTAUT model, which consists of five dimensions that can used as predictors of the intention to adopt and use technology, as well as validation criteria.

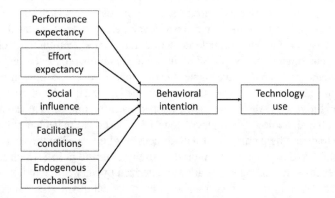

Fig. 1. Diagram representing the UTAUT, which incorporates several user experience measures as drivers for the behavioral intention to buy a product or adopt a technology.

Performance Expectancy. This aspect refers to the belief that the use of a particular technology will enhance the performance of an individual or will produce some advantage in realizing a task. This is particularly relevant to non-conventional wearable technology, as signals acquired from the body can be utilized to seamlessly control interaction, make it more efficient, or improve the range or accuracy of commands.

Effort Expectancy. This is a two-fold measure: on the one hand, it refers to the perceived skillset required to be able to utilize a system and the expected learning curve (human-machine co-evolution). Simultaneously, it relates to the extent of convenience perceived in using the system. Innovative wearable devices that recognize gestures, brain signals, or gaze simplify input, though mastering their use requires training.

Social Influence. This component refers to user's perception of beliefs or impressions that the product will generate in others (their milieu, their social group, or the external community). This includes the ability of a device to improve the social status of an individual or to create a desired social image. Moreover, this measure involves social acceptance of technology in a given context of reference. This dimension is particularly relevant in wearables, especially if they are visible or intrusive: they could be perceived as innovative gadgets or associate the user with some medical condition and, depending on the social context of use, pose risks to users' safety (e.g., theft).

Facilitating Conditions. Extrinsic factors, such as, battery life, device compatibility and availability of product accessories and features that render the product more versatile might be a driver for adoption. Also, presence of technical support and user's guide might increase the likelihood of acquiring products. Moreover, individuals might consider expected product longevity, which, in turn, is influenced by the presence of a number of applications, features that demonstrate other options for potential use, or an active community or marketplace.

Endogenous Mechanisms. Intrinsic factors that are not related to product experience are associated with individuals' conditions or beliefs, social background, and education. As this often is a multifaceted aspect, we included open-ended questions to elicit participants' impressions.

3.1 Technology

Three devices were chosen for this study, involving diverse types of innovative wearable technology. Products were selected among non-conventional interfaces that are representative of a niche of wearables involving more sophisticated human-computer interaction dynamics compared to fitness bands and smartwatches. Although the devices had diverse levels of maturity, all of them were non-mainstream technology. This was to avoid any confounding effect introduced by individuals' familiarity with brand. Figure 2 shows the three products.

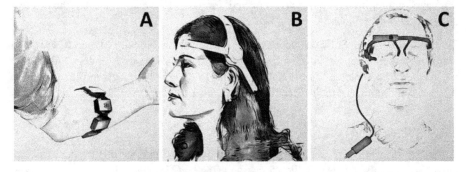

Fig. 2. The products utilized in the study: Gesture controller (A), a Brain-Computer Interface device (B), and eyewear integrating gaze tracking (C).

Gesture Controller. Myo is a wristband equipped with electromyographic sensors [12] that convert electrical activity in the muscles of a user's forearm into gestures for controlling computers and other devices. Different types of gesture controllers have been embedded in wearables [13] and utilized for enabling more realistic control especially in tasks that involve manipulation [14]. Although sales have been discontinued, documentation and videos are available on the Internet.

Brain-Computer Interface. Mindwave [15] is an affordable commercial EEG device for non-medical use that utilizes one electrode placed on the forehead to sense the electrical activity of the occipital area of the brain and detect attention and concentration. It is one of the several attempts to introduce BCI in scenarios that primarily are associated with clinical conditions [16].

Eye Tracking. Pupil is an eyewear that incorporates high speed cameras that capture movements of the pupil, so that the position of the eyes can be calculated and represented over the video stream acquired using a front camera, which represents user's perspective [17]. Typically, the wearable eye-tracking devices are utilized in the evaluation of immersive and physically-immersive experiences [18, 19].

3.2 Protocol

The experimental protocol was divided in three phases based on three stages of user experience: indirect, direct, and early use.

Task 1 – Indirect Use. Participants were provided with the description of the three products and a survey that asked them questions that assessed different items in the five dimensions of user experience defined in UTAUT. Respondents were given one week to complete the questionnaire and evaluate the devices based on indirect use, that is, by experiencing the technology through information available on the Internet (e.g., product website, tutorials, reviews).

Task 2 – Direct Use. Subjects were provided with the three devices: they were asked to install and use each of the products for 20 min, and to rate them by filling a questionnaire similar to the one they received for task 1, which captured their direct experience with products.

Task 3 – Early Use. Subjects were provided with the three devices for four days, and they were asked to realize three simple projects each involving a technology. Also, they were asked to complete a questionnaire similar to the one they received for task 1, so that we could compare their direct experience over a longer period.

3.3 Participants

A total of 14 subjects (7 females and 7 males) were recruited for this study, 14 (43%) aged 18–24, 11 (36%) were in the 25–34 range, and 3 (21%) were in the 35–44 bracket. None of them had previous experience with the devices utilized in this study, though all of them were familiar with information technology and high-tech gadgets. All of them had basic programming skills (i.e., they utilized Arduino). Also, 3 of them (21%, 2 males and 1 females) reported themselves as early adopters, whereas the others (79%) showed different patterns in regard to the adoption of new technology.

4   Results and Discussion

The objective of our study was not to evaluate the specific products or technologies, as preferences can be influenced by several factors that were not under our control. Moreover, as products in each category come in different versions and have diverse characteristics and advantages, we did not compare individual devices with one another. Conversely, our objective was to evaluate how user experience of different types of innovative wearable technology changes over time and, specifically, between indirect and direct use. To this end, we calculated the behavioral intention to use as a compound measure of the five items in the UTAUT inventory, in addition to assessing its individual dimensions separately, to identify their role at different stages of user-product experience. Figure 3 describes users' behavior after indirect use.

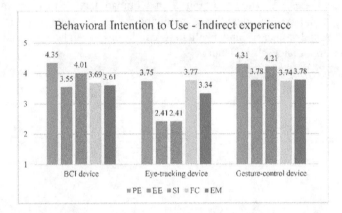

Fig. 3. Behavioral intention to use the technology based on indirect user-product experiences. Performance expectancy (PE), effort expectancy (EE), social influence (SI), facilitating conditions (FC), and endogenous mechanisms (EM) were considered for each technology.

Participants considered performance expectancy as their most important drivers (see Fig. 1). As users were familiar with technology gadgets, they described the eye-tracking device as a less user-friendly system. However, this was mainly due to the fact that they recognized that it is mostly available to researchers in the field of gaze tracking, whereas the other products appeared specifically geared towards consumers. Therefore, subjects expected the eyewear to be more difficult to operate and less impactful from a social standpoint. Moreover, the form factor of the wearable eye tracker resulted more intrusive than other products and less suitable for social interactions.

Despite specific differences due to the design of the products and their documentations, all the wearables considered in the study were well received by participants during indirect user-product experiences. In this phase, the compound measure of the behavioral intention to adopt a technology aligns with the actual willingness to buy stated by subjects, as shown in Fig. 4. The eye tracking device represents an exception for the reasons discussed earlier in this Section. In the second part of our study, subjects received the devices and tried to install and use them. In this phase, effort expectancy had a significant impact on the results. Consequently, as the eye-tracking device is tailored to a technical audience, its setup time affected users' willingness to adopt it. Finally, data from early direct user-product interaction show an overall decrease in the likelihood to adopt the three products. In this case, the actual and compound measure of the willingness to adopt are not statistically comparable, though the data from the eye tracking device might introduce some bias in the outcome.

Furthermore, our findings show that users' perceived importance of the aspects described in UTAUT changes in the three phases of product use. Specifically, performance and effort are considered key factors during indirect use, whereas effort is considered as a crucial aspect during the first phase of direct experience. As users become more familiar with the product, the importance of performance and facilitating conditions increases. As shown in Fig. 5, the human factors that are involved in indirect use could be considered as a predictor for performance expectancy, effort expectancy, social influence, and facilitating conditions, though users tend to underestimate the importance of the latter.

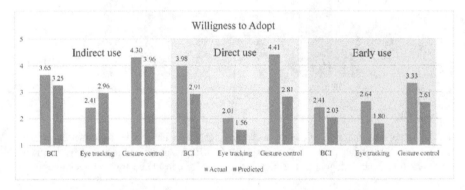

Fig. 4. Individuals' willingness to Adopt a technology based on user experience in indirect, direct, and early user-product interaction. The predicted value is calculated as a compound measure of the five dimensions of UTAUT, whereas the actual value represents subjects' response to a question that explicitly asked their willingness to buy the product.

Finally, as users became more proficient with the use of technology, they mentioned lack of applications as one of the factors related to facilitating conditions that affected user experience in the early adoption of all the devices. Although companies offer software development kits that enable creating new opportunities to use the technology, users perceived the three wearables as very situation-specific, which contrasted with the performance expectancy fostered experienced during indirect interaction.

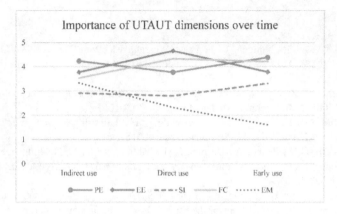

Fig. 5. Perceived importance of each of the five UTAUT dimensions in the three stages of product use recorded in the study.

5 Conclusion

In this paper, we analyzed user experience in user-product interaction and compared three categories of innovative wearable interfaces, with the objective of understanding dynamics of user adoption in indirect and direct use. Assessing the criteria that guide choice and abandonment of technology is relevant especially considering that the offer of alternative, non-mainstream input and output interfaces is expected to grow in the next years. In our study, we utilized the Unified Theory of Acceptance and Use of Technology to model and capture the diverse human factors that occur at different stages of user-product experience. From our findings, we can conclude that endogenous mechanisms played a secondary role in indirect and direct use; also, social influence was a minor factor, though it might be due to individuals' low awareness of the types of technology assessed in our study. Furthermore, our results show that the dimensions of the behavioral intention to buy have different weights depending on the stage: users have high performance expectations when they indirectly interact with wearables, whereas effort expectancy is considered the main factor in direct user-product experience. Moreover, facilitating conditions play an increasing role as users become more proficient with the product and, simultaneously, discover the lack of available applications, which results one of the main motivations for potential product discontinuation. In conclusion, the main long-term challenge in the introduction of novel interfaces for pervasive interaction consists in increasing product longevity by providing users with more applications and usage scenarios that align their direct experience with performance expectancy and product expectations generated in indirect interaction.

References

1. Antin, A., Atwal, R., Nguyen, T.: Forecast: Wearable Electronic Devices, Worldwide, 2018. Gartner (2018). https://www.gartner.com/doc/3891988/forecast-wearable-electronic-devices-worldwide
2. Venkatesh, V., Zhang, X.: Unified theory of acceptance and use of technology: US vs. China. J. Glob. Inf. Technol. Manag. 13(1), 5–27 (2010)
3. Hassenzahl, M.: The thing and I: understanding the relationship between user and product. In: Funology 2, pp. 301–313. Springer, Cham (2018)
4. Schifferstein, H.N., Fenko, A., Desmet, P.M., Labbe, D., Martin, N.: Influence of package design on the dynamics of multisensory and emotional food experience. Food Qual. Preference 27(1), 18–25 (2013)
5. Nenonen, S., Rasila, H., Junnonen, J.M., Kärnä, S.: Customer Journey–a method to investigate user experience. In: Proceedings of the Euro FM Conference Manchester, pp. 54–63 (2008)
6. Karapanos, E., Zimmerman, J., Forlizzi, J., Martens, J.B.: User experience over time: an initial framework. In: Proceedings of the SIGCHI Conference on Human Factors in Computing Systems, pp. 729–738. ACM (2009)
7. Fenko, A., Schifferstein, H.N., Hekkert, P.: Shifts in sensory dominance between various stages of user–product interactions. Appl. Ergon. 41(1), 34–40 (2010)
8. Hamilton, R.W., Thompson, D.V.: Is there a substitute for direct experience? Comparing consumers' preferences after direct and indirect product experiences. J. Consumer Res. 34 (4), 546–555 (2007)
9. Zhu, F., Zhang, X.: Impact of online consumer reviews on sales: The moderating role of product and consumer characteristics. J. Market. 74(2), 133–148 (2010)
10. Füller, J., Matzler, K.: Virtual product experience and customer participation—A chance for customer-centred, really new products. Technovation 27(6–7), 378–387 (2007)
11. Papagiannidis, S., See-To, E., Bourlakis, M.: Virtual test-driving: the impact of simulated products on purchase intention. J. Retail. Consum. Serv. 21(5), 877–887 (2014)
12. Rawat, S., Vats, S., Kumar, P.: Evaluating and exploring the MYO ARMBAND. In: 2016 International Conference System Modeling & Advancement in Research Trends (SMART), pp. 115–120. IEEE (2016)
13. Caporusso, N., Biasi, L., Cinquepalmi, G., Trotta, G.F., Brunetti, A., Bevilacqua, V.: A wearable device supporting multiple touch-and gesture-based languages for the deaf-blind. In: International Conference on Applied Human Factors and Ergonomics, pp. 32–41. Springer, Cham (2017). https://doi.org/10.1007/978-3-319-60639-2_4
14. Caporusso, N., Biasi, L., Cinquepalmi, G., Bevilacqua, V.: An immersive environment for experiential training and remote control in hazardous industrial tasks. In: International Conference on Applied Human Factors and Ergonomics, pp. 88–97. Springer, Cham (2018). https://doi.org/10.1007/978-3-319-94619-1_9
15. Sałabun, W.: Processing and spectral analysis of the raw EEG signal from the MindWave. Prz. Elektrotechniczny 90(2), 169–174 (2014)
16. Cincotti, F., Kauhanen, L., Aloise, F., Palomäki, T., Caporusso, N., Jylänki, P., Mattia, D., Babiloni, F., Vanacker, G., Nuttin, M., Marciani, M.G.: Vibrotactile feedback for brain-computer interface operation. Comput. Intell. Neurosci. (2007). https://doi.org/10.1155/2007/48937
17. Kassner, M., Patera, W., Bulling, A.: Pupil: an open source platform for pervasive eye tracking and mobile gaze-based interaction. In: Proceedings of the 2014 ACM International Joint Conference on Pervasive and Ubiquitous Computing: Adjunct Publication, pp. 1151–1160. ACM (2014)

18. Caporusso, N., Ding, M., Clarke, M., Carlson, G., Bevilacqua, V., Trotta, G.F.: Analysis of the relationship between content and interaction in the usability design of 360° videos. In: International Conference on Applied Human Factors and Ergonomics, pp. 593–602. Springer, Cham (2018). https://doi.org/10.1007/978-3-319-94947-5_60
19. Carlson, G., Caporusso, N.: A physically immersive platform for training emergency responders and law enforcement officers. In: International Conference on Applied Human Factors and Ergonomics, pp. 108–116. Springer, Cham (2018). https://doi.org/10.1007/978-3-319-93882-0_11

A Novel Joint Angle Measurement System to Monitor Hip Movement in Children with Hip Diseases

Donato G. Leo[1,2(✉)], Badr M. Abdullah[2], Daniel C. Perry[3], and Helen Jones[1]

[1] School of Sport and Exercise Sciences,
Liverpool John Moores University, Liverpool, UK
d.g.leo@2016.ljmu.ac.uk
[2] Built Environment and Sustainable Technologies (BEST) Research Institute,
Liverpool John Moores University, Liverpool, UK
[3] Nuffield Department of Orthopaedic, Rheumatology and Musculoskeletal
Sciences, University of Oxford, Oxford, UK

Abstract. Children's hip diseases are an umbrella term to define different conditions (e.g. Perthes' disease; hip dysplasia) that affect the hip bone during the first months or years after birth. Assessing the degree of hip stiffness is important in the management of the disease, but to date there is no system able to continuously monitor hip angle in children. We aimed to characterize a novel wearable joint angle monitoring system able to collect data during the day in everyday life to assess hip mobility in children with hip diseases. We developed a flexible sensor embedded in a microcontroller based device, including an external SD card to store data. Preliminary data collected by the sensor shows its feasibility into monitor hip flexion/extension (SEM of $\pm0.20°$) during daily tasks. The preliminary results support moving forward with the prototype and improving its wearability, validating it in a wider study.

Keywords: Optical flexible sensor · Hip diseases · Wearable technology · Hip mobility

1 Introduction

Childhood hip diseases, including Perthes' disease and hip dysplasia, affect the femoral acetabular joint [1, 2] during the first months or years after birth with different grades of severity and symptoms. The two main characteristics of these hip conditions are pain and changes to normal range of motion (ROM) at the hip joint [1, 2]. The conditions induce stiffness of the hip joint, which causes difficulty in walking and affects normal daily life activities (e.g. climbing stairs or standing up from bed). Treatments for these conditions include surgery or conservative approaches, but common targets of the treatments are to manage the pain and to restore the normal hip mobility allowing a normal life in the affected children [1, 2].

The usual assessment of the impact of reduced mobility of daily life of these children is via quality of life questionnaires [3, 4], which indicates that hip stiffness

© Springer Nature Switzerland AG 2020
T. Ahram (Ed.): AHFE 2019, AISC 973, pp. 14–19, 2020.
https://doi.org/10.1007/978-3-030-20476-1_2

reduces the ability to perform the daily tasks (i.e. limping and functional impairments during walking). However, there is no objective tool to measure functional joint mobility during daily living. In order to objectively assess the impact and extent of hip stiffness on the child's life a dynamic measurement instrument is required. This device could also be useful in monitoring disease progression and rehabilitation.

Nowadays, wearable technology is an emerging field in the health and medical sector (i.e. heart rate monitoring; body temperature measurement) [5, 6]. Despite this, no wearable instrumentation is available to monitor hip stiffness during daily life. Existing devices are only able to obtain hip ROM in a laboratory (i.e. electronic goniometers) or in clinical (i.e. manual goniometer) environment [7].

The aim of our study was to report the development of a wearable prototype for real-time wireless, continuous monitoring of hip ROM during everyday life, to be used as a monitoring tool in childhood hip diseases.

2 Developing of the Joint Angle Measurement Device

2.1 The Device

We developed a wireless device, with a core microcontroller (ATMEL ATME-GA328P), with 1 optical flexible sensor, to detect changes in hip motion (flexion/extension) (Fig. 1).

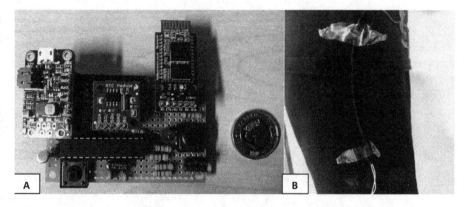

Fig. 1. Device prototype board (A) and optical flexible sensor attached over the clothes to the hip joint (B).

A Bluetooth interface (Tronixlabs HC-06) was implemented to send data for real time acquisition (Lab setting) to a computer. A local SD module (Hobby Components, HCARDU0008) for local data storage when the device is outside the laboratory environment; a real time clock microchip (Maxim Integrated, DS1307) for time and date recording; and a tilt ball rolling switch to detect changes in body position (person in standing or lying in down position), were also implemented.

The device runs at 5 V and it is supplied by a 3.7 V lithium battery 2000 mAh (Adafruit), connected to a power booster (Adafruit PowerBoost500) to reach the running voltage.

The optical flexible sensor was structured as a variable resistor embedded in a voltage divider design. The sensor implements a light-emitting diode (LED) to one side of a plastic optic fibre (POF), and a light-dependent resistor (LDR) to the other side. The POF was isolated by external light interferences through an external coating made of black shrinking tubes. When the optical flexible sensor is bent, the changes in angle reflection of the light from the LED through the POF changes the amount of light received by the LDR. This induce changes in resistance, read by the device, allowing conversion of the resistance value to a change in angle (degrees).

The bending of the optical flexible sensor induces macro-bending loss of the light that causes the change in the amount of light received by the LDR. Kim and colleagues [8] report that when the angle (Θ) of incidence light in a POF is greater than its critical angle (Θc), the light is transmitted to the end of the POF through the total internal reflection. The critical angle is the incidence angle (Θi) when the reflective angle (Θr) of the light is at 90° of bending. Θi as showed in Eq. (2) can be directly obtained from Eq. (1):

$$\frac{n_1}{n_2} = \frac{\sin \theta_2}{\sin \theta_1} = \frac{\sin \theta_r}{\sin \theta_i} \quad (\theta_r - 90^\circ) \tag{1}$$

$$\theta_i = \sin^{-1}\left(\frac{n_2}{n_2}\right) = \theta_c \tag{2}$$

The light leak in the bent area when a POF is bended makes the angle Θ smaller than the Θc, inducing changes in light reflection through the POF and less light exposure to the LDR.

The changes in light exposure to the LDR (R1) increases it resistance, changing the output voltage (Vout) of the voltage divider connected to the micro-controller (with R2 as fix resistor) which reads the different output and convert it in different joint angle degrees, following Eq. (3):

$$V_{out} = V_{in}\left(\frac{R2}{R1 + R2}\right) \quad V_{in} = 5V \tag{3}$$

We set the value of R2 as a middle value between the minimum and the maximum value reached by R1 (in Ω).

In order to fit the subjective variation in hip mobility among subjects, the device self-calibrates itself in the first 15 s of recording. This is performed by the subject extending the joint in the 0° position (neutral hip flexion) and in the 90° flexion position.

2.2   Microcontroller's Code

Example pseudo code implementation of the voltage divider data acquisition from the microcontroller shown below (based on the example code made for flexible sensors implementations by Cates, Barton and Takahashi [9]):

```
#define flexion_PIN = *Input pin of the flex
                                     sensor*;
const float VCC = 4.98;
const float R_DIV = *R1 VALUE*;
const float STRAIGHT_RESISTANCE = *R1 resistance
                        when POF is straight*;
const float BEND_RESISTANCE = *R1 resistance when
                        POF is bended*;
int flexADC = analogRead(FLEX_PIN);
float flexV = flexADC * VCC / 1023.0;
float flexR = R_DIV * (VCC / flexV - 1.0);
```

3   Methods

Data obtained by the device were compared with a manual goniometer examine the accuracy of the measurements. The device and the manual goniometer were positioned statically at 0°, 45° and 90° of flexion. Measurements from the optical flex sensor were taken at a sample rate of 1 ms, for a period of 10 s for each angulation and then averaged.

Additional data were recorded using the device during dynamic movements that simulated daily activities such as sitting. The device was attached to the hip of the participant though medical tape (see Fig. 1B) while the participant performed sit to stand maneuvers on a chair.

4   Results and Discussion

4.1   Optical Flexible Sensor Response

The optical flex sensor demonstrates linear relationship between the changes in LDR resistance made by the POF bending and the changes in angle detected by the device (Fig. 2).

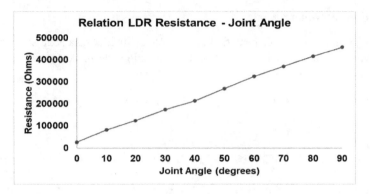

Fig. 2. Linear relationship between changes in LDR resistance and device angle detection.

During the measurements taken in static position, the optical flex sensor shown good agreement with the manual goniometer at 0°, 45° and 90° of flexion (Table 1).

Table 1. Agreement in measurement in both devices at 0°, 45° and 90°.

Manual goniometer angle	Optical flexible sensor angle mean (±SEM)
0°	1° (±0.20°)
45°	44° (±0.20°)
90°	89° (±0.20°)

4.2   Example Tests Results

During 5 repeated trials the device was able to detect the expected hip flexion response while performing sit to stand maneuvers (Fig. 3).

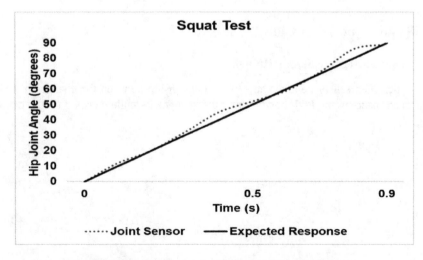

Fig. 3. Changes in hip joint angle during body weight squat.

Taken together, this preliminary data show the device's ability to detect the changes in angle of the hip joint which reflect flexion and extension (relating to sitting and standing). The data were successfully transmitted to a computer/laptop (through the Bluetooth interface) or were stored on the SD card included in the device.

5 Conclusion

The aim of our study was to report the development of a wearable prototype for real-time, wireless, continuous monitoring of hip ROM during everyday life.

Our device has shown good preliminary results in the simulated daily activity tests performed, showing fast and accurate reading of the changes in the POF bending angle during flexion/extension of the hip joint. The preliminary data have shown the concept is feasible, In order to make the sensor suitable for implementation in clinical practice, further miniaturization and testing in ambulatory environments are required. Further modification will seek to improve the current prototype, improving its features and its wearability to fit the population of interest. Additional tests of reliability will be performed which include longer durations of data collection (i.e. 24 h/7 day period) using a larger sample size and including children.

Acknowledgments. The authors want to thank the Liverpool John Moores University for the PhD scholarship with which this work has been funded. Authors want also to thank Dr Alex Mason (Animalia AS, Process and Product, Oslo – NO) for his support and suggestions during the device development.

References

1. Perry, D., Bruce, C.: Hip disorders in childhood. Surgery **29**(4), 1–7 (2011)
2. Zucker, E., Lee, E., Restrepo, R., Eisenberg, R.: Hip disorsders in children. Am. J. Roentgenol. **201**(6), W776–W796 (2013)
3. Hailer, Y., Haag, A., Nilsson, O.: Legg-Calvé-Perthes' disease: quality of life, physical activity, and behaviour pattern. J. Pediatr. Orthop. **34**(5), 8 (2014)
4. Malheiros, C., Lisle, L., Castelar, M., Sá, K., Matos, M.: Hip dysfunction and quality of life in patients with sickle cell disease. Clin. Paediatr. **54**(14), 5 (2015)
5. Zheng, Y., Ding, X., Poon, C., Lo, B., Zhang, H., Zhou, X., Yang, G., Zhao, N., Zhang, Y.: Unobtrusive sensing and wearable devices for health informatics. IEEE Trans. Biomed. Eng. **61**(5), 1538–1554 (2014)
6. Bonato, P.: Advances in wearable technology and applications in physical medicine and rehabilitation. J. NeuroEng. Rehabil. **2**(2), 1–4 (2005)
7. Owen, J., Stephens, D., Wright, J.: Reliability of hip range of motion using goniometry in pediatric femur shaft fractures. Can. J. Surg. **50**(4), 251–255 (2007)
8. Kim, S., Jang, K., Yoo, W., Shin, S., Cho, S., Lee, B.: Feasibility study on fiber-optic goniometer for measuring knee joint angle. Opt. Rev. **21**(5), 694–697 (2014)
9. Cates, J., Barton, R., Takahashi, N.: Flex sensor glove (2017). https://github.com/JonathanCates/Flex_Sensor_Glove

Nudging vs. Budging – Users' Acceptance of Nudging for More Physical Activity

Chantal Lidynia[(✉)], Julia Offermann-van Heek, and Martina Ziefle

Human-Computer Interaction Center, RWTH Aachen University,
Campus-Boulevard 57, 52074 Aachen, Germany
{lidynia, vanheek, ziefle}@comm.rwth-aachen.de

Abstract. Life-logging devices do not only record data such as daily steps, distances traveled, or energy used. Instead these systems also offer ready-made analyses of recorded data. Furthermore, they enable to give the user information and recommendations (nudges) that predefined goals are within reach. While other studies have proven technical accuracy and positive health-related effects of life-logging, the use of such devices is still not wide-spread. The present study aims at examining acceptance of life-logging in general and nudging in particular. To do so, a questionnaire study was conducted (N = 190). It was found that active users and non-users of life-logging technologies differ in their evaluation of benefits and barriers of these technologies and their acceptance of nudging to increase their daily or weekly physical activity. Experienced life-logging users were significantly more positive in their evaluations than non-users who rather rejected future life-logging technology use. Finally, nudging was more accepted by already experienced life-logging users than by non-users. The study's insights provide a deeper understanding of diverse users' requirements and needs regarding life-logging technologies and enable to derive guidelines for user-specific life-logging technology development.

Keywords: Technology acceptance · Life-logging · Privacy · Nudging

1 Introduction

The technological advancements have brought forth many advantages and a lot of physical labor can now be done by machines, especially in the Western World. However, this also means that a lot of work nowadays happens while sitting at a desk and working on a computer, oftentimes for long, uninterrupted periods of time [1, 2]. In addition, even when not at work, people rather tend to stay in reclined or seated positions [3]. This prolonged sitting or sedentary behavior can have negative effects on people's health, be it short term with postural problems [4], or even long term with increased weight and the risks connected to that [5]. Most people do not meet the recommended amount of physical activity prescribed by the World Health Organization [6, 7]. Despite knowing the positive effects of physical activity, not only on the body with weight control, increased mobility, and decreased risks of cardiovascular disease [8–10] but also on the mind with improved memory, mood, and less bouts of depression [11–13]. However, the ubiquity of mobile computing and new technology

© Springer Nature Switzerland AG 2020
T. Ahram (Ed.): AHFE 2019, AISC 973, pp. 20–33, 2020.
https://doi.org/10.1007/978-3-030-20476-1_3

developments also offer a possible solution, namely life-logging technologies. These includes smartphone apps but also specialized devices such as wearables or activity trackers. These do not only record data such as daily steps, distances traveled, or energy used. Instead these systems also offer ready-made analyses of the recorded data: they can be (easily) programed or already are programed to give the user information and recommendations (nudges) that predefined goals are within reach. For example, some brands either remind the wearer each hour that not even 250 steps have been taken, which would require a break from sitting down in front of a computer. Or, should the daily goal be a total of 10,000 steps, then once about 80% of that goal have been reached, the user is informed and encouraged to fulfill that goal. While studies have shown these devices do have the potential to increase physical activity and help with weight management, e.g., [14, 15], their use, especially in Germany, is still rather low [16]. The present study aims at finding out why, if the health benefits of physical activity have been proven, these technologies are not well accepted.

2 Related Work

Life-Logging can be used to record data from daily life and use this to facilitate behavior changes [17, 18], such as more activity or less nutritional intake. The best-known data to be tracked are typically daily steps, energy expenditure, energy intake, i.e., nutrition, distances traveled and the like [19]. Many studies concerning life-logging and especially the devices used to do so, deal with hardware properties such as sensor accuracy [14, 20] and how this is reliant on the activity [21], where it is attached to the body [22], or the brand and model [14, 23, 24]. However, an ever-increasing number of research is devoted to user interaction with these technologies. Oftentimes, to understand influencing factors on the perception of and interaction with these devices, technology acceptance models are used as starting point. These models, such as the Technology Acceptance Model [25, 26] or the Unified Theory of Acceptance and Use of Technology [27, 28], with their respective extensions, could all show that factors such as performance expectancy or social influence have an influence on the intention to use a device. This is usually a rather good indicator of actual adoption of a technology [29]. However, with such sensitive data as heartrate or GPS, both of which can be tracked by life-logging apps and devices, it is also important to include privacy into the study of acceptance of these technologies. The so-called privacy calculus is an attempt to understand when data are made available online and when the service is declined in favor of keeping the data from others [30]. In the case of life-logging and wearables, the impact of privacy has been shown in multiple studies and for different contexts, e.g., [31–33]. The present study also investigates another feature of these fitness apps and wearables, namely that of nudging. Nudges are well-timed reminders or recommendations meant to induce wanted behavior or try to prevent unwanted behavior [34].

3 Empirical Approach

This section presents the study's empirical approach, describing the design of the online questionnaire as well as introducing the sample's characteristics.

3.1 Online Questionnaire

Based on previous qualitative studies and literature reviews, an online questionnaire was conceptualized focusing on evaluations of benefits and barriers of life-logging technology usage and on the relationship between nudging and technology acceptance.

In the first part of the questionnaire, the participants were asked for demographic information such as age, gender, educational level etc. In addition, the participants evaluated their current (perceived) health status. In order to get an idea on the importance of maintaining a healthy lifestyle, the participants were asked for different factors that have been proven to impact health, e.g., avoiding stress, eating healthy, getting enough exercise, avoiding unnecessary medication, etc. As life-logging and physical activity is often linked to sports, the participants were further asked to estimate hours per week they spend doing sports. In order to address relevant aspects in the context of life-logging, the participants' *Motivation for Physical Activity Measure (MPAM)* was applied using eleven items based on [35] including different dimensions of motives being physically active: fitness, appearance, competence, enjoyment, and a social dimension. Subsequently, actual use of life-logging technologies was queried.

The next part of the questionnaire focused on additional individual attitudes of the participants. Based on [36], the participants' *Affinity for Technology Interaction (ATI)* surveyed using a reduced number of five items focusing on interest in as well as self-efficacy when interacting with technology. As privacy-related aspects have proven to be relevant for life-logging acceptance in literature reviews as well as previous studies, it was necessary to consider the participants' needs for privacy and potential concerns with regard to privacy in the present survey. Thereby, *Need for Privacy (NfP)* was measured using four items taken from different validated studies focusing on the general need to limit access to one's person. *Privacy Concern* on the other hand, was measured with three items, and addressed concerns that stem from the use of online services (including life-logging apps as well as devices).

Within the next part of the questionnaire, relevant usage factors were addressed that are of interest in the context of life-logging technology acceptance. This section started with a short introduction of life-logging technology examples. Then, different usage motives (using eleven items; $\alpha = .864$) as well as usage barriers (using sixteen items; $\alpha = .894$) were examined. All these items are based on previously conducted studies and can be looked at the results Sects. 4.2 and 4.3, e.g., [37]. As an important point from previous studies, the participants were finally asked to evaluate their acceptability of nudging, based on a self-developed item collection. The so-called *Nudging Acceptability Scale* consisted of six items ($\alpha = .864$) which can be found within the results section (see 4.4). Unless specified otherwise, all items were evaluated using six-point Likert scales (min = 1: "I strongly disagree"; max = 6: "I strongly agree").

3.2   Sample of the Study

A total of N = 190 German participants completed the questionnaire and were included in the following analyses. 107 (56.3%) participants were female and 83 (43.7%) male. The age of the sample ranged from 15 to 91 years, with a mean age of 36.9 years (SD = 15.4). As described above, usage of Life-Logging technology was determined by *usage frequency*: the higher the value, the higher the frequency of use, which ties in with expertise: 69 participants (36.3%) indicated to have not ever used life-logging devices, 34 (17.9%) to use Life-Logging technology very rarely, 18 participants (9.5%) to use it a few times a month, 9 (4.7%) once a week, another 9 (4.7%) a few times per week, 31 participants (16.3%), and 20 (10.5%) even around the clock. Considering individual user factors for the whole sample, the participants' *Affinity for Technology Interaction (ATI)* was rather average (M = 3.8; SD = 1.1; min = 1; max = 6), whereas both *Need for Privacy* (M = 4.4; SD = 1.0) and *Privacy Concern* (M = 4.6; SD = 1.0) were rather high. Overall, the participants indicated to be somewhat *motivated to engage in physical activity* (M = 4.2; SD = 0.8). Correlation analyses revealed significant relationships between some of the demographic and user-related variables. Gender was related with the participants' *ATI*, indicating men to have a higher affinity than women (r = 0.422; p < 0.01). Age correlated with the participants' *ATI*, indicating that younger people have a higher affinity than older people (r = −0.156; p < 0.05). A high positive correlation between the participants' *Need for Privacy* and *Privacy Concerns* was found (r = 0.757; p < 0.01). Finally, the *motivation for physical activity* was negatively related with age, indicating that younger people reported a higher motivation to engage in physical activity than older people (r = −0.206; p < 0.01). With regard to user factors, no further correlations were found.

4   Results

Within this section, the results of the study are presented. First, identified life-logging user groups and their specific characteristics are described. Afterwards, the evaluations of motives and barriers of life-logging technology usage are detailed differentiating between non-users, sporadic users, and frequent users of life-logging technology. In a last step, it is analyzed if and to what extent the acceptability of nudging is related with life-logging technology usage. Besides descriptive and correlation analyses, non-parametric tests (Kruskal-Wallis) were applied to analyze group differences with post-hoc-tests using Bonferroni correction. Level of significance was set at 5% (p < 0.05). Within the results section the following notation is used to differentiate between levels of significance: *: p < 0.05; **: p < 0.01; ***: p < 0.001.

4.1   Defining Life-Logging User Groups

Based on the frequency of life-logging technology usage, the sample was divided into three user groups: *non-users*, who do not use life-logging devices (n = 69), *sporadic users*, who have some experience but do not regularly use life-logging (barely once a week) (n = 61), *frequent users*, who use the device daily or around the clock (n = 60).

The characteristics of the three groups were summarized in Table 1. The first group, the **non-users**, group included 59.4% female and 40.6% male participants. The mean age of this group was 38.0 years. Regarding *their motivation for physical activity*, the participants of this group were slightly above average (M = 3.9; SD = 0.8). The group was rather average in their *ATI* (M = 3.4; SD = 1.0). The privacy-related attitudes, *NfP* (M = 4.6; SD = 1.0) and *Privacy Concern* (M = 4.6; SD = 1.0), were both comparably high. In contrast, this group's *acceptability of nudging* was not high (M = 3.2; SD = 1.0).

Table 1. Overview of group characteristics.

	"Non-Users" (n = 69)	"Sporadic Users" (n = 61)	"Frequent Users" (n = 60)	Significance a = non- to sporadic users b = non- to frequent users
Motivation for physical activity *(M, (SD))*	3.9 (0.8)	4.3 (0.6)	4.2 (0.7)	H(2) = 10.352, p = 0.006 a: U = 1471.5, p = 0.003, r = 0.259 b: U = 1536.5, p = 0.012, r = 0.222
ATI *(M, (SD))*	3.4 (1.0)	4.1 (0.9)	4.1 (1.1)	H(2) = 14.372, p = 0.001 a: U = 1348.5, p < 0.001, r = 0.310 b: U = 1447.0, p = 0.003, r = 0.259
NfP *(M, (SD))*	4.6 (1.0)	4.4 (1.0)	4.1 (1.1)	H(2) = 6.992, p = 0.03 a: U = 1798.5, p = 0.152, n.s. b: U = 1530.5, p = 0.011, r = 0.224
Nudging Acceptability *(M, (SD))*	3.2 (1.0)	4.0 (0.8)	4.3 (0.7)	H(2) = 45.333, p < 0.001 a: U = 1103.0, p < 0.001, r = 0.0410 b: U = 718.0, p < 0.001, r = 0.562

The second group, the **sporadic users**, also consisted of more women than men, with 57.4% female and 42.6% male participants and the mean age of this group was 34.1 years. Their *motivation for physical activity* was comparably higher (M = 4.3; SD = 0.6). This group had on average a higher *ATI* (M = 4.1; SD = 0.9). Considering privacy-related aspects, the participants of this group had a comparably high *NfP* (M = 4.4; SD = 1.0) and *Privacy Concern* (M = 4.5; SD = 0.9). In contrast to the non-users, the sporadic users reported a higher *acceptability of nudging* (M = 4.0; SD = 0.8).

The third group, the **frequent users**, showed rather similar evaluations to the sporadic users and differed also from the non-users. In this group, gender was almost equally spread with 51.7% women and 48.3% men. The mean age of this group was 38.4 years (SD = 15.0) with a slightly smaller age range from 20 to 69 years. This group reported a rather high *motivation for physical activity* (M = 4.2; SD = 0.7). As the sporadic users, this group revealed to have a higher *ATI* (M = 4.1; SD = 1.1) compared to the non-users. In this group, *NfP* was rather high (M = 4.1; SD = 1.1) and the participants' *Privacy Concern* was even higher (M = 4.5; SD = 1.1). This group showed the highest *acceptability of nudging* (M = 4.3; SD = 0.7). Analyses of differences between the user groups revealed that frequent users and sporadic users did not differ significantly in their characteristics. In contrast (see Table 1), non-users differed significantly from the other groups concerning their *motivation for physical* *activity, ATI, NfP,* and *nudging acceptability,* while age, gender, and *Privacy Concern* differed not significantly.

4.2 Motives of Life-Logging Usage

Except of *comparison with others* (M = 3.4; SD = 1.6) and *avoiding stigmatization by using unobtrusive technology* (M = 3.2; SD = 1.3), all other potential motives of using life-logging technology were seen as such and received agreement (means > 3.5), although to varying degrees (see Fig. 1). Referring to the whole sample, the best evaluations were given to more control (M = 4.3; SD = 1.4), *improvement of health* (M = 4.2; SD = 1.3), *improvement of physical fitness* (M = 4.1; SD = 1.2), and *fast access to data* (M = 4.1; SD = 1.3). Compared to that, *quick help in emergencies* (M = 3.8; SD = 1.4), *preservation of mobility* (M = 3.7; SD = 1.3), and *relief in everyday life* (M = 3.7; SD = 1.3) were clearly lower evaluated, while *increase of life quality* (M = 3.6; SD = 1.4) and *comfort* (M = 3.6; SD = 1.3) received almost neutral evaluations.

Analyzing the user groups' evaluations of potential motives of using life-logging revealed significant differences with regard to seven motives (see Table 2). On the one hand, all participants confirmed the benefits of life-logging *to improve health* and *physical fitness, have more control, have fast access to their data* (H(2) = 5.914, p = 0.06, n.s.), and also *get quick help in case of emergencies* (H(2) = 3.276, p = 0.194, n.s.).

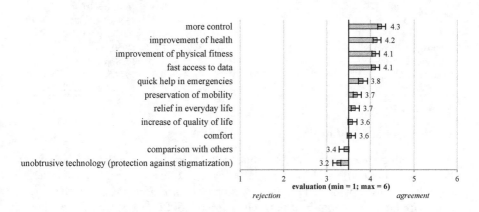

Fig. 1. Evaluations (means, standard errors) of life-logging motives (whole sample).

On the other hand, the user groups differed sometimes in the polarization of the agreement or disagreement: i.e., both sporadic and frequent users evaluated life-logging as beneficial to *increase the quality of life, comfort, preservation of mobility,* and *bringing relief to daily life* (H(2) = 5.956, p = 0.051, n.s.); however, non-users did not ascribe those qualities to life-logging. In addition, only sporadic users perceived and confirmed *comparison with others* (H(2) = 5.339, p = .069, n.s.) as a benefit of life-logging.

Table 2. Overview of significant group differences in the evaluations of motives.

	"Non-Users", n = 69 (M, (SD))	"Sporadic Users", n = 61 (M, (SD))	"Frequent Users", n = 60 (M, (SD))	Significance a = non- to sporadic users b = non- to frequent users c = sporadic to frequent users
Improvement of health ***	3.6 (1.4)	4.3 (1.1)	4.7 (1.0)	H(2) = 19.528, p < 0.001 a: U = 1543, p = 0.007, r = 0.230 b: U = 1202, p < 0.001, r = 0.361 c:1487, p = 0.062, n.s.
More control *	3.9 (1.5)	4.3 (1.1)	4.6 (0.9)	H(2) = 7.197, p = 0.027 a: U = 1797.5, p = 0.139, n.s. b: U = 1532, p = 0.008, r = 0.224 c: U = 1609, p = 0.227, n.s.
Improvement of physical fitness ***	3.6 (1.3)	4.3 (1.1)	4.6 (1.0)	H(2) = 25.001, p < 0.001 a: U = 1417, p = 0.001, r = 0.281 b: U = 1096, p < 0.001, r = 0.405 c: U = 1531, p = 0.099, n.s.
Increase of quality of life ***	3.1 (1.4)	3.6 (1.2)	4.2 (1.3)	H(2) = 21.477, p < 0.001 a: U = 1617, p = 0.019, n.s. b: U = 1151.5, p < 0.001, r = 0.382 c: U = 1336, p = 0.008, r = 0.223
Comfort ***	3.1 (1.4)	3.6 (1.1)	4.1 (1.1)	H(2) = 19.607, p < 0.001 a: U = 1661.5, p = 0.034, n.s. b: U = 1189.5, p < 0.001, r = 0.366 c: U = 1352, p = 0.01, r = 0.225
Preservation of mobility **	3.2 (1.4)	3.9 (1.1)	4.1 (1.2)	H(2) = 14.574, p = 0.001 a: U = 1508, p = 0.004, r = 0.244 b: U = 1350, p < 0.001, r = 0.319 c: U = 1642.5, p = 0.312, n.s.
Unobtrusive technology (protection against stigmatization) *	2.9 (1.3)	3.3 (1.1)	3.6 (1.3)	H(2) = 8.930, p < 0.05 a: U = 1728, p = 0.07, n.s. b: U = 1477.5, p < 0.01, r = 0.246 c: U = 1585.5, p = 0.192, n.s.

4.3   Barriers of Life-Logging Usage

Complementing the evaluations of benefits, the participants were asked to rate sixteen different potential barriers of life-logging technology usage (see Fig. 2). Considering the whole sample of participants, *misuse of data by a third party* (M = 4.7; SD = 1.3), *unauthorized access of data by a third party* (M = 4.6; SD = 1.3), and forwarding of data (M = 4.6; SD = 1.3) represented the barriers receiving the highest approval. In contrast, *dealing with new technology* (M = 3.0; SD = 1.4), *easier to get addicted to sports* (M = 2.7; SD = 1.3), and *missing aesthetics of the app or device* (M = 2.7; SD = 1.4) represented the barriers that were slightly rejected to be barriers of using lifelogging technology.

With regard to the evaluations of the three user groups, the results did not reveal significant differences for six of the evaluated barriers: *surveillance through technology* (H(2) = 2.502, p = 0.286, n.s.), *transmission of false information* (H(2) = 2.089; p = 0.352, n.s.), *likelihood of setting unrealistic goals* (H(2) = 0.661; p = 0.719, n.s.), *dealing with new technology* (H(2) = 3.635; p = 0.162, n.s.), *easier to get addicted to*

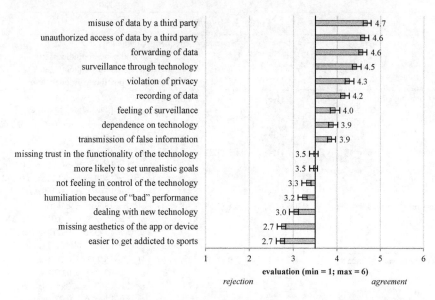

Fig. 2. Evaluations (means, standard errors) of life-logging barriers (whole sample).

sports (H(2) = 2.257; p = 0.323, n.s.), and *a lack of aesthetic concerning app or device design* (H(2) = 3.792; p = 0.150, n.s.). In contrast, significant differences between the frequent users and the non-users were found for ten of the evaluated barriers (Table 3). In addition, differences were sometimes also found between the sporadic users and the non-users. However, there was also a significant difference between frequent and sporadic users in case of *a possible humiliation based on supposedly bad performance results*. For the present sample, the evaluation pattern was found that almost all barriers are seen as more influential by non-users, followed by sporadic users, while frequent users are less concerned with the presented barriers than the two other groups. A pattern applying to all groups was that barriers dealing with privacy and the violation thereof were the most important barriers of life-logging technology usage.

Table 3. Overview of significant group differences in the evaluations of barriers.

	"Non-Users" n = 69 (M, (SD))	"Sporadic Users" n = 61 (M, (SD))	"Frequent Users" n = 60 (M, (SD))	*Significance* a = non- to sporadic users b = non- to frequent users c = sporadic to frequent users
Misuse of data by a third party *	5.0 (1.1)	4.7 (1.2)	4.4 (1.4)	H(2) = 8.450, p < 0.05 a: U = 1739.5, p = 0.075, n.s. b: U = 1492, p < 0.001, r = 0.159 c: U = 1616.5, p = 0.253, n.s.
Unauthorized access of data by a third party *	4.9 (1.1)	4.6 (1.1)	4.3 (1.5)	H(2) = 7.043, p < 0.05 a: U = 1766, p = 0.100, n.s. b: U = 1543, p < 0.01, r = 0.230 c: U = 1636.5, p = 0.301, n.s.

(continued)

Table 3. (*continued*)

	"Non-Users" n = 69 (M, (SD))	"Sporadic Users" n = 61 (M, (SD))	"Frequent Users" n = 60 (M, (SD))	*Significance* a = non- to sporadic users b = non- to frequent users c = sporadic to frequent users
Forwarding of data *	4.8 (1.3)	4.7 (1.1)	4.3 (1.4)	H(2) = 6.223, p < 0.05 a: U = 1886, p = 0.288, n.s. b: U = 1568.5, p < 0.016, r = 0.212 c: U = 1588, p = 0.146, n.s.
Violation of privacy **	4.7 (1.4)	4.1 (1.3)	4.0 (1.3)	H(2) = 10.727, p < 0.01 a: U = 1576.5, p < 0.05, r = 0.216 b: U = 1439.5, p < 0.01, r = 0.272 c: U = 1774.5, p = 0.769, n.s.
Recording of data **	4.6 (1.3)	4.1 (1.3)	3.8 (1.4)	H(2) = 13.808, p < 0.01 a: U = 1567.5, p < 0.05, r = 0.220 b: U = 1343, p < 0.001, r = 0.230 c: U = 1604.5, p = 0.231, n.s.
Feeling of surveillance ***	4.5 (1.4)	3.9 (1.5)	3.4 (1.6)	H(2) = 15.433, p < 0.001 a: U = 1643, p = 0.028, n.s. b: U = 1267, p < 0.001, r = 0.265 c: U = 1496, p = 0.078, n.s.
Dependence on technology **	4.4 (1.4)	3.8 (1.3)	3.5 (1.5)	H(2) = 14.618, p < 0. 01 a: U = 1525.5, p < 0.01, r = 0.237 b: U = 1330, p < 0.001, r = 0.236 c: U = 1627, p = 0.283, n.s.
Not feeling in control of the technology ***	4.0 (1.5)	3.0 (1.2)	2.7 (1.3)	H(2) = 27.511, p < 0.001 a: U = 1283.5, p < 0.001, r = 0.336 b: U = 1071.5, p < 0.001, r =0.358 c: U = 1556, p = 0.253, n.s.
Missing trust in the functionality of the technology **	3.9 (1.4)	3.2 (1.1)	3.2 (1.4)	H(2) = 10.842, p < 0.01 a: U = 1488, p < 0.01, r = 0.252 b: U = 1516, p < 0.01, r = 0.148 c: U = 1822.5, p = 0.968, n.s.
Humiliation because of "bad" performance *	3.4 (1.4)	3.4 (1.3)	2.8 (1.4)	H(2) = 8.804; p < 0.05 a: U = 1988, p = 575, n.s. b: U = 1536.5, p < 0.01, r = 0.138 c: U = 1342.5, p < 0.016, r = 0.222

4.4 Nudging and Its Relationship with Life-Logging Usage?

Slightly positive evaluations referring to the nudging acceptability scale for the whole sample indicated that participants were at least not against giving nudging a try (M = 3.8; SD = 1.0). In more detail, the *opting out of the suggestion made by life-logging technology* (M = 4.6; SD = 1.1) received the highest approval, suggesting that participants are aware a nudge is not a decree but merely a timed recommendation. Further, the results showed that the participants were overall slightly willing *to try nudging by life-logging technology* (M = 3.9; SD = 1.5). In contrast, the results showed that the participants did *not like taking instructions from a machine* (M = 3.9; SD = 1.5) very much. Other aspects such as the expectation that *nudging would make it easier to meet personal goals* (M = 3.4; SD = 1.4) or the statement that n*udging is* *like a navigation system* (M = 3.6; SD = 1.3) were evaluated almost neutrally. In contrast, the idea that *behavior suggestions from a lifeless machine scare the participants* was rather rejected (M = 2.7; SD = 1.4). In addition to these descriptive results, correlation analyses revealed a medium strong correlation between the acceptability of nudging and the frequency of use life-logging technology (ρ = 0.489, p < 0.01). Hence, the user groups' evaluations were also analyzed with regard to nudging acceptability.

As shown in Fig. 3, the three groups differed significantly in their evaluations of all individual statements. Thereby, non-users differed clearly from sporadic as well as frequent users with regard to all statements except of *not taking instructions from a machine*: here, non-users and sporadic users did not show significant differences in their evaluations (U = 1660.5, p = 0.034, n.s.). Significant differences between sporadic and frequent users were only found for the evaluation of their *willingness to try out nudging* (U = 1381, p = 0.016, r = 0.212; M_{spo} = 4.3, SD_{spo} = 1.0; M_{fre} = 4.8, SD_{fre} = 1.1) and the expectation that *nudging would make it easier to reach personal goals* (U = 1350, p = 0.03, r = 0.226; M_{spo} = 3.5, SD_{spo} = 1.2; M_{fre} = 4.0; SD_{fre} = 1.2). It is unsurprising that both actual user groups (sporadic and frequent users) report that they would *try nudging*. Non-users, however, rather reject the notion (M_{Non} = 2.7, SD_{Non} = 1.5), as they are also adamant to *not take instructions from a machine* (M_{Non} = 4.3, SD_{Non} = 1.5).

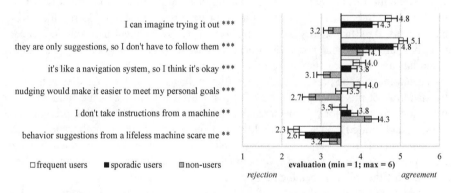

Fig. 3. Group-dependent evaluations (means, standard errors) of nudging acceptability.

5   Discussion

This study provided insights into life-logging users' and non-users' perception of benefits and barriers of life-logging technology usage as well as their acceptability of nudging by life-logging. The study revealed that most of the considered potential benefits of life-logging technologies were seen as such by the participants, whereas improving one's health and fitness were the best evaluated benefits and fast access to one's data was also seen as beneficial. Compared to that, the preservation of mobility and increase of quality of life were less regarded as real benefits of life-logging, although authorities and experts, e.g., [38], strongly recommend keeping active and research in this regard has proven that usage of life-logging technology has the potential to increase physical activity, e.g., [39]. Possible explanations can be that these benefits either are not communicated sufficiently to the potential users or not seen as beneficial enough to trump perceived barriers of using life-logging technology.

Most of the evaluated barriers focused on privacy-related issues and represented the perceived barriers receiving the highest approval. This indicates that privacy is of high relevance in the context of life-logging usage and that the potential risks might dominate perceived benefits of using life-logging technology. In this regard, clear and transparent communication is needed, and it should be reflected if possibilities can be ensured not to share data online or not to make data accessible to third parties. Ensuring these options would perhaps lead to life-logging benefits dominating concerns for privacy. Another interesting aspect meets the fact that competition as motivation element was largely disregarded within this sample, while most life-logging concepts employ gamification as a central element. In the present study, frequent as well as non-users of life-logging technology (slightly) rejected comparison with others to be motivating, whereas only the group of sporadic users (i.e., using life-logging technology sometimes but not every week) see an added value in compare their activity with others. This directly leads to another focus of the current study – the acceptance of nudging during life-logging technology usage. Here, the results show that current non-users disliked the concept of nudging indicated by an unwillingness to take suggestions from a machine, even though non-users are not scare by those suggestions. One possible explanation lies in non-mandatory nature of nudges [34], that is probably – while obvious to all three user groups – not as clear to current non-users. Here, nudges are only perceived as facilitators of reaching one's goal by frequent users while sporadic users are undecided and non-users reject it. This may be due to an expertise effect which has already been shown for other technologies, e.g., [40].

To summarize, the present study has shown in line with previous studies in the field that handling of privacy represents the biggest barrier for the acceptance and adoption of life-logging technology. Transparency and defined models of handling privacy, data security, and limited data access should thereby be considered as guidelines to increase the adoption of life-logging technology and therewith also the physical activity of users. For the future, this could also help to improve health and well-being and could thus support the relief of healthcare systems.

Besides new insights, there are of course also some limitations of the present study which should be considered for future work. Although the sample had an adequate size and was heterogenous according to age and gender distribution, future studies should aim for a validation of the results trying to reach an even larger sample: thereby a focus on older people suffering from chronic diseases or in need of care would give the opportunity for analyses of more specific health-related life-logging applications. In addition, the current study was conducted in Germany and represents a country- and culture-specific perspective. As in particular regulations and handling of privacy as potential barrier of technology usage are diverse in different countries, cross-national and -cultural comparisons seem to be very useful for future studies.

Acknowledgments. The authors thank all participants for their openness to share opinions on lifelogging. This work has been funded by the German Federal Ministry of Education and Research projects MyneData (KIS1DSD045) and PAAL (6SV7955).

References

1. Thorp, A.A., Healy, G.N., Winkler, E., Clark, B.K., Gardiner, P.A., et al.: Prolonged sedentary time and physical activity in workplace and non-work contexts: a cross-sectional study of office, customer service and call centre employees. Int. J. Behav. Nutr. Phys. Act. **9**, 128 (2012)
2. Bennie, J.A., Chau, J.Y., van der Ploeg, H.P., Stamatakis, E., Do, A., Bauman, A.: The prevalence and correlates of sitting in european adults—a comparison of 32 eurobarometer-participating countries. Int. J. Behav. Nutr. Phys. Act. **10**, 107 (2013)
3. Kirk, M.A., Rhodes, R.E.: Occupation correlates of adults' participation in leisure-time physical activity. A systematic review. Am. J. Prev. Med. **40**, 476–485 (2011)
4. Zemp, R., Fliesser, M., Wippert, P.M., Taylor, W.R., Lorenzetti, S.: Occupational sitting behaviour and its relationship with back Pain-A pilot study. Appl. Ergon. **56**, 84–91 (2016)
5. Biswas, A., Oh, P.I., Faulkner, G.E., Bajaj, R.R., Silver, M.A., et al.: Sedentary time and its association with risk for disease incidence, mortality, and hospitalization in adults. A systematic review and meta-analysis. Ann. Intern. Med. **162**, 123–132 (2015)
6. Hallal, P.C., Andersen, L.B., Bull, F.C., Guthold, R., Haskell, W., et al.: Global physical activity levels: surveillance progress, pitfalls, and prospects. Lancet. **380**, 247–257 (2012)
7. Grieco, A.: Sitting posture: an old problem and a new one. Ergonomics **29**, 345–362 (1986)
8. Goldberg, J.H., King, A.C.: Physical activity and weight management across the lifespan. Annu. Rev. Public Health. **28**, 145–170 (2007)
9. Wahid, A., Manek, N., Nichols, M., Kelly, P., Foster, C., et al.: Quantifying the association between physical activity and cardiovascular disease and diabetes: a systematic review and meta-analysis. J. Am. Heart Assoc. **5**, e002495 (2016)
10. World Cancer Research Fund: American Institute for Cancer Research: Food, Nutrition, Physical Activity, and the Prevention of Cancer: a Global Perspective. AICR, Washington DC (2007)
11. Stroth, S., Hille, K., Spitzer, M., Reinhardt, R.: Aerobic endurance exercise benefits memory and affect in young adults. Neuropsychol. Rehabil. **19**, 223–243 (2009)
12. Cooney, G.M., Dwan, K., Greig, C.A., Lawlor, D.A., Rimer, J., Waugh, F.R., et al.: Exercise for depression. Cochrane Database Syst. Rev. **9** (2013)
13. Thayer, R.E., Newman, J.R., McClain, T.M.: Self-regulation of mood: strategies for changing a bad mood, raising energy, and reducing tension. J. Pers. Soc. Psychol. **67**, 910–925 (1994)
14. El-Amrawy, F., Nounou, M.I., McGrath, M., Scanaill, C., McGrath, M., et al.: Are currently available wearable devices for activity tracking and heart rate monitoring accurate, precise, and medically beneficial? Healthc. Inform. Res. **21**, 315 (2015)
15. Coughlin, S.S., Whitehead, M., Sheats, J.Q., Mastromonico, J., Smith, S.: A review of smartphone applications for promoting physical activity. J Commun. Med. **2** (2016)
16. Bitkom: Zukunft der Consumer Technology—2018. Marktentwicklung, Trends, Mediennutzung, Technologien, Geschäftsmodelle. Berlin (2018)
17. Bice, M.R., Ball, J.W., McClaran, S.: Technology and physical activity motivation. Int. J. Sport Exerc. Psychol. **14**, 295–304 (2016)

18. Gualtieri, L., Rosenbluth, S., Phillips, J.: Can a free wearable activity tracker change behavior? The impact of trackers on adults in a physician-led wellness group. JMIR Res Protoc. **5** (2016)
19. Lupton, D.: The Quantified Self. Polity Press, Cambridge (2016)
20. Battenberg, A.K., Donohoe, S., Robertson, N., Schmalzried, T.P.: The accuracy of personal activity monitoring devices. Semin. Arthroplasty. **28**, 71–75 (2017)
21. Sasaki, J.E., Hickey, A., Mavilia, M., Tedesco, J., John, D., Keadle, S.K.: Validation of the fitbit wireless activity tracker for prediction of energy expenditure. J. Phys. Act. Heal. **12**, 149–154 (2015)
22. Chow, J.J., Thom, J.M., Wewege, M.A., Ward, R.E., Parmenter, B.J.: Accuracy of step count measured by physical activity monitors: the effect of gait speed and anatomical placement site. Gait Posture. **57**, 199–203 (2017)
23. Kaewkannate, K., Kim, S.: A comparison of wearable fitness devices. BMC Public Health. **16**, 433 (2016)
24. O'Connell, S., Ólaighin, G., Kelly, L., Murphy, E., Beirne, S., Burke, N., et al.: These shoes are made for walking: sensitivity performance evaluation of commercial activity monitors under the expected conditions and circumstances required to achieve the international daily step goal of 10,000 steps. PLoSOne **11**, e0154956 (2016)
25. Davis, F.: Perceived usefulness, perceived ease of use, and user acceptance of information technology. MIS Q. **13**, 319–340 (1989)
26. Venkatesh, V., Davis, F.D.: A theoretical extension of the technology acceptance model: four longitudinal field studies. Manage. Sci. **46**, 186–204 (2000)
27. Venkatesh, V., Morris, M.G., Davis, G.B., Davis, F.D.: User acceptance of information technology: toward a unified view. MIS Q. **27**, 425–478 (2003)
28. Venkatesh, V., Walton, S.M., Thong, J.Y.L., Xu, X.: Consumer acceptance and use of information technology: extending the unified theory of acceptance and use of technology. MIS Q. **36**, 157–178 (2012)
29. Venkatesh, V., Thong, J.Y.L., Xu, X.: Unified theory of acceptance and use of technology: a synthesis and the road ahead. J. Assoc. Inf. Syst. **17**, 328–376 (2016)
30. Smith, H.J., Dinev, T., Xu, H.: Information privacy research: an interdisciplinary review. MIS Q. **35**, 989–1015 (2011)
31. Lidynia, C., Brauner, P., Ziefle, M.: A step in the right direction—understanding privacy concerns and perceived sensitivity of fitness trackers. In: Ahram, T., Falcão, C. (eds.) Advances in Human Factors in Wearable Technologies and Game Design, Advances in Intelligent Systems and Computing 608, pp. 42–53. Springer International Publishing, Cham (2018)
32. Lidynia, C., Brauner, P., Burbach, L., Ziefle, M.: Step by step—users and non-users of life-logging technologies. In: Ahram, T.Z. (ed.) Advances in Intelligent Systems and Computing, pp. 123–134. Springer International Publishing AG, Cham, Switzerland (2019)
33. Schomakers, E.-M., Lidynia, C., Ziefle, M.: Exploring the acceptance of mhealth applications—do acceptance patterns vary depending on context? In: Ahram, T.Z. (ed.) Advances in Intelligent Systems and Computing, pp. 53–64. Springer International Publishing, Cham, Switzerland (2019)
34. Thaler, R.H., Sunstein, C.R.: Nudge. Improving Decisions About Health, Wealth, and Happiness. Yale University Press, New Haven & London (2008)
35. Ryan, R.M., Frederick, C.M., Lepes, D., Rubio, N., Sheldon, K.M.: Intrinsic motivation and exercise adherence. Int. J. Sport Psychol. **28**, 335–354 (1997)
36. Franke, T., Attig, C., Wessel, D.: A personal resource for technology interaction: development and validation of the affinity for technology interaction (ATI) scale. Int. J. Hum. Comput. Interact. 1–12 (2018)

37. Lidynia, C., Schomakers, E.-M., Ziefle, M.: What are you waiting for?—perceived barriers to the adoption of fitness-applications and wearables. In: Ahram, T.Z. (ed.) Advances in Intelligent Systems and Computing, pp. 41–52. Springer International Publishing AG, Cham, Switzerland (2019)

38. World Health Organization: Global Recommendations on Physical Activity for Health. WHO Press, Geneva, Switzerland (2010)

39. Bravata, D.M., Smith-Spangler, C., Sundaram, V., Gienger, A.L., Lin, N., Lewis, R., Stave, C.D., Olkin, I., Sirard, J.R.: Using pedometers to increase physical activity and improve health: a systematic review. JAMA **298**, 2296–2304 (2007)

40. van Heek, J., Arning, K., Ziefle, M.: Reduce, reuse, recycle: acceptance of co2-utilization for plastic products. Energy Policy **105**, 53–66 (2017)

Evaluation of Suitable Rest Times for the Use of Optical Head-Mounted Displays

Chih-Yu Hsiao[1(✉)], Yi-An Liu[1], and Mao-Jiun Wang[2]

[1] Department of Industrial Engineering and Engineering Management,
National Tsing Hua University, NO. 101, Sec. 2, Kuang Fu Road,
Hsinchu 30013, Taiwan, R.O.C.
{cyhsiao,yaliu}@ie.nthu.edu.tw
[2] Department of Industrial Engineering and Enterprise Information,
Tunghai University, NO. 1727, Sec. 4, Taiwan Boulevard, Xitun District,
Taichung 40704, Taiwan, R.O.C.
mjwang@ie.nthu.edu.tw

Abstract. This study aimed to determine suitable rest times for visual tasks of different lengths by evaluating visual fatigue and visually induced motion sickness while 20 healthy participants were using an optical head-mounted display. The independent variables tested were participant gender, visual task condition (30 min and 60 min), and length of rest time (5 min, 10 min, and 15 min). The critical fusion frequency (CFF) value and simulator sickness questionnaire (SSQ) scores were used as dependent measures. Gender showed no significant effect on CFF values or SSQ scores. Greater CFF values and SSQ scores were produced by the 60-min visual task than the 30-min visual task. The results of this study suggest that after completing a 30-min visual task, a rest time of at least 10 min is necessary for participants to recover, and a rest time of more than 15 min needs to follow a 60-min visual task.

Keywords: Optical head-mounted display · Visual fatigue ·
Visually induced motion sickness · Critical fusion frequency ·
Simulator sickness questionnaire

1 Introduction

With the advancement of wearable technology in recent years, head-mounted displays (HMDs) and smart eyewear devices have become more and more popular. HMDs consist of a helmet or goggles with a wide-range cathode ray tube monitor or a liquid crystal display (LCD) monitor [1]. Recently, several light-weight and easy-control optical HMDs (OHMDs) have been developed. OHMDs integrate augmented reality (AR) and computing techniques into eyeglasses [2] and have brought about many benefits in a variety of applications, such as surgery [3] and maintenance [4].

OHMDs are effective tools used in many fields and are already influencing user experience. There is therefore an increasing need for an investigation into the ergonomic issues of OHMD setups. Shibata et al. [5] reported that the stereo display devices result in vergence-accommodation conflicts and thus cause visual discomfort.

© Springer Nature Switzerland AG 2020
T. Ahram (Ed.): AHFE 2019, AISC 973, pp. 34–42, 2020.
https://doi.org/10.1007/978-3-030-20476-1_4

Even when HMDs or OHMDs are used in non-stereo display mode, users will experience a phoria on the eye, which results in eyestrain [6]. Moreover, OHMD devices lack screen size adjusting and cannot be used in combination with prescription sizes, which may lead to visual fatigue [7].

Apart from visual fatigue, Stanney et al. [8] reported that sickness symptoms (e.g., nausea or oculomotor and disorientation symptoms) become more severe as the time of using eyewear devices increases. Kennedy et al. [9] indicated that these symptoms occur during and after using some types of display devices. Even in the absence of physical motion, these sickness symptoms may occur due to visually perceived motion and are therefore classified as visually induced motion sickness (VIMS). Moreover, images displayed on OHMDs are fixed to the devices, which will result in the suppression of vestibule-ocular reflexes; the images are also unnatural and can therefore lead to nausea [6, 10].

Previous studies indicated that long-time use of OHMDs leads to visual fatigue as well as visually induced motion sickness. However, studies on suitable rest times for OHMDs that allow users to recover from physical discomfort are still lacking. Hence, this study aims to determine suitable rest times for visual tasks of different lengths by evaluating visual fatigue and visually induced motion sickness in subjects using an OHMD.

2 Method

2.1 Participants

Twenty healthy participants (10 males and 10 females) voluntarily participated in this study. Their age ranged from 20 to 30 years (mean = 21.61, SD = 1.35). The participants all had normal or corrected-to-normal vision (20/20).

2.2 Apparatus

An Epson Moverio BT-200 (Epson Inc., CA, USA) augmented reality OHMD was used as the binocular OHMD for this study. The field of vision was similar to a 40-inch high definition screen viewed from about 2.5 m away. The display projector had a 960×540 resolution. Participants wore the BT-200 during the experimental sessions and could adjust the nose pad for comfort. The critical flicker fusion (CFF) test (instrument model no. 501c, Takei Kiki Kogyo Co., Japan) was used to evaluate visual fatigue.

2.3 Experimental Design

A within-subject factorial design was employed using three factors: gender, visual task condition, and length of rest time. Participants were requested to watch a movie using the BT-200. The experimental procedure comprised two visual task conditions, that is, a 30-min and a 60-min movie session. Each participant completed one condition per day. After the video session, a 15-min rest was provided. The CFF value and simulator sickness questionnaire (SSQ) score were recorded every 5 min (i.e., three times) during the 15-min rest period. Each participant completed each visual task condition three times (the experiment was conducted on 6 different days), in random order.

Independent Variables. Gender, visual task condition, and rest time length were used as independent variables. There were two visual task conditions (30 min and 60 min) and three rest time lengths (5 min, 10 min, and 15 min).

Dependent Variables. Prior to each visual task condition, the CFF value and SSQ score were measured as a baseline. The same measurements were recorded during and after the rest period (see above).

The CFF value was defined as the minimum frequency at which an intermittent light stimulus appears fused to an observer. We used the average value of three flicker-to-fusion trials and three fusion-to-flicker trials [12]. A drop in the CFF value reflects an increase in visual fatigue [13].

Kennedy et al. [14] developed the SSQ, which is composed of three symptom subscale scores (i.e., nausea as well as oculomotor and disorientation symptoms) and the total score. A sixteen-item symptoms checklist on a 4-point scale (0 = not at all, 1 = slight, 2 = moderate, 3 = severe) was used to reflect the discomfort level for each symptom. The total simulator sickness score was calculated by multiplying the respective subscale's coefficient after summing the three subscales scores [14].

2.4 Experimental Procedure

The experimental environment was standardized. The visual acuity check ensured that each participant had normal vision. A 5-min training session was provided to help participants familiarize themselves with the BT-200. After completing the training session, a 5-min rest was provided. Then, the CFF value and SSQ score baseline measurements were taken.

At the beginning of visual task, the movie was displayed on the BT-200 screen. After one visual task condition was executed, the CFF value and SSQ score were measured. Then, a 15-min rest was scheduled, the CFF value and SSQ score for participants were taken every 5-min during a 15-min rest period. The sequence of visual task conditions for each participant was arranged randomly. Each participant executed each visual task condition three times.

2.5 Statistical Analysis

The CFF values and SSQ scores were analyzed using repeated measures analyses of variance (ANOVA), to evaluate how gender, visual task condition, and length of rest time affected the measurements. Fisher's LSD test was conducted as a post hoc test. The adopted statistical significance level was $\alpha = 0.05$. All statistical analyses were performed using SPSS Statistics software version 22.

3 Results

The average CFF values and SSQ scores for the two visual task conditions are shown in Tables 1 and 2. The ANOVA results indicate that gender had no significant effect on CFF values ($[F\ 1, 9] = 1.19$, $p = 0.31$) or SSQ scores ($[F\ 1, 9] = 0.17$, $p = 0.69$). Visual task condition had a significant effect on CFF values ($[F\ 1, 9] = 6.43$, $p = 0.03$) and SSQ scores ($[F\ 1, 9] = 28.62$, $p < 0.01$). Rest time duration had a significant effect on CFF values ($[F\ 1.99, 17.88] = 26.59$, $p < 0.01$) and SSQ scores ($[F\ 1.81, 16.31] = 52.24$, $p < 0.01$).

Table 1. Means (SDs) and post hoc test results for CFF values for each visual task condition at different measurement times

CFF value (Hz)	Baseline	Post-experiment	Rest 5-min	Rest 10-min	Rest 15-min
	Mean (SD)	Mean (SD)	Mean (SD)	Mean (SD)	Mean (SD)
30-min	35.08 (2.44)[a]	33.86 (2.45)[b]	34.33 (2.66)[c]	34.74 (2.81)[a]	34.85 (2.74)[a]
60-min	35.38 (2.58)[a]	31.76 (2.30)[b]	32.73 (2.56)[c]	33.01 (2.64)[c]	33.21 (2.58)[c, d]

Note: [a, b,] and [c] means, followed by the same letter, are not significantly different, as determined by ANOVA followed by Fisher's LSD test at the significant level $\alpha = 0.05$.

Table 2. Means and post hoc test results for SSQ scores for each visual task condition at different measurement times

SSQ score	Baseline	Post-experiment	Rest 5-min	Rest 10-min	Rest 15-min
	Mean (SD)	Mean (SD)	Mean (SD)	Mean (SD)	Mean (SD)
N (30-min)	0.95 (2.86)[a]	5.72 (6.98)[b]	1.91 (3.39)[a]	1.43 (3.62)[a]	1.43 (3.34)[a]
O (30-min)	3.41 (4.40)[a]	21.22 (14.20)[b]	9.48 (9.44)[c]	4.93 (4.17)[a]	4.20 (4.40)[a]
D (30-min)	2.78 (5.57)[a]	19.49 (23.44)[b]	4.18 (7.07)[c]	3.86 (8.66)[a]	2.78 (8.63)[a]
T (30-min)	3.53 (3.79)[a]	17.07 (13.38)[b]	6.92 (4.14)[c]	4.13 (3.41)[a]	3.55 (2.70)[a]
N (60-min)	1.43 (3.34)[a]	14.79 (7.24)[b]	10.02 (6.87)[c]	7.63 (6.54)[d]	7.16 (5.40)[d]
O (60-min)	2.99 (3.50)[a]	32.59 (14.08)[b]	22.74 (9.78)[c]	17.81 (10.71)[d]	14.02 (9.93)[d]
D (60-min)	2.78 (5.57)[a]	29.23 (19.14)[b]	18.10 (17.96)[c]	13.92 (12.64)[d]	6.24 (5.50)[e]
T (60-min)	3.93 (4.16)[a]	30.11 (12.69)[b]	21.69 (12.12)[c]	17.20 (11.11)[d]	14.96 (9.33)[d]

Note: (1) [a, b,] and [c] means, followed by the same letter, are not significantly different, as determined by ANOVA followed by Fisher's LSD tests at the significant level $\alpha = 0.05$. (2) N, O, D, and T indicate nausea (N), oculomotor symptoms (O), disorientation (D), and total (T).

An interaction effect of visual task condition and rest time length was found for CFF values ([F 1.60, 14.37] = 6.07, p = 0.02), as shown in Fig. 1. Generally, the 60-min visual task condition produced significantly lower CFF values than the 30-min visual task condition. The CFF value was the lowest at the end of the experimental session, compared to all rest-time measurements. The data trend over time showed that CFF values increased with increased rest time in both visual task conditions. CFF values measured after the 30-min visual task followed by a 10-min rest were not significantly different from baseline values. However, CFF values measured after the 60-min visual task followed by a 15-min rest were significantly different from baseline CFF values.

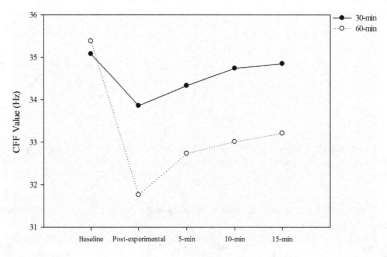

Fig. 1. Interactions between visual task conditions and length of rest time for CFF values

A visual task condition × rest time length interaction effect was found for SSQ scores ([F 4, 36] = 5.98, p < 0.01), as shown in Fig. 2a–d. The 60-min visual task condition produced significantly greater SSQ scores than the 30-min visual task condition. SSQ scores taken after the 30-min visual task followed by a 10-min rest were not significantly different from baseline. However, SSQ scores taken after the 60-min visual task followed by a 15-min rest were significantly different from baseline scores.

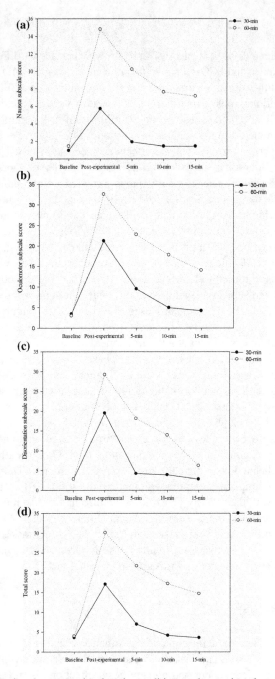

Fig. 2. (a). Interaction between visual task condition and rest time length for SSQ-Nausea subscale scores (b). Interaction between visual task condition and rest time length for SSQ-Oculomotor subscale scores (c). Interaction between visual task condition and rest time length for SSQ-Disorientation subscale scores. (d). Interaction between visual task condition and rest time length for SSQ-Total scores.

4   Discussion

This study investigated the suitable rest time for visual tasks of different lengths by evaluating CFF values and SSQ scores while subjects used an OHMD. Both the visual task condition and the rest time duration had a significant effect on CFF values. In both visual task conditions, post-experiment CFF values were significantly decreased compared to baseline: The decrease in the 30-min visual task condition was 1.22 Hz and the decrease in the 60-min visual task condition was 3.62 Hz. Several studies have indicated that a decrease in CFF values reflects an increase in visual fatigue [13, 15]. Moreover, visual fatigue has been defined as a drop of 1 Hz or more in CFF values after a visual task [16]. Hence, the participants experienced a significant increase in visual fatigue while using binocular OHMDs in both visual task conditions; the 60-min condition led to a higher level of visual fatigue than the 30-min condition.

The data trend over time shows that the level of visual fatigue decreased with increased rest time in both visual task conditions, as indicated in Fig. 1: After the 30-min visual task and a 10-min rest, the subjects' level of visual fatigue had significantly dropped and almost returned to baseline levels. However, after the 60-min visual task and a 15-min rest, the subjects' level of visual fatigue was significantly above baseline level; this shows that when participants used OHMDs in the longer visual task condition, they needed more rest time to recover from the task-induced visual fatigue.

Both the visual task condition and the rest time duration had a significant effect on SSQ scores. In both conditions, the SSQ score peaked at the end of the experimental session. Oculomotor symptoms scores were the highest, and nausea symptoms scores were the lowest overall, as shown in Fig. 2a–b. The results of this study are similar with to a report by Häkkinen et al. [17] that oculomotor symptoms were the most severe symptoms in each visual task condition. Kennedy et al.'s (1994) study also indicated that using eyewear devices for a long time leads to oculomotor symptoms.

In both visual task conditions, the subscale and total SSQ scores after the 5-min rest period were significant lower than the post-experiment scores. This shows that the severity of VIMS symptoms decreased with increased rest time. After the 30-min visual task and a 10-min rest, the severity of VIMS symptoms had almost returned to baseline levels, while, in contrast, there were significant differences in SSQ scores between baseline and the end of the 15-min rest period after the 60-min task condition. It thus seems that a 15-min rest is necessary to allow task-induced VIMS symptoms to return to baseline levels after subjects perform a 60-min visual task.

5   Conclusions

This study aimed to determine the suitable rest time for visual tasks of different lengths by evaluating visual fatigue and visually induced motion sickness while subjects were using an OHMD. Subject gender had no significant effect on visual fatigue and VIMS symptoms, while visual task condition and rest time duration showed significant effects on visual fatigue and VIMS symptoms: Watching a movie using the OHMD in a 60-min visual task resulted in increased visual fatigue and greater VIMS symptoms in our participants, reflected by significant changes in CFF values and SSQ scores.

The current results also show that levels of visual fatigue and VIMS symptoms decreased with increased rest time. After the 30-min visual task and a 10-min rest, levels of visual fatigue and VIMS symptoms returned to baseline levels. However, after the 60-min visual task and a 15-min rest, the significant differences in visual fatigue and VIMS symptoms between baseline and the end of the 15-min rest period remained. It seems that providing a 15-min rest might not be sufficient for participants to recover their visual fatigue and VIMS symptoms after completing a 60-min visual task. Considering the results of this study, when people use OHMDs to watch movies, a rest period of at least 10 min needs to follow a 30-min movie session, and a rest period of more than 15 min should follow a 60-min movie session. The results of this study provide useful information for the use of OHMDs with regards to suitable resting times.

References

1. Hsiao, C.Y., Wang, M.J., Lu, Y.T., Chang, C.C.: Usability evaluation of monocular optical head-mounted displays on reading tasks. J. Ambient Intell. Humaniz. Comput., 1–10 (2018)
2. Hua, H., Javidi, B.: A 3D integral imaging optical see-through head-mounted display. Opt. Express **22**(11), 13484–13491 (2014)
3. Wu, T.S., Dameff, C.J., Tully, J.L.: Ultrasound-guided central venous access using Google Glass. J. Emerg. Med. **47**(6), 668–675 (2014)
4. Robertson, T., Bischof, J., Geyman, M., Lise, E.: Reducing maintenance error with wearable technology. In: 2018 Annual Reliability and Maintainability Symposium (RAMS), pp. 1–6. IEEE Press (2018)
5. Shibata, T., Kim, J., Hoffman, D. M., Banks, M. S.: Visual discomfort with stereo displays: effects of viewing distance and direction of vergence-accommodation conflict. In: Woods, A. J., Holliman, N.S., Dodgson, N.A. (Eds.), Proceedings of SPIE: Stereoscopic Displays and Applications XXII, vol. 7863, pp. 1–9 (2011)
6. Ukai, K., Howarth, P.A.: Visual fatigue caused by viewing stereoscopic motion images: background, theories, and observations. Displays **29**(2), 106–116 (2008)
7. Du, X., Arya, A.: Design and evaluation of a learning assistant system with optical head-mounted display (OHMD). In: Learning and Collaboration Technologies, pp. 75–86 (2015)
8. Stanney, K.M., Hale, K.S., Nahmens, I., Kennedy, R.S.: What to expect from immersive virtual environment exposure: influences of gender, body mass index, and past experience. Hum. Factors **45**(3), 504–520 (2003)
9. Kennedy, R.S., Drexler, J., Kennedy, R.C.: Research in visually induced motion sickness. Appl. Ergon. **41**(4), 494–503 (2010)
10. Ukai, K., Kibe, A.: Counterroll torsional eye movement in users of head-mounted displays. Displays **24**(2), 59–63 (2003)
11. Moss, J.D., Muth, E.R.: Characteristics of head-mounted displays and their effects on simulator sickness. Hum. Factors: J. Hum. Factors Ergon. Soc. **53**(3), 308–319 (2011)
12. Chi, C.F., Lin, F.T.: A comparison of seven visual fatigue assessment techniques in three data-acquisition VDT tasks. Hum. Factors **40**(4), 577–590 (1998)
13. Kang, Y.Y., Wang, M.J.J., Lin, R.: Usability evaluation of e-books. Displays **30**(2), 49–52 (2009)

14. Kennedy, R.S., Lane, N.E., Berbaum, K.S., Lilienthal, M.G.: Simulator sickness questionnaire: an enhanced method for quantifying simulator sickness. Int. J. Aviat. Psychol. 3(3), 203–220 (1993)
15. Gunnarsson, E., Söderberg, I.: Eye strain resulting from VDT work at the Swedish telecommunications administration. Appl. Ergon. 14(1), 61–69 (1983)
16. Wu, H.C.: Visual fatigue and performances for the 40-min. mixed visual work with a projected screen. Ergon. Open J. 5, 10–18 (2012)
17. Häkkinen, J., Vuori, T., Paakka, M.: Postural stability and sickness symptoms after HMD use. In: IEEE International Conference on Systems, Man and Cybernetics, pp. 147–152 (2002)

Development of Ergonomic Swimming Goggles Based on Benchmarking and 3D Facial Scan

Jangwoon Park[✉], Waranya Khamnasak, Jared Crawford,
Justin Bonds, Elsy Hernandez, and Mehrube Mehrubeoglu

Department of Engineering, Texas A&M University-Corpus Christi,
Corpus Christi, TX, USA
jangwoon.park@tamucc.edu

Abstract. Swimming goggles are important tools for professional swimmers, as well as the general public, to have clear visibility, fit, and wearing comfort under water; however, commercialized swimming goggles have been designed as one-size-fits-all, which may cause several issues to the wearers who have different facial dimensions; Typical issues include improper fit which causes water leakage, and sometimes exceeded pressure that might occur on the eyes. The present study is intended to design and validate ergonomic swimming goggles that are specially designed toward a wearer's 3D facial shape. Before designing novel swimming goggles, 26 existing swimming goggles were investigated in terms of the design features of their lens, strap, and gasket. In addition, dimensions of lens, nose bridge, frame, and gasket were measured by using a caliper. To design the shape of a gasket, one male's 3D facial scan data was constructed by using multiple mobile phone 2D images. The shape of the gasket is being designed based on the facial structure of the subject using computer aided design (CAD). Detailed design and validation results will be presented at the conference.

Keywords: Swimming goggles · Ergonomics in design ·
2D & 3D facial scan · Prototyping

1 Introduction

Swimming goggles are important equipment for professional swimmers and all swimmers in general. Swimming goggles are designed to provide clear visibility under water and protect swimmers' eyes from any harmful irritants; however, the existing typical swimming goggles are designed as one-size-fits-all, which does not allow a swimmer to choose proper size and shape of goggles for his/her facial structure.

A few studies have been conducted to design customized swimming goggles based on a swimmer's 3D facial scan data to improve fit. Coleman et al. [2] introduced a design protocol for customized swimming goggles by using a wearer's 3D facial scan data. The group validated the protocol by creating a prototype of the customized swimming goggles (Fig. 1a) and testing water leakage and wearing comfort under water (Fig. 1b). Park et al. [1] conducted a similar study to Coleman et al., but the former group used a typical camera instead of an expensive 3D scanner, and acquired

© Springer Nature Switzerland AG 2020
T. Ahram (Ed.): AHFE 2019, AISC 973, pp. 43–48, 2020.
https://doi.org/10.1007/978-3-030-20476-1_5

similar quality 3D facial data. The group's described reconstruction process requires several selfie images. The quality of a reconstructed data is good enough to design customized swimming goggles using images from a camera (Fig. 2).

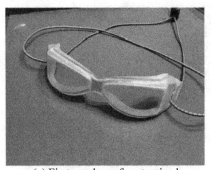
(a) First mockup of customized swimming goggles

(b) Validation of customized swimming goggles at a swimming pool

Fig. 1. Reconstructed 3D facial data based on multiple 2D images at different angles [2].

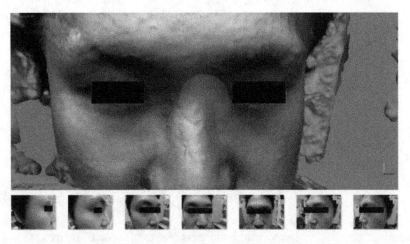

Fig. 2. Reconstructed 3D facial data based on multiple 2D images at different angles [1]

Although previous studies have provided potential progressive ergonomic advantages and cost-effective production process for customized swimming goggles, research in extensively understanding design characteristics of commercialized swimming goggles in terms of shape and material of lens and strap is still limited. In addition, there is lack of research studies to quantitatively evaluate the effects of lens shape on hydrodynamics under water as well as quantitative measurements of wearing pressure of swimming goggles on eyes.

The objectives of this study include (1) conducting extensive reviews on the existing swimming goggles, (2) identifying effects of shape of lens on hydrodynamics using a quantitative measure, and (3) measuring and comparing wearing pressure of customized swimming goggles on the user using a pressure film.

2 Reviews of Existing Swimming Goggles

In this step, 26 commercialized swimming goggles were analyzed in terms of their design features and dimensions. Figure 3 shows the summary results of overall design features. Out of the 26 swimming goggles analyzed, 21 goggles (81%) have an anti-fog feature and 18 goggles (69%) have ultraviolet (UV) protection. On the other hand, only a few of the goggles have the anti-scratch (3 out of 26, 12%), hypo-allergenic (4 out of 26, 15%), and hydrodynamic design features (8 out of 26, 31%). Specifically, 75% of those eight hydrodynamic goggles have flat lens, 50% have symmetric lens, and 75% have double head straps.

Figure 4 shows the review results regarding design characteristics and materials of a head strap. Based on the review on the 26 existing goggles, all the head straps were designed to be adjustable, 69% of the goggles used double straps, 31% of the straps were made of a silicone. Silicone are being used to replace a rubber which contains a latex that might cause an allergy to skin [3].

Figure 5 shows the review results regarding lens of the existing swimming goggles. 73% of the goggle lens were designed as an asymmetric shape and 62% of the lens were designed as flat, 38% of the lens were designed as curved shape.

Lastly, Fig. 6 shows the design characteristics of nose-bridges. 50% of the goggles have an exchangeable nose-bridge. Most of the exchangeable nose-bridges (12 out of 13) were made of plastic that do not allow any adjusting while the only once sample is made of a rubber which can provide better flexibility toward a user's preference.

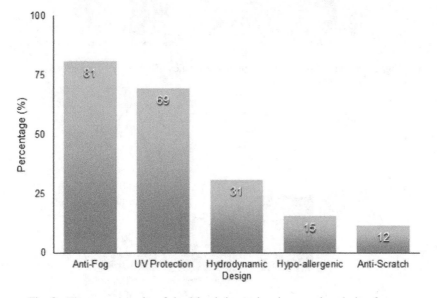

Fig. 3. The survey results of the 26 existing swimming goggles: design features

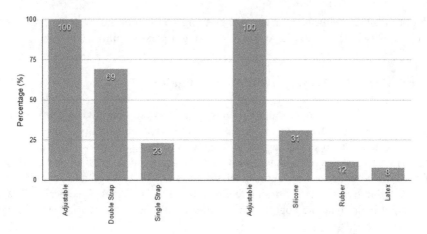

Fig. 4. Review results: design characteristics of head strap

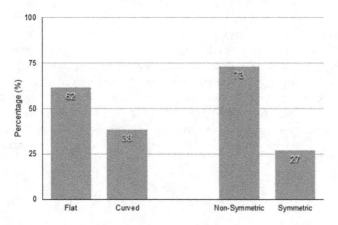

Fig. 5. Review results: the design characteristics of the existing goggles lens

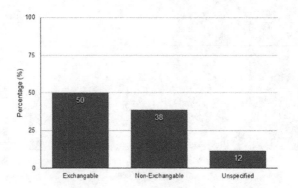

Fig. 6. Review results: design characteristics of nose-bridges

The dimensions (length, width, and depth) of the five existing swimming goggles were measured for each component (nose-bridge, frame, and gasket; see Fig. 7 and Table 1) and summarized (see Table 1). The vertical average width of the lens measured from top to bottom in y-direction was 31.6 mm. The horizontal average length measured from left to right of the lens in x-direction was 50.3 mm. Similarly, to the dimension of frame, the average width was 143.2 mm, and the average length of the frame was 40.1 mm. The nose-bridge was measured for its length and width as shown in Fig. 7. The average of the nose-bridge was 26.5 mm for length, and 7.9 mm for width. The gasket was measured for its depth as shown in Fig. 7 and its average value was found as 10.4 mm. The dimensional information was used to design new swimming goggles in CAD software in this project.

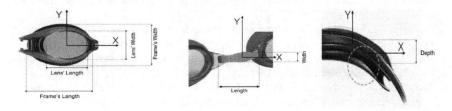

Fig. 7. Dimensions of lens (left), nose-bridge (center), and gasket (right)

Table 1. Measurements of the six existing swimming goggles (unit: mm)

Existing swimming goggles (model names)	Lens		Nose bridge		Frame		Gasket
	W	L	W	L	W	L	D
Speedo	27.0	40.1	26.0	7.9	135.6	38.7	5.8
Zionor G5	35.8	56.8	25.0	7.0	147.2	43.0	13.8
Outdoor Master OM-2	36.4	57.5	28.0	6.8	137.7	43.0	13.8
AEGENO	27.5	53.1	31.6	12.2	161.0	37.2	9.0
ProSWIMS	31.4	43.8	22.0	5.7	134.3	38.6	7.8
Average	31.6	50.3	26.5	7.9	143.2	40.1	10.0
Standard deviation	4.4	7.9	3.6	2.5	11.2	2.7	3.6

Note: W = width, L = length, D = depth.

3   Conclusions and Future Work

In the present study, extensive reviews were conducted on the existing swimming goggles in terms of design, materials, and dimensions. Significant variations among different models were identified. The identified design features of the existing swimming goggles will be useful when designing customized swimming goggles and optimizing parameters for usability and minimal eye pressure. Experiments involving human subjects will be conducted to evaluate the customized swimming goggles in terms of water leakage, wearing pressure, and wearing comfort. The research team is preparing an experimental design for approval from institutional review board (IRB) before experiments involving human subjects can commence.

Acknowledgments. This project was supported by a Research Enhancement Grant from Division of Research, Commercialization and Outreach at Texas A&M University-Corpus Christi. Prototyping was supported by the Department of Engineering Capstone Projects funds.

References

1. Park, J., Mehrubeoglu, M., Baca, J., Rylance, K., Almoosa, H., Salazar, G., Franks, C., Falahati, S.: Development of a design protocol for customized swimming goggles using 2D facial image data. In: International Conference on Applied Human Factors and Ergonomics, pp. 151–155. Springer, Cham (2018)
2. Coleman, J., Hernandez, C., Hernandez, J., Hubenak, S., McBride, A., Mehrubeoglu, M., Park, J.: Development of a design protocol for customized swimming goggles using 3D facial scan data. In: International Conference on Applied Human Factors and Ergonomics, pp. 1017–1021. Springer, Cham. Los Angeles, USA (2017)
3. Tomazic, V.J., Withrow, T.J., Fisher, B.R., Dillard, S.F.: Latex-associated allergies and anaphylactic reactions. Clin. Immunol. Immunopathol. **64**(2), 89–97 (1992)

Advancement in Orthopedics Through an Accessible Wearable Device

Troy Kelly Jr.[1]([⊠]), Behzad Mottahed[2], and Ryan Integlia[1]

[1] Florida Polytechnic University, Lakeland, FL 33805, USA
tkelly5126@floridapoly.edu, integlia@ieee.org
[2] Stevens Institute of Technology, Hoboken, NJ 07030, USA

Abstract. Wearable technology holds great potential in transforming health-care. This technology already has applications in medical education and in enhanced accessibility in healthcare. Wearable technology allows medical professionals to access data more efficiently while monitoring patient behavior, thus improving patient care as a whole. The implementation of wearable technology will increase the accuracy of the diagnostic process and decrease cases of medical malpractice, due to improved accuracy. Proposed is a wearable device concept, utilizing wrist-band technology to assist in the development and maintenance of orthopedic health, while avoiding other ailments through enhanced training.

Keywords: Healthcare · Migraines · Orthopedics · Training · Wearable technology

1 Introduction

In this paper, the positive impact of "autonomous wearable" technology is explored. An "autonomous wearable" is, in part, a wearable computer. The first of which was made in the 1960s by mathematics professor Edward Thorp [1]. Thorp created a timing device that helped to predict where a ball would land on a roulette table. Since that time, there have been several devices that have modernized the scope of wearable technology. In the 1980s, the wearable technology improved considerably with the release of the first ever set of digital hearing aids [1]. From the years of 2006 and 2013, newer fitness wearable technology devices were released, including Fitbit, Google Glass, and Nike+.

The world of wearable technology has unlimited opportunities. Fitness and healthcare are two separate applications in wearable technology, which may be combined as a single wearable device. The prospective work of a wearable device in the form of a wrist band is proposed to assist in the improvement and maintenance of orthopedic health, which would potentially assist in avoidance of other ailments through training and acclimation.

T. Ahram (Ed.): AHFE 2019, AISC 973, pp. 49–56, 2020.
https://doi.org/10.1007/978-3-030-20476-1_6

2 Contemporary Topic and Importance

In terms of the proposed autonomous wristband, it will have body sensors in it that would be used to monitor diet and to assist with strength performance in training. The prospective wristband is an energy wristband, or "NRG", and has many key functions. One of which would be the way it would be used to monitor performance in strength training and weight-bearing exercises. In this case, the NRG would be connected to a computer using a micro-USB or Bluetooth, which would allow users to input their age, gender, and pre-existing health conditions. This information would allow for the device to administer a weight-bearing routine for the user. The device would be specific to each individual in recording their data, as a secure profile. Some studies have highlighted that children, including those that have type 1 diabetes have found that weight-bearing activities and high impact exercise increase the amount of bone created during the years of peak bone growth [2, 3]. Strength training and weight-bearing exercises have also been reported as beneficial in preventing bone loss in older adults [4]. Strength training exercise is not only beneficial for increasing muscle mass, but may also help protect against bone loss in younger and older women, including those with osteoporosis, osteopenia or breast cancer [4].

The NRG wristband would play a role in individual's dietary habits by advising individuals caloric intake based their profile. The profile would be an implementation of machine learning tools, practitioner training and set of sensors. Some of the sensors include pressure, temperature, impedance sensor impacted by the skin of the user and an accelerometer. The calculation of caloric intake, in this context, may be useful for improvement of orthopedic health. Guidance could be provided based on calculations of the user's energy levels, and how much energy a given user would need before participating in a fitness activity. The NRG wristband first would collect the data via the sensors based on the individual's performance in training and their dietary habits. This collected data would be securely transmitted back to a registered smartphone through Bluetooth or a virtual private network application login, which would inform individuals on their glucose levels - whether too high, too low, or if it is normal for their profile. A minimum safe range caloric intake would be established. A caloric minimum is significant, given that studies have shown that diets providing fewer than 1,000 calories per day can lead to a lower bone density in normal-weight, and overweight individuals [5–7].

2.1 Migraine Alleviation Through Training

The NRG will also be equipped with a setting for individuals that suffer from migraines or severe headaches. As studies have shown that there is an increased migraine risk in osteoporosis patients [8]. Osteoporosis is a disease that thins and weakens the bones, causing them to become fragile and to break easily, especially the bones in the hip, spine, and wrist [9]. Unless the individual has a rare condition, exercise would act as a potential aid to manage headaches. It is reported that exercise may cause migraines and headaches, but it is understood that it is not the physical activity itself that causes headaches [10]. Headaches and migraines that arise during exercise are more likely due to deficient energy levels, such as exercising on an empty stomach or not remaining hydrated while exercising [10].

3 Data Collection and Accessibility

There is a significant digital divide between the ages of 16 to 65, most notably in the United States of America [11]. The integration of the proposed NRG wristband has the potential to bridge the digital divide within healthcare by addressing the health literacy gap. Based on a recent study conducted in 2016, it has been found that low health literacy is associated with lower usage rates for consumer health technology, such as devices that can be compared to the NRG [12].

During the set-up period of the device to a computer, and related technical resources, an assessment test will be provided to users, to assess their literacy of wearable technology. Based on user information put into the device upon set up of the device, and the assessment, the NRG will send messages throughout the day per the user inputting at least 5 times to receive messages, one early in the morning, one in the afternoon, one in the evening and two at random times throughout the day. The device would vibrate the users arm, through the use of haptic technology, reminding them of activities they may need to fulfill for the day or helpful information about any pre-existing conditions they may have or any goals the individual may want to reach. All medical information provided securely to users in these messaged, would be facilitated by a healthcare provider, as it would be valuable for users to have the most accurate and timely information possible to aid in developing a sense of trust from users. It is understood that trust is an important detail with this device as trust and privacy concerns are major, especially with individual's health information. Studies show that 46 percent of adults are distrustful of information they put into a computerized system, due to the worry of individuals accessing unauthorized information [13]. This will not be a problem for the NRG as the system will utilize a virtual private network (VPN) from the supplier's end, and also an Advance Encryption Standard (AES). Encryption would be sought out to be used with modifications for tighter security [14].

One of the unique features of the NRG wristband would be its accessibility. It is important to remember the feature of accessibility, because most devices are not composed of features that are equipped with strong accessibility features for users nor features that would consider health complications and disability. There will be a feature in the NRG wristband entitled, "Personal Assistant". This would be a capability functioning primarily through a voice interface with support from haptics, sensors, camera and related components in the wristband. A feature of this system, would allow individuals to have a growing set of response to possible questions they may ask or any problems they may have encountered through continuous training of the personal assistant. In situations where individuals may be unable to feel or touch while being unable to speak, the NRG will be programmed with an artificial intelligence based algorithm that counts eye blinks, as an extension of the personal assistant. With this, individuals with amputated limbs where they may not have a wrist on one hand may still have accessibility using the NRG's blink algorithm as commands. For instance, a possibility could be a user blinking one time for yes, two times for no and the user would also be prompted with an "are you sure?" prompt for cases of miscommunication.

The information received from patients using the NRG will enable anonymized analysis and future diagnosis for patients facilitated by the practitioner. One of the outstanding features of the NRG wristband would be that it prompts users after exercises of any kind. Asking users how their workout was and if they experienced any pain through the process. This information would be researched to gain a better understanding of why the individual would have experienced this pain along with giving patients alternatives to allow them to carry on with the strengthening of their bones. Wearable technology keeps patients engaged, encourages proactive health care, benefits healthcare providers and employers, and monitors vulnerable patients [15]. The purpose of this wristband would be to assist individuals, most notably patients, in increasing their bone strength through strong discipline practices comprised of strength training and dietary wellness. With this being acknowledged it is understood that feedback and analysis will be pivotal in this process.

4   Current Development

The NRG wristband will consist of a pressure sensor, impedance sensor and accelerometer; along with a camera, speaker and microphone. Sensors within the NRG wristband would be used to collect raw data from measurements, which would be a resource for assessing performance, exercise activity, and most importantly monitoring the health of users and enable ongoing practitioner training. The pressure sensor would be able to measure an individual's blood pulse rate in real time. This information would be useful, allowing the NRG wristband to prompt users if they should slow down or increase their performance in their workout. The impedance sensor would be used in measuring the amount of electrical resistance within the skin. Sweat produced through exercising and exposed to the impedance sensor will allow additional information to be relayed to the software on user activity. The accelerometer will be measuring user movement in a complimentary manner to the other sensors. These three sensors will work together, recommending levels of exertions in the exercises being conducted, and the measurements being taken based on practitioner guidance. These sensors will also work to assist in measuring caloric intake through the skin and related behavior [16]. The NRG wristband will also be equipped with a speaker, microphone, and camera. The speaker and microphone will allow interaction between user and device through speech, in cases for preference or disability. This may also allow users to communicate with practitioners through the NRG wristband. The camera within the device allows for larger number of applications to be possible and enable service to a broader range of disabilities. This gives users, in this case, the option of "blink responses" as the development of this wearable device progresses.

4.1   Figures

In the image below, to the left is the microphone and speakerphone in one, with the display screen to the right of that reading "NRG". To the very right, the camera can be seen (Fig. 1).

Fig. 1. Front View of NRG wristband

Figure 2 depicts the external view of NRG wristband prototype, where the pressure sensor, impedance sensor, and accelerometer can be seen. The accelerometer can be seen to the left and the impedance and pressure sensor can be seen to the right.

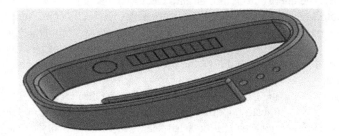

Fig. 2. Rear View of NRG wristband

4.2   Future Development

Future studies in the field of ergonomics will be crucial in the progression of wearable technology. As for future development of the "NRG", one consideration would be color preference of the device. As it is a known fact that color influences the reaction of users, the color blue is a popular color among hospitals and healthcare facilities as it is recognized as a cool calming color. Further studies may suggest otherwise, with another color being more suitable for the integration of the "NRG" wristband and an ability to change the color of the exterior through selectable LED illumination color might be more suitable, based on the detected state of the user. Future ergonomic studies can also influence the idea of future features of the device, in terms of user usability, and managing a wide range of usability across demographics. For instance, the size of buttons or the size of the screen of the device could be enlarged to serve a wider range of accessibility needs.

5   Practitioner Use and Acclimation

It is clear that the integration of this new technology may need additional work and support for full adoption. It may also be expected that health practitioners who adopt these types of products may need to teach their patients how to use it which can lead to a process flow for acclimation to be put in place. There would be a group of modules available for first time users to provide awareness on the wearable technology device and guide them on how to use it, this will allow users to progress from acclimation to more advanced uses. There would be a procedure for training with the device allowing users to understand what the device is doing while they are exercising and training, how to input data, and accessing their data for their personal analysis of their performance. There will be a module entailing how the product works and how each sensor is important, while emphasizing that the device should not be tampered with. The inputting information feature will be important for users to understand as it would be only through this, that the device can be made better. Analysis from this device can lead to a new future device catering to needs noted. It is important for the NRG wristband to keep in mind human user factors. Thus, the idea of all the sensors of the device being made very small allows the wristband to be very light in weight for comfortability. The NRG wristband will be made of silicone and energy harness smart fabrics, which would allow the device to harvest its energy from human motion. It would convert the energy from motion into electricity using fiber-based solar cells and nano-generators, then it's stored as chemical energy in fiber-shaped super capacitors. This textile would be woven into fabrics that will be in the design of the NRG. With the proprietorship of an NRG wristband, users would be informed of what is needed for effective use of the product, and how the product is both comfortable and beneficial at the same time.

The results from the NRG wristband would assist healthcare practitioners as a gateway for more healthcare opportunities being available. Results and feedback from the device would also give insight on what seems to be a necessity of the device and what could be added to a future device. The results would also give feedback for the implementation of research to be conducted for future studies of bone development and sustainability. Upon development of the NRG wristband, physicians can associate a physician profile with the NRG online web where they would be able to access approved data once given consent by their patients. One study indicated over 50% of physicians that gave their input in a survey taken have stated that online visits can replace over 10% of all office visits because it allows physicians to communicate with their patients easier especially with problems that may be very minimal and may not require physician-patient face to face interaction [17]. It is also important to note that more than 60% of physicians spend their online time seeking information on diagnoses, diseases, pathogens, treatment methods, and other topics [17]. This data may indicate industry awareness to the trend as it is obtained from CLODOC, which is an example of a company claiming to use technology to streamline healthcare processes [17]. The integration of the NRG would play a great part in assisting practitioners in understanding new and emerging treatment methods. In cases where certain methods are being sought out to increase bone growth, the feedback from the patient to the NRG wristband would be securely sent to servers to ensure that the data is being analyzed and that alternative options are being administered to patients until further tests have been conducted.

6 Conclusion

Wearable technology holds great potential to transform the healthcare industry through exercise and training. The integration of the NRG wristband with personalized training resources and wireless networks may be useful in enhancing the adoption of wearable technology. The proposed device is presented as a tool to assist users through a regimented exercise routine that is influenced by the individual's biometrics and progression under continual practitioner guidance. The NRG wristband will also contain machine learning based features used to identify contrary performance and prompt the use to correct their performance. The NRG does not only show potential value for individuals trying to develop and sustain strong bones, but also highlights significance for practitioners in collecting data for their patients and enhancing the potential value through training. This implementation may bridge the digital divide between technology and healthcare by combining consumer electronic wearable applications of activity tracking with healthcare applications related to injury reduction and fitness. The implementation of this device, with the related modules, would be a great tool for users and healthcare practitioners about new emerging technology.

References

1. "The Past, Present and Future of Wearable Technology." Grace College, 17 November 2016. https://online.grace.edu/news/business/the-past-present-future-of-wearable-technology/. Accessed 21 Jan 2019
2. Meiring, R.M., Micklesfield, L.K., Avidon, I., McVeigh, J.A.: Osteogenic effects of a physical activity intervention in South African black children. Curr. Neurol. Neurosci. Rep. September 2014. https://www.ncbi.nlm.nih.gov/pubmed/25198222. Accessed 21 Jan 2019
3. Maggio, A.B., Rizzoli, R.R., Marchand, L.M., Ferrari, S., Beghetti, M., Farpour-Lambert, N.J.: Physical activity increases bone mineral density in children with type 1 diabetes. Curr. Neurol. Neurosci. Rep. (2012). https://www.ncbi.nlm.nih.gov/pubmed/22246217. Accessed 21 Jan 2019
4. Metcalfe, L., Lohman, T., Going, S., Houtkooper, L., Ferriera, D., Flint-Wagner, H., Guido, T., Martin, J., Wright, J., Cussler, E.: Post-menopausal women and exercise for prevention of osteoporosis, May/June 2001. https://cals.arizona.edu/cpan/files/Metcalfe%20ACSM%20final%20article.pdf. Accessed 15 Feb 2019
5. Villareal, D.T., Fontana, L., Das, S.K., Redman, L., Smith, S.R., Saltzman, E., Bales, C., Rochon, J., Pieper, C., Huang, M., Lewis, M., Schwartz, A.V., Calerie, G.R.: Effect of two-year caloric restriction on bone metabolism and bone mineral density in non-obese younger adults: a randomized clinical trial. Curr. Neurol. Neurosci. Rep. (2016). https://www.ncbi.nlm.nih.gov/pubmed/26332798. Accessed 21 Jan 2019

6. Redman, L.M., Rood, J., Anton, S.D., Champagne, C., Smith, S.R., Ravussin, E., Pennington, T.E.: Calorie restriction and bone health in young, overweight individuals. Curr. Neurol. Neurosci. Rep. **22** (2008). https://www.ncbi.nlm.nih.gov/pubmed/18809812. Accessed 21 Jan 2019

7. Compston, J.E., Laskey, M.A., Croucher, P.I., Coxon, A., Kreitzman, S.: Effect of diet-induced weight loss on total body bone mass. Curr. Neurol. Neurosci. Rep. (1992). https://www.ncbi.nlm.nih.gov/pubmed/1315653. Accessed 21 Jan 2019

8. Wu, C.-H., Zhang, Z.-H., Wu, M.-K., Wang, C.-H., Lu, Y.-Y., Lin, C.-L.: Increased migraine risk in osteoporosis patients: a nationwide population-based study, 22 August 2016. https://www.ncbi.nlm.nih.gov/pmc/articles/PMC4993742/#. Accessed 21 Jan 2019

9. "Osteoporosis." MedlinePlus, 19 December 2018. https://medlineplus.gov/osteoporosis.html . Accessed 21 Jan 2019

10. "Exercise Can Help You Beat Headaches." Stroke Center - EverydayHealth.com. 20 June 2014. https://www.everydayhealth.com/headache-and-migraine/exercise-to-beat-headaches.aspx. Accessed 21 Jan 2019

11. "Healthcare's opportunity to bridge the digital divide – and boost health literacy." Chilmark Research, 15 January 2018. https://www.chilmarkresearch.com/healthcares-opportunity-bridge-digital-divide-boost-health-literacy/. Accessed 10 Feb 2019

12. Health literacy and health information technology adoption: the potential for a new digital divide. J. Med. Internet Res. https://www.jmir.org/2016/10/e264. Accessed 10 Feb 2019

13. Brodie, M., Flournoy, R.E., Altman, D.E., Blendon, R.J., Benson, J.M., Rosenbaum, M.D.: Health information, the Internet, and the digital divide. https://www.healthaffairs.org/doi/pdf/10.1377/hlthaff.19.6.255.Accessed 9 Feb 2019

14. Singh, G., Supriya, S.: A study of encryption algorithms (RSA, DES, 3DES and AES) for information security. Int. J. Comput. Appl. **67**(19), 33–38 (2013). https://doi.org/10.5120/11507-7224

15. Haghi, M., Thurow, K., Stoll, R.: Wearable devices in medical Internet of things: scientific research and commercially available devices. Healthc. Inf. Res. **23**(1), 4 (2017). https://doi.org/10.4258/hir.2017.23.1.4

16. Patel, S., Park, H., Bonato, P., Chan, L., Rodgers, M.: A review of wearable sensors and systems with application in rehabilitation. J. NeuroEng. Rehabil. **9**(1), 21 (2012). https://doi.org/10.1186/1743-0003-9-21

17. CLODOC: Uses of Wearable Technology in Healthcare ~ CLODOC, 13 December 2017. https://www.clodoc.com/blog/uses-of-wearable-technology-in-healthcare/. Accessed 10 Feb 2019

Physiological Evaluation of a Non-invasive Wearable Vagus Nerve Stimulation (VNS) Device

Se Jin Park[1,2(✉)], Seunghee Hong[1], Damee Kim[1], Iqram Hussain[1,2], Young Seo[1], and Min Kyu Kim[3]

[1] Korea Research Institute of Standards and Science, Daejeon, South Korea
{sjpark,hsh82622,dameeing,iqram,young2da}@kriss.re.kr
[2] University of Science & Technology, Daejeon, South Korea
[3] AMO Lab, Seoul, South Korea
min@amo-lab.com

Abstract. Recent clinical studies suggest that Vagus nerve stimulation (VNS) may be a safe and potentially effective for disorder related to heart. In the rise of wearable medical devices, non-invasive wearable VNS devices are available to treatment of disorders related to the central nervous system. The goal of this study was to investigate the effects of VNS devices on physiological parameters of human. AMO+ is a wearable VNS device that was worn to the neck as a necklace. ECG (Electrocardiogram) and EEG (Electroencephalogram) of 30 healthy persons are measured in Center for Medical Metrology, Korea Research Institute of Standards and Science, Daejeon, South Korea. Wireless Bionomadix (Biopac Systems, Goleta, CA) RSPEC-R used for measuring ECG and EEG2-R as used to measure EEG. EEG electrodes are placed in F3 and F4 position of head as per 10/20 System. Acqknowledge Ver. 14 used for data analysis. EEG α, β, θ, δ power spectrums are extracted from raw EEG signals. HR (Heart Rate), HF (High Frequency), LF (Low Frequency) features are extracted from raw ECG data. Data was taken before using VNS deice and after one hour of using wearable AMO VNS device. All data was taken in resting state. Statistical analysis was done by IBM SPSS Ver. 23.0. No significant difference of HR is found between cases of before using VNS device and after using VNS device. A significant difference found in case of HF power of ECG. As ECG HF power reflects the parasympathetic nervous system activity, HF power increase indicates improvement of parasympathetic nervous system activity. No significant difference of EEG α, β, θ power spectrums except δ power is found between cases of before using VNS device and after using VNS device. Wearable AMO+ VNS stimulation is expected to improve the parasympathetic nervous system activity.

Keywords: Balance seat · Vibration comfort · Human vibration

© Springer Nature Switzerland AG 2020
T. Ahram (Ed.): AHFE 2019, AISC 973, pp. 57–62, 2020.
https://doi.org/10.1007/978-3-030-20476-1_7

1 Introduction

Vegas nerve is a one of most significant nerve connecting autonomic nervous system and cardiovascular, respiratory, gastrointestinal, immune, and endocrine systems and has been called the "great wandering protector" [1]. This nerve involves in large number of senses such as; pain, temperature, inflammation. Vegas nerve controls respiration, airway diameter, heart rate, blood pressure, vascular resistance, and feeding [2].

In several studies, it is found that electrical stimulation effects brain electrical activity. As the initial study, an increased electrical potential is observed on the contralateral orbitofrontal cortex during VNS stimulation (24–50 Hz) [3]. Global cortical desynchronization and sleep spindle blockage was found in "encephale isole" cats with strychnine-induced seizures during VNS (2 volts, 50 Hz, 0.5 ms pulse) [4]. EEG synchronization or desynchronization was produced during stimulated the nucleus of the solitary tract at low (1–16 Hz) or high (> 30 Hz) frequencies [5].

James Corning firstly used Vagus nerve stimulation (VNS) to treat epilepsy in the late 19th century [6]. Researchers are trying to develop the neuroprosthetic devices and neural stimulators for treatment of a variety of neurological diseases (e.g., Parkinson's disease, epilepsy, Atrial Fibrillation, and depression) [6–9]. Among those studies, it is found that epilepsy and depression is most treatable diseases by VNS therapy [9–12]. The US Food and Drug administration (FDA) approved VNS device for managing epilepsy (AspireSR®, SenTiva®) and for depression (NET-2000) [13]. GammaCore© (electroCore LLC, Basking Ridge, NJ, USA), a nVNS device received FDA approval for the acute treatment of episodic cluster headache and for the acute treatment of migraine [13].

This study focused on evaluating physiological responses of a non-invasive wearable Vagus nerve Stimulation (VNS) Device measuring ECG and EEG.

2 Methodology

2.1 Vagus Nerve Stimulation (VNS) Device

AMO+ is a wearable VNS device containing low-frequency electromagnetic field (LF-EMF) signal supposed to reducing stress and stabilizing heart rate. AMO+ is a necklace type wearable device.

2.2 Experimental Setup

Those who participated in the experiment were male students and their average age was 24 ± 1.3 years. All the participants were performed two tasks (resting) with simultaneous recording of physiological parameters. The participants had no clinical history for mental diseases. ECG (Electrocardiogram) and EEG (Electroencephalogram) of 30 healthy persons are measured in Center for Medical Metrology, Korea Research Institute of Standards and Science, Daejeon, South Korea. Wireless Bionomadix (Biopac Systems, Goleta, CA) RSPEC-R used for measuring ECG and EEG2-R as used to measure EEG. An EEG test was conducted in order to study the activity of the central nervous system. EEG electrodes are placed in F3 and F4 position (frontal lobe) of head as per 10/20 System. Room temperature was kept on 24 °C and 43% RH (Fig. 1).

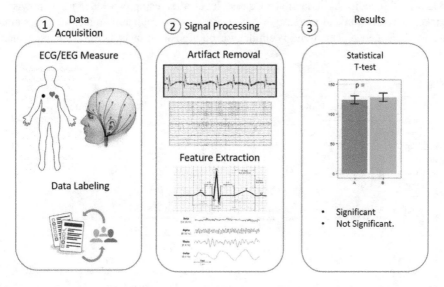

Fig. 1. Multi-axis noiseless BSR exciter simulator (*top*), Vibration Analysis ToolSet, VATS™ (bottom).

2.3 Analysis Methods

Acqknowledge Ver. 14 used for data analysis. EEG α, β, θ, δ power spectrums are extracted from raw EEG signals. HR (Heart Rate), HF (High Frequency), LF (Low Frequency) features are extracted from raw ECG data. Data was taken before using VNS deice and after one hour of using wearable AMO VNS device. All data was taken in resting state. Statistical analysis was done by IBM SPSS Ver. 23.0.

2.4 Feature Extraction

The spectral features are most commonly extracted features from EEG; divided into five frequency bands namely delta, δ (0–4 Hz), theta, θ (5–7 Hz), alpha, α (8–12 Hz), beta, β (13–30 Hz). For processing the EEG, data were converted using A/D transformation, and then low-pass filtered. Frequency analysis was executed using an FFT transform for signals from EEG position of F_3 and F_4. Spectral features of α, β, θ, δ were extracted. Mean Power, mean frequency were calculated for all five spectral bands and used as EEG features in different driving status. HR (Heart Rate), HF (High Frequency), LF (Low Frequency) features are extracted from raw ECG data.

3   Results and Discussion

In EEG features, it is found that α, β, θ power increased and δ power has decreased after using VNS device as shown in Fig. 2. β spectrum is representative of stress and decline of β indicates reduction of stress level after using VNS device. δ power is maker of relaxation and increase of δ power represents relaxation effect increases after using VNS device. This trend is similar for both position of EEG electrode placement (F_3, F_4).

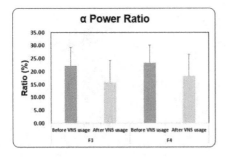
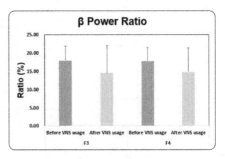
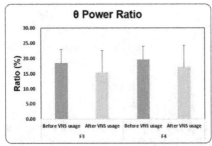
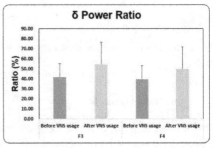

Fig. 2. Comparison of result of α, β, θ, δ power between EEG acquisition before VNS and after VNS device usage.

No significant difference of HR is found between cases of before using VNS device and after using VNS device. LF oscillation represents to vasomotor activity, which indicates activity and stress. On the other hand, the HF oscillation corresponds to respiratory activity, which represents relaxation. A significant difference found in case of HF power of ECG. ECG HF component is representation of vagal activity [14]. ECG HF power reflects the parasympathetic nervous system activity. HF power increase indicates improvement of parasympathetic nervous system activity. Rise of High frequency (HF) component indicates improved vagal nerve activity which in observed in Fig. 3 after using VNS device. There is a significantly different HF ratio between ECG acquisition before VNS and after VNS device usage.

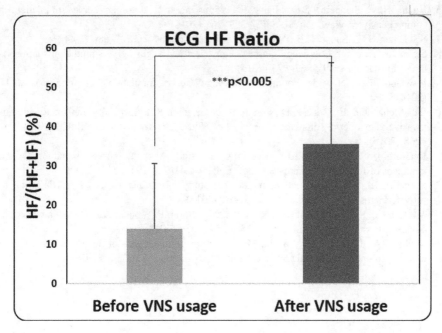

Fig. 3. Comparison of result of HF ratio between ECG acquisition before VNS and after VNS device usage.

4 Conclusion

The present study demonstrates physiological response of the Vegas nerve stimulation as limited as HF component for ECG and delta power for EEG. More broad experiment needed to check effects other factors.

Acknowledgments. The AMO Lab (www.amo-lab.com) supported this work.

References

1. Berthoud, H.R., Neuhuber, W.L.: Functional and chemical anatomy of the afferent vagal system (in eng). Auton. Neurosci. **85**(1–3), 1–17 (2000)
2. Yuan, H., Silberstein, S.D.: Vagus nerve and vagus nerve stimulation, a comprehensive review: Part I **56**(1), 71–78 (2016)
3. Bailey, P., Bremer, F.: A sensory cortical representation of the vagus nerve: with a note on the effects of low blood pressure on the cortical electrogram. **1**(5), 405–412 (1938)
4. Zanchetti, A., Wang, S.C., Moruzzi, G.: The effect of vagal afferent stimulation on the EEG pattern of the cat (in eng). Electroencephalogr. Clin. Neurophysiol. **4**(3), 357–361 (1952)
5. Magnes, J., Moruzzi, G., Pompeiano, O.: Synchronization of the EEG produced by low-frequncy electrical stimulation of the region of the solitary tract. **99**(1), 33–67 (1961)

6. Yuan, H., Silberstein, S.D., Pain, F.: Vagus nerve and vagus nerve stimulation, a comprehensive review: Part II. **56**(2), 259–266 (2016)
7. Schachter, S.C., Saper, C.B.: Vagus nerve stimulation. **39**(7), 677–686 (1998)
8. Stavrakis, S., et al.: Low-level transcutaneous electrical vagus nerve stimulation suppresses atrial fibrillation. **65**(9), 867–875 (2015)
9. Howland, R.H.: New developments with vagus nerve stimulation therapy. **52**(3), 11–14 (2014)
10. Ben-Menachem, E., Revesz, D., Simon, B., Silberstein, S.J.: Surgically implanted and non-invasive vagus nerve stimulation: a review of efficacy, safety and tolerability. **22**(9), 1260–1268 (2015)
11. DeGiorgio, C.M., et al.: Prospective long-term study of vagus nerve stimulation for the treatment of refractory seizures (in eng). Epilepsia **41**(9), 1195–1200 (2000)
12. Boon, P., et al.: A prospective, multicenter study of cardiac-based seizure detection to activate vagus nerve stimulation. **32**, 52–61 (2015)
13. Mertens, A., Raedt, R., Gadeyne, S., Carrette, E., Boon, P., Vonck, K.J.: Recent advances in devices for vagus nerve stimulation. **15**(8), 527–539 (2018)
14. Malliani, A., Lombardi, F., Pagani, M.: Power spectrum analysis of heart rate variability: a tool to explore neural regulatory mechanisms. Br. Heart J. **71**(1), 1–2 (1994)

The Ergonomic Design of Wearable Robot Based on the Shoulder Kinematic Analysis by Walking Speed

Seung-Min Mo[1(✉)], Jaejin Hwang[2], Jae Ho Kim[3],
and Myung-Chul Jung[3]

[1] Department of Industrial and Chemical Engineering, Suncheon Jeil College,
Suncheon 57997, Republic of Korea
smmo@suncheon.ac.kr
[2] Department of Indusrtial and Systems Engineering,
Northern Illinois University, Dekalb, IL 60115, USA
jhwang3@niu.edu
[3] Department of Indusrtial Engineering, Ajou University, Suwon 16499,
Republic of Korea
{jaeho82,mcjung}@ajou.ac.kr

Abstract. The purpose of this study was to suggest the design of wearable robots based on the shoulder kinematics including the range of motion, angular velocity, and angular acceleration during walking. A treadmill system was used to measure the kinematics data of the shoulder joint during different walking speed. The independent variables of this study were walking speed. Walking speed was set as four levels including 3.6, 5.4, 7.2, and preferred walking speed (PWS) km/h. The subject walked according to the randomized walking speed during 90 s on the treadmill. Twenty gait cycles of motion capture data from each experimental condition of each subject were extracted. Data was analyzed by one-way repeated measures analysis of variance. There were significant differences of minimum joint angle, mean of ROM, maximum joint angular velocity, minimum joint angular velocity, maximum joint angular acceleration and minimum joint angular acceleration. There was no significant difference of maximum joint angle. The kinematics data of ROM, angular velocity, and angular acceleration revealed an increasing trend up to walking speed of 7.2 km/h. It indicates that the arm swinging was sufficiently performed to maintain the walking stability. The maximum angular acceleration increased as the walking speed increased, which meant the instantaneous velocity of the shoulder joint increased. It indicated that the load of the shoulder joint increased with the increase of the walking speed. Hence, this study suggests that the ergonomic threshold for walking speed of the wearable robot could be limited to 5.4 km/h.

Keywords: Wearable robot · Human-robot interaction · Shoulder · Kinematics · Walking speed

© Springer Nature Switzerland AG 2020
T. Ahram (Ed.): AHFE 2019, AISC 973, pp. 63–69, 2020.
https://doi.org/10.1007/978-3-030-20476-1_8

1 Introduction

The robot technology is developing to improve industry productivity and convenience in human life. Wearable robots have been designed and developed to improve humans' strength, speed, and endurance. They have been applied to various tasks such as walking, running, lifting, lowering, pushing, and pulling. A wearable robot can be seen as a technology that extends, complements, substitutes human function and capability or replaces [1].

Despite of the rapidly grown technology, wearable robots still face several challenges. Previous studies still have focused on improving the mechanical performance of wearable robots [2–4]. Users could experience the resistance of their movement while walking and feel discomfort on their body regions if the joints of the wearable robots are not properly aligned [5]. In addition, adding extra mass to the human body could increase the metabolic expenditure and lead to the early fatigue during walking [6].

In order to overcome these limitations of the wearable robot, understanding natural biomechanics of walking could be essential. Natural rhythm of movements in the upper extremity while walking could be useful to determine the functional limit (e.g., joint torque and range of motion) of the wearable robot to promote the safety level and comfort of users. And also, the key distinctive aspect in wearable robots is their human-robot interaction. A human-robot interaction is a hardware and software link that connects to both human and robot systems [7].

The purpose of this study was to suggest the design of wearable robots based on the shoulder kinematics including the range of motion, angular velocity, and angular acceleration during walking.

2 Methods

2.1 Subjects

Ten male subjects in their 20s participated in this study. All the subjects gave written informed consent form to the experiment. Subjects had no history of musculoskeletal disorders or injuries in the past 12 months. Table 1 shows the basic information including age, height, and weight of the subjects.

Table 1. Anthropometry of the subjects.

Age (year)	Height (mm)	Weight (kg)
25.3 (±1.9)	1756.3 (±62.8)	73.0 (±9.1)

2.2 Apparatus

A treadmill system was used to measure the joint angle, angular velocity and angular acceleration of the shoulder joint during different walking speed. Eight 1.3-megapixel

Flex 13 cameras were used to capture the walking motion (Natural Point, Inc. Optitrack, USA).

The marker set for calculating the kinematics values of the human body during walking was Plugin Gait Full Body Model and 39 markers were attached to the entire body of the subject.

2.3 Experimental Design

The independent variables of this study were walking speed. Walking speed was set as four levels including 3.6, 5.4, 7.2, and preferred walking speed (PWS) km/h. The preferred walking speed (PWS) was determined as a comfortable walking speed of each subject while using a treadmill. The mean and standard deviation of PWS was 4.3 ($\pm$ 0.5) km/h. The level of walking speed was referenced criteria to slow, normal and fast from previous studies [8, 9]. As the dependent variables, the kinematic data were analyzed maximum joint angle [deg], minimum joint angle [deg], range of motion (ROM) [deg], maximum joint angular velocity [deg/sec], minimum joint angular velocity [deg/sec], maximum joint angular acceleration [deg/sec^2] and minimum joint angular acceleration [deg/sec^2].

A within-subject design with repeated measures on the cycles was used. All experiment conditions were randomized across subjects.

2.4 Procedure

Thirty nine retro-reflective skin markers were placed on the head, arms, torso, legs, and feet of subject. All subject exercised different walking speeds on the treadmill. Then PWS was measured for each subject. The subject walked according to the randomized walking speed during 90 s on the treadmill. To minimize the fatigue, subjects were given at least 5 min break between experimental conditions.

2.5 Analysis

The period for the walking was analyzed based on one gait cycle. One gait cycle was defined from the time when the heel reached the floor to the time when the heel reached the floor again [10]. Twenty gait cycles of motion capture data from each experimental condition of each subject were extracted. Then, these data were post-processed using MOTIVE version 2.0.3 (Natural Point, Inc. Optitrack, USA), and the joint angle, angular velocity and angular acceleration of the shoulder joint was calculated using the Visual 3D v6 (C-Motion, Inc., USA). The obtained motion capture data was filtered using a lowpass filter at a cut off frequency of 6 Hz. The joint angle, angular velocity and angular acceleration of the shoulder joint of right side were calculated, and only the sagittal plane was analyzed (flexion-extension) as seen in Fig. 1.

Data was analyzed by one-way repeated measures analysis of variance (ANOVA) with Tukey's test. A significance level of 0.05 was used all statistical analysis which were performed using SAS software 9.4 (SAS Institute, USA).

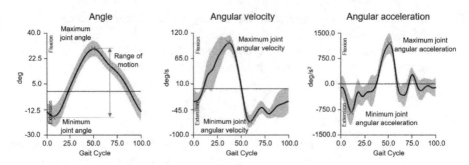

Fig. 1. Plot showing the dependent variables were calculated kinematic data.

3 Result

As a result of ANOVA, there were significant differences of minimum joint angle (F = 12.72 $p < 0.0001$), mean of ROM (F = 3.65 $p = 0.0248$), maximum joint angular velocity (F = 11.58 $p < 0.0001$), minimum joint angular velocity (F = 12.91 $p < 0.0001$), maximum joint angular acceleration (F = 10.46 $p < 0.0001$) and minimum joint angular acceleration (F = 30.81 $p < 0.0001$). There was no significant difference was found dependent variable for maximum joint angle (F = 0.52 $p = 0.6749$).

As shown in Fig. 2, the mean of ROM for walking speed 7.2 km/h was significantly higher than other walking speeds. The result of Tukey's test shows that there were no significant differences in walking speed of 3.6, 5.4, and 7.2 km/h. The mean of ROM was analyzed to the lowest at PWS of 4.3 km/h.

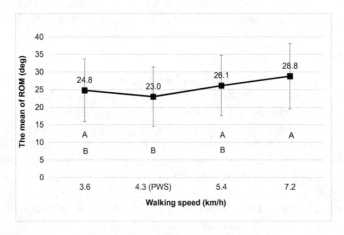

Fig. 2. The mean and standard deviation of range of motion (ROM) of shoulder joint for the different walking speeds. * Alphabetical order represents the Tukey's post-hoc test.

As shown in Fig. 3, the mean of maximum angular velocity for walking speed of 7.2 km/h was significantly higher than other walking speeds. Similarly, the mean of minimum angular velocity for walking speed of 7.2 km/h was analyzed the lowest. The mean of maximum angular velocity was analyzed to the lowest at PWS of 4.3 km/h. Also, the mean of minimum angular velocity was analyzed to the highest at PWS of 4.3 km/h.

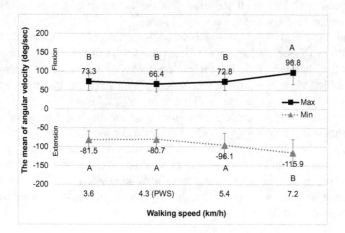

Fig. 3. The mean and standard deviation of angular velocity of shoulder joint for the different walking speeds. * Alphabetical order represents the Tukey's post-hoc test.

The mean of maximum angular acceleration also showed a similar result of angular velocity as seen in Fig. 4. There was an increasing trend up to walking speed of 7.2 km/h. Similarly, the mean of minimum angular acceleration was a decreasing trend up to walking speed of 7.2 km/h.

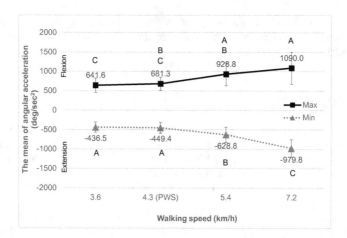

Fig. 4. The mean and standard deviation of angular acceleration of shoulder joint for the different walking speeds. * Alphabetical order represents the Tukey's post-hoc test.

4   Discussion

This study analyzed the tendency of kinematic variables of the shoulder joint by different walking speeds in the repetitive gait pattern.

The mean of ROM, angular velocity, and angular acceleration were the highest at walking speed of 7.2 km/h. It indicates that the arm swinging was sufficiently performed to maintain the walking stability [11–13].

The walking speed of 7.2 km/h was set as the fastest walking speed. At this speed, the gait pattern of one cycle was too short due to the fast walking. Accordingly, it considered that the time of motion for sufficiently swinging the arm was short. Hence, the mean of ROM, maximum angular velocity, and maximum angular acceleration analyzed the highest value during walking speed of 7.2 km/h.

The maximum angular acceleration increased as the walking speed increased, which meant the instantaneous velocity of the shoulder joint increased. It indicated that the load of the shoulder joint increased with the increase of the walking speed. Based on the results of this study, it is possible to have relatively higher biomechanical load and discomfort in the shoulder joint during a walking speed of 7.2 km/h than other speeds. Hence, this study suggests that the ergonomic threshold for walking speed of the wearable robot could be limited to 5.4 km/h.

5   Conclusion

Based on the result of this study, we suggest that the ergonomic threshold of the walking speed in the wearable robot could be limited to 5.4 km/h. In addition, the fast walking speed could cause higher biomechanical load and discomfort in the arm movement. The limitation of this study was the small size of the sample, resulting in the fact that no general conclusions could be drawn. Further study should consider to validate these results, including more samples.

Acknowledgments. This study was supported by Basic Science Research Program through the National Research Foundation of Korea (NRF) funded by the Ministry of Education (NRF-2017R1D1A3B03035407).

References

1. Alami, R., Albu-Schaeffer, A., Bicchi, A., Bischoff, R., Chatila, R., De Luca, A., De Santis, A., Giralt, G., Guiochet, J., Hirzinger, G., Ingrand, F., Lippiello, V., Mattone, R., Powell, D., Sen, S., Siciliano, B., Tonietti, G., Villani, L.: Safe and dependable physical human-robot interaction in anthropic domains: state of the art and challenges. In: 2006 IEEE/RSJ International Conference on Intelligent Robots and Systems. https://doi.org/10.1109/iros.2006.6936985 (2006)
2. Riani, A., Madani, T., Benallegue, A., Djouani, K.: Adaptive integral terminal sliding mode control for upper-limb rehabilitation exoskeleton. Control. Eng. Pract. **75**, 108–117 (2018)
3. Bergamasco, M., Salsedo, F., Marcheschi, S.: A novel actuator for wearable robots with improved torque density and mechanical efficiency. Adv. Robot. **24**, 2019–2041 (2012)

4. Copaci, D., Cano, E., Moreno, L., Blanco, D.: New design of a soft robotics wearable elbow exoskeleton based on shape memory alloy wire actuators. Appl. Bionics. Biomech. https://doi.org/10.1155/2017/1605101 (2017)
5. Asbeck, A.T., De Rossi, S.M., Galiana, I., Ding, Y., Walsh, C.J.: Stronger, smarter, softer: next-generation wearable robots. IEEE Robot. Autom. Mag. **21**, 22–33 (2014)
6. Browning, R.C., Modica, J.R., Kram, R., Goswami, A.: The effects of adding mass to the legs on the energetics and biomechanics of walking. Med. Sci. Sports. Exerc. **39**, 515–525 (2007)
7. Pons, J.L.: Wearable Robots: Biomechatronic Exoskeletons. Wiley, UK (2008)
8. Bilney. B., Morris. M., Webster. K.: Concurrent related validity of the GAITRite® walkway system for quantification of the spatial and temporal parameters of gait. Gait. Posture. **17**, 68–74 (2003)
9. Kang, H.G., Dingwell, J.B.: Separating the effects of age and walking speed on gait variability. Gait. Posture. **27**, 572–577 (2008)
10. Zeni Jr, J.A., Richards, J.G., Higginson, J.S.: Two simple methods for determining gait events during treadmill and overground walking using kinematic data. Gait. Posture. **27**, 710–714 (2008)
11. Ortega, J.D., Fehlman, L.A., Farley, C.T.: Effects of aging and arm swing on the metabolic cost of stability in human walking. J. Biomech. **41**, 3303–3308 (2008)
12. Umberger, B.R.: Effects of suppressing arm swing on kinematics, kinetics, and energetics of human walking. J. Biomech. **41**, 2575–2580 (2008)
13. Bruijn, S.M., Meijer, E.G., Beek, P.J., van Dieën, J.H.: The effects of arm swing on human gait stability. J. Exp. Biol. **213**, 3945–3952 (2010)

Improvements in a Wearable Device for Sign Language Translation

Francesco Pezzuoli[⊠], Dario Corona, and Maria Letizia Corradini

School of Science and Technology, Mathematics Division,
University of Camerino, Via Madonna Delle Carceri 9, 62032 Camerino, Italy
{francesco.pezzuoli, dario.corona,
letizia.corradini}@unicam.it

Abstract. Nowadays a commercial product for sign language translation is still not available. This paper presents our latest results towards this goal, presenting a functional prototype called Talking Hands. Talking Hands uses a data-glove to detect the hand movements of the user, and a smartphone application to gather all the data and translates them into voice, using a speech synthesizer. Talking Hands implements the most suitable solutions for a massive production without penalizing its reliability. This paper presents the improvements of the last prototype in terms of hardware, software and design, together with a preliminary analysis for the translation of dynamic gestures through this device.

Keywords: Sign Language Recognition · Deaf · Data-glove · Gesture recognition

1 Introduction

According to the World Federation of the Deaf, 70 million people are deaf-mute in the world.[1] The majority of them communicate using sign languages (SL), and there are more than 300 sign languages in the world. Two people can establish a communication only if they know the same SL. Moreover, SLs are very unfamiliar for those who are not deaf and mute, thus a conversation between disabled and able individuals is very difficult. For this reason, deaf people face many difficulties to connect with the society, and this leads to social and employment limitations [1].

The technology can play a central role to help deaf people in their ordinary life, ensuring the two way of communication: a translation from SL to spoken languages and vice versa. In both cases, the progress of mobile technology has been a break-through. There are some cross-platforms applications to translate spoken languages into SL. They have still some limitations, like the number of spoken languages and SL they are able to translates and the extensions of their vocabularies, but they are currently on the market, and some for free, like ProDeaf.[2] On the contrary, nowadays there is not a commercial product able to translate a SL into a spoken one, and our study aims to fill this gap.

---

[1] https://wfdeaf.org/.

[2] https://play.google.com/store/apps/details?id=com.Proativa.ProDeafMovel&hl=en.

© Springer Nature Switzerland AG 2020
T. Ahram (Ed.): AHFE 2019, AISC 973, pp. 70–81, 2020.
https://doi.org/10.1007/978-3-030-20476-1_9

1.1   State of Art

The Sign Language Recognition (SLR) field is strictly connected with the Gesture Recognition (GR) one. The difference is slightly but essential. A gesture is a bodily motion, configuration or a combination of them, typically of hands, arms and head. Gestures can be used for many proposes, such as Human Computer or Human Robot Interaction (HCI – HRI), gaming, proxemics, real-time pose estimation [2]. A sign is instead a *meaningful gesture*, i.e., a gesture that expresses a word, a concept or a particular grammar structure. This difference is crucial for a correct implementation of a sign language translation system.

There are many limitations in gesture recognition systems for sign languages translation. The first challenging tasks is the collection of movement's data. A sign language is primarily expressed through the hands, but even facial expressions and full-body movements play a central role, so no one of these aspects can be neglected to achieve a perfect translation. At the same time, data acquisition has to be simple enough to realize a portable system. Second, the translation must be conducted in real time to truly support deaf people. A real time translation within a very large set of signs needs a heavy computation that requires powerful hardware and software systems. Third, the signs can be static or dynamic. A sign is static if it does not involve motion, like signs for numbers and some letters in most of SLs. On the contrary, a dynamic sign involves the motion, which can differentiate two signs with the same posture. Last, a sign language translation system has to reconstruct the grammar structure of the phrases, because sign languages are different from their respective spoken languages.

Different papers and studies of last decades face with these challenging tasks. An extensive survey about the gesture recognition works from 2009 to 2017 focusing on SL is given in [3]. The majority of these works use cameras for data acquisition (e.g. [4–7]). The drawback of these solutions is system portability: the user has to be stand in front of the camera and to rearrange the camera location each time they move from a place to another. Moreover, most of the papers deals with a simple background and this is a very strong restriction in a real-life application.

As a consequence, some other researches use a special hardware. Among others, the most suitable and widely studied are data-gloves, which are gloves equipped with sensors and an electronic circuit. The major advantage is that gloves can acquire data directly (degree of bend, wrist orientation, hand motion, etc.) thus eliminating the need to process raw data into meaningful values. Furthermore, this approach is not subject to environmental influences, for example, the location of the individual or the background conditions and lighting effects, and the collected data are accurate. An extensive and detailed survey on system-based sensory gloves for SLR from 2007 to 2017, together with their advantages and drawbacks, is presented in [8].

In [9] an applicative hand glove architecture is described. It has five flex sensors, three contact sensors and one three-axis accelerometer that serves as an input channel. It can translate American SL (ASL) in form of alphabet and it can transmit data to a mobile phone, and the overall translation efficiency is of 83%. Similar projects are described in [10–14]. The hardware is almost the same, and they can vary for some details of implementation and for mathematical techniques for signs detection and translation. Their classification accuracy is between 85% and 99%. The major draw-backs of these solutions are:

- the user must wear a cumbersome data glove to capture hand and finger movements;
- reliable segmentation methods should be developed to assist in continuous gestures recognition;
- the inability to obtain meaningful data, complementary to gestures, giving the full meaning of the conversation, such as facial expressions, eye movements, and lip-perusing.

To the best of our knowledge, these issues are still unsolved and a commercial product for an effective support of deaf people in their ordinary life is still lacking.

1.2   Main Contribution

This paper describes the development of *Talking Hands*, a wearable device for sign language translation that is oriented to the deaf-hearing support in ordinary life. Even if it does not achieve a complete translation of SL, Talking Hands is a user-friendly device with a great portability, which allows deaf people a basic interaction with everyone. These goals are achieved through software and design solutions that simplify the different tasks. There are several advantages of Talking Hands with respect to other similar projects [9–14] in terms of hardware, software and product design:

- a customized board to reduce the dimensions and optimize the performances;
- two IMUs for a more accurate detection of hand movements and orientation;
- a scenario-based translation approach, that achieves a high-level of translation between a large set of static signs;

This paper highlights our last improvements, in terms of hardware, software and design, with respect to the previous prototype, presented in [15]. The actual prototype is shown in Fig. 1. It is less bulky, its translation is more reliable and it can be mass produced. However, it does not translate dynamic signs yet, and this paper presents also a preliminary analysis to overcome this issue.

1.3   Outline

The next section presents the last prototype of Talking Hands, with its hardware, software and design solutions. Section 3 describes a preliminary analysis to translate dynamic signs using Talking Hands, with some interesting and promising results. Section 4 contains our discussion and the future directions of the project.

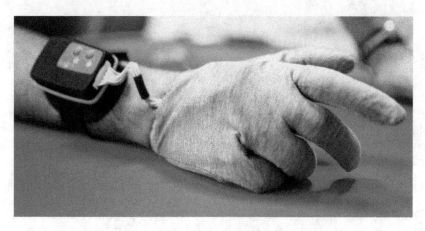

Fig. 1. Last prototype of talking hands

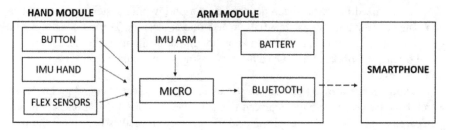

Fig. 2. Scheme architecture of talking hands

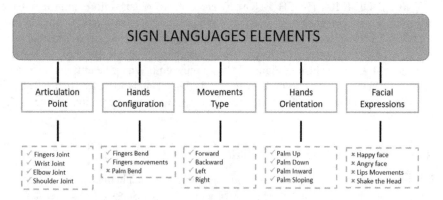

Fig. 3. Sign languages elements acquired by talking hands

2 Talking Hands

2.1 Hardware

Talking Hands is composed by one data glove connected with an ordinary smartphone through bluetooth. Even if the sign languages are mainly expressed with two hands, Talking Hands uses only one glove in the perspective of a commercial product, reducing the final prize for the user. Moreover, Talking Hands does not use a commercial data-glove, but a dedicated glove is realized to obtain a low cost device. The hardware is divided into two main modules (Fig. 4):

- Hand Module: 10 flex sensor (Spectra Symbol) to detect fingers position; one IMU (BOSCH BNO055) to detect hand orientation; one button to initialize the system.
- Arm Module: one custom board with a 32-bit microprocessor (Cortex M3 Atmel SAM3X8E) and one IMU to detect arm orientation; a led RGB to check the system status; mini-usb connection; a battery and a charge module; Bluetooth module (Microchip RN42).

A customized board has been developed as a prototyping system for the application. The board's dimensions are 3.05 mm × 5 mm × 1.5 mm. The initialization of the system through a button is required to know the initial position of the user through the initial orientation of the IMUs. Once the system knows the initial position, the user can start signing.

Talking Hands uses 10 flex sensors to acquire data of finger bent. One flex sensor which have two sensible parts is placed in each finger to collect the information about metacarpus and proximal inter phalangeal joints. In some studies (e.g. [16]) the position of the fingers are detected using EMG sensors. This is an elegant solution that implements only one sensor, but it does not recognize the positions of single fingers, so it is not enough for a sign language translation application.

Two 9 DOF IMUs (BOSH BNO055) are used to obtain information about the forearm and hand orientation. These IMUs require an initial calibration, which it last about 30 s, but they achieve a high-level of reliability even for prolonged use, thanks to the built-in functions of sensor fusion. The bluetooth is a simple BT module to communicate with the smartphone. Figure 3 shows the sign languages elements introduced in [8], together with the possibilities of the Talking Hands to detect them. Talking Hands acquires data for almost all the elements, with the exception of the facial expressions.

2.2 Software and Translation Algorithm

Most of the works in gesture recognition use advanced mathematical tools, such as Neural Networks [17], Hidden Markov Models [7, 18], Support Vectors Machines [16], Fuzzy C-Means Clustering [19], Multi-layer Perceptron [20]. Talking Hands uses a simpler solution, based on a distance function defined on the space of sensors data, to translate static signs. The most similar approaches to our solution are given in [11, 13]. In these papers, the tests show some promising results, with an accuracy from 78.33% to 95%. However, the tests have been conducted on 9 and 5 signs respectively, so they

Fig. 4. Previous prototype of talking hands

are very far from a real-life application. Our approach reaches a higher level of translation with respect to both recognition rate and the number of recognized signs, and can guarantee a satisfying communication experience to the final user.

The main feature of our solution is a two-stage computation. The firmware pre-processes the sensors data and establishes if the user is performing a gesture. Only in that case, it sends data by the Bluetooth connection, and this reduces the communication load. The smartphone receives data from the glove, performs the translation algorithm to link the gesture with its eventual meaning and uses the speech synthesizer to talk. This leads to three important advantages. Firstly, the translation algorithm is disjoint from the feature extraction and the glove of Talking Hands can be applied in all the other application fields of the gesture recognition. Secondly, the dataset of signs is stored in the smartphone, without having issues of memory capacity. Moreover, the recording phase of new signs does not require a re-initialization of the firmware.

Another remarkable improvement of Talking Hands is its translation through scenarios. We define a *scenario* as a set of signs that Talking Hands can translate in a single session. Hence, the system can translate the signs of one scenario at a time. The user can switch among the scenarios on-line, i.e. during the usage without the need of re-initializing. This approach leads to some important advantages. Across the world, there are many sign languages, like the spoken ones, and some of them have different dialects. For this reason, it is almost infeasible to realize a pre-build universal system. In a scenario approach, the user can easily record the signs through the smartphone application and can assign the sign to one or more scenarios. This approach enlarges the set of signs that the system can translate, without losing reliability. In fact, similar signs would not be misunderstood if they were not in the same scenarios.

The translation algorithm is based on a particular distance function on the space of vectors that represent the gestures, both to establish if the user is performing a sign and to link the sign to its translation. This distance is both accurate and robust, i.e., if the data-glove has one or two broken sensors, the translation still works. A particular aspect is the usage of a proper distance function for the quaternions that coming from the IMUs. Following the results in [21], we define the distance of two quaternions with unit Euclidean norm as

$$\mathbf{d}(\mathbf{h}_1, \mathbf{h}_2) = \frac{2}{\pi}\cos^{-1}(|\mathbf{h}_1 \cdot \mathbf{h}_2|) \in [0, 1]. \tag{1}$$

Then we define the distance between two distinct vectors coming from the hardware as the sum of the distances between the values of each sensors. With this definition, the comparison between the coming values of the sensors and the stored values for the signs is trivial and fast, and if the distance is below a certain threshold, the translation occurs. A more detailed explanation of all these software solutions is given in [15].

2.3   Design

The design of Talking Hands is very mild, light and comfortable, solving the main problem of usual cumbersome data-gloves highlighted in [8]. The design of the last prototype has significantly improved with respect to the previous one.

Figure 4 shows the previous prototype of Talking Hands, which has a bulky structure and a complex wiring system. This bulky structure allowed the flexion of the fingers in a comfortable way, without having problem with flex sensors returning in initial position after fingers flexion. Without this wiring/mechanical system, flex sensors jammed themselves with glove structure. The last prototype, shown in Fig. 5. Talking Hands Design, is easier and more comfortable to wear, especially about the hand part. This last prototype simplifies the wiring/mechanical system. Thanks to hardware complexity reduction, we designed a special textile glove to embed the F-PCB, which has inside the same hardware architecture of previous prototype and adds a micro motor for haptic feedback. This special glove ensures an excellent wearing, transpiration, comfort, and it can be easily mass-produced. Now the hand system has a maximum height of 6.90 mm, caused by hand-arm connector, and a minimum height of 0.32 mm in all the flexible PCB area. Moreover, the Flexible PCB can be removed to wash the glove.

3   Preliminary Study for Dynamic Signs Translation

he current prototype does not handle with dynamic gestures and this is the major drawback. To overcome this important issue in the near future, a preliminary study has been conducted to test some ideas and methods in a simple framework. Firstly, we consider only the data coming from one IMUs, i.e., a quaternion vector representing the orientation of the hand. Secondly, we perform our study on only three dynamic signs, which are "Deaf", "Hard" and "Hearing" from the American SL.[3] The whole study is performed off-line; we log and save all the data coming from the IMU during the

---

[3] https://www.youtube.com/watch?v=Raa0vBXA8OQ.

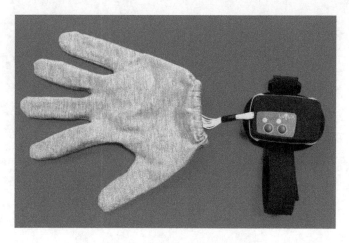

Fig. 5. Talking hands design

performances of the gestures and then conduct our algorithm on a pc. In four different logs, the three signs have been performed consecutively for five times, for a total of 20 occurrences of each sign. Figure 6 shows the data of one log, collected with a frequency of 50 Hz.

There are two main different tasks: the segmentation and the translation. The segmentation distinguishes two consecutives but different gestures. The translation eventually links a gesture to a meaning, according to a database previously created.

We use the velocity in the space of quaternions to perform the segmentation. To compute the velocity of the quaternions, one can use the Eq. (1) for the distance over the sampling period, thus

$$v(t) = \frac{d(h(t), h(t - T))}{T}. \tag{2}$$

However, the quaternions coming from the IMUs could have some computational errors, so they could slightly differ from the unit norm. In that case, the distance in (1) is not reliable for near quaternions. This can be seen in Fig. 8: even if the system is still, a little velocity is detected. Since we are dealing with two consecutive quaternions, they can differ only for a small amount and the Euclidean distance could lead to better result. The velocity computed with the Euclidean distance is reported in Fig. 7. The different gestures are isolated setting a threshold on the velocity signal.

The translation phase requires a preliminary feature extraction from the signal of the gestures, both for the creation of a database and for the effective translation during usage. To this purpose, we use a polynomial interpolation. For each of the four components of a quaternion, a fitting polynomial of seventh degree is construct, thus a gesture can be described by $4 \cdot (7 + 1) = 32$ numbers. Figure 9 shows a signal of a component of the quaternion for a gesture (in blue), together with its fitting polynomial (in red). Since the interpolation error is almost zero, the loss of information during this process is irrelevant, but the benefit in terms of storage data is notable. For example, the

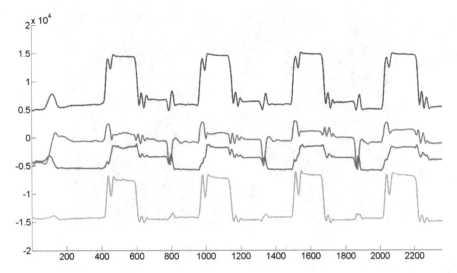

Fig. 6. One log of the three signs performed consecutively for five times.

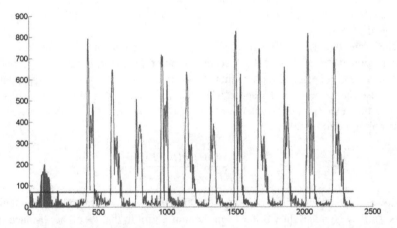

Fig. 7. Velocity computed with standard Euclidean distance and threshold for segmentation

polynomial in Fig. 9 requires eight numbers to be stored, while the whole sequence requires about seventy numbers. Moreover, a comparison between the coefficients of different polynomial can be conducted even if they come from signals of different length, and this is a very remarkable advantage.

After this feature extraction, we used different techniques to translate the signs, namely nearest neighbors, linear SVM, neural networks, naïve bayes and decision tree methods. As usual in these cases, we divided our data in a training set and a test set. The training set is used for the creation of the database, while the test set is used to verify the efficacy of the methods. Table 1 reports the methods applied together with their translation accuracy on the test set. The results are promising, with a 100% of

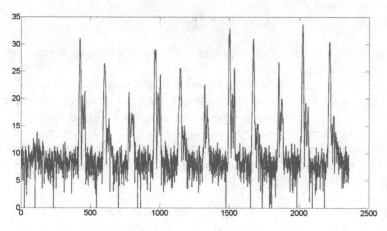

Fig. 8. Velocity computed with distance (1)

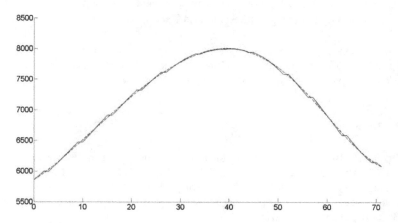

Fig. 9. A component signal of a quaternion (blue) together with its fitting polynomial (red)

translation rate for the Linear Support Vector Machine, Naïve Bayes and Decision Tree techniques. Hence, a deeper analysis and an online implementation of the proposed scheme will be the next steps towards the translation of dynamic signs.

Table 1. Classification techniques and their translation accuracies

Technique	Nearest neighbors	Linear SVM	Neural networks	Naïve Bayes	Decision tree
Accuracy	93%	100%	33%	100%	100%

4 Discussion and Future Development

This work presented Talking Hands, a functional prototype for a commercial device for sign language translation.

In [8] some recommendations are provided for developers and we took in consideration them all. In particular, Talking Hands will be a low-cost device (under 1,000 USD) that will ensure an effective support to deaf people to overcome the difficulty in communicating. Both high-quality sensors and robust algorithms guarantee a reliable and real time translation. The design embeds the hardware and provides a comfortable dress of the entire system. The preliminary analysis to translate dynamic signs lead to promising results, even with simple techniques.

As explained in [22], successful recognition of lexical signs is not sufficient for a full understanding of sign language communication. Nevertheless, we think that Talking Hands will support deaf people in real-life. Thus, extensive studies will be conducted together with deaf people to test the overall system usability and effectiveness in the near future. Moreover, we will continue the study proposed in Sect. 4 to full implement a dynamic signs translation on Talking Hands.

Acknowledgments. This work is supported by Limix S.r.l. (www.limix.it). Limix is an Italian start-up and spin-off of the University of Camerino. The intellectual property of Talking Hands and its different parts (hardware, software, design) is of Limix S.r.l.

References

1. Perkins, R., Battle, T., Edgerton, J., Mcneill, J.: A survey of barriers to employment for individuals who are deaf. J. Am. Deaf. Rehabil. Assoc. **49**(1), 66–85 (2015)
2. Kim, H., Lee, S., Lee, D., Choi, S., Ju, J., Myung, H.: Real-time human pose estimation and gesture recognition from depth images using superpixels and SVM classifier. Sensors (Switzerland) **15**(6), 2410–12427 (2015)
3. Hirafuji Neiva, D., Zanchettin, C.: Gesture recognition: a review focusing on sign language in a mobile context. Expert Syst. Appl. **103**, 159–183 (2018)
4. Cooper, H., Pugeault, N., Bowden, R.: Reading the signs: a video based sign dictionary. In: IEEE International Conference Computer Vision workshops, ICCV 2011, Barcelona (2011)
5. Starner, T., Weaver, J., Pentland, A.: Real time american sign language recognition using desk and wearable computer based video. IEEE Trans. Pattern Anal. Mach. Intell. **20**(12) (1998)
6. Kelly, D., McDonald, J., Markham, C.: A person independent system for recognition of hand postures used in sign language. Pattern Recognit. Lett. **31**, 1359–1368 (2010)
7. Yoon, H.S., Soh, J., Bae, Y.J., Seung Yang, H.: Hand gesture recognition using combined features of location, angle and velocity. Pattern Recognit. **37**(4), 1491–1501 (2001)
8. Ahmed, M.A., Zaidan, B.B., Zaidan, A.A., Salih, M.M., Bin Lakulu, M.M.: A review on systems-based sensory gloves for sign language recognition state of the art between 2007 and 2017. Sensors (Switzerland) **18**(7) (2018)
9. Bajpai, D., Porov, U., Srivastav, G., Sachan, N.: Two way wireless data communication and American sign language translator glove for images text and speech display on mobile phone. In: Proceedings 2015 5th International Conference on Communication Systems and Network Technologies. CSNT 2015, pp. 578–585 (2015)

10. Bukhari, J., Rehman, M., Malik, S.I., Kamboh, A.M., Salman, A.: American sign language translation through sensory glove: SignSpeak. Int. J. u- e-Serv. Sci. Technol. **8**, 131–142 (2015)
11. Shukor, A.Z., Miskon, M.F., Jamaluddin, M.H., Bin Ali Ibrahim, F., Asyraf, M.F., Bin Bahar, M.B.: A new data glove approach for malaysian sign language detection. In: 2015 IEEE International Symposium on Robotics and Intelligent Sensors (IRIS), vol. 76, pp. 60–67 (2015)
12. Seymour, M., Tsoeu, M.: A mobile application for South African Sign Language (SASL) recognition, pp. 1–5 (2015)
13. Kau, L.J., Su, W.L., Yu, P.J., Wei S.J.: A real-time portable sign language translation system. In: 2015 IEEE 58th International Midwest Symposium on Circuits and Systems (MWSCAS), pp. 1–4 (2015)
14. Devi, S., Deb, S.: Low cost tangible glove for translating sign gestures to speech and text in Hindi language. In: 3rd International Conference on Computational Intelligence & Communication Technology (CICT), pp. 1–5 (2017)
15. Pezzuoli, F., Corona, D., Corradini, M.L., Cristofaro, A.: Development of a wearable device for sign language translation. In: International Workshop on Human-Friendly Robotics (HFR2017), pp. 115–126 (2017)
16. Akhmadeev, K., Rampone, E., Yu, T., Aoustin, Y., Le Carpentier E.: A testing system for a real-time gesture classification using surface EMG. In: 20th IFAC World Congress, pp. 11498–11503 (2017)
17. Kouichi, M., Hitomi, T.: Gesture recognition using recurrent neural networks. In: ACM Conference on Human factors in computing systems: reaching through technology (1999)
18. Vogler, C.: American sign language recognition: reducing the complexity of the task with phoneme-based modeling and parallel hidden markov models. University of Pennsylvania (2003)
19. Li, X.: Gesture recognition based on fuzzy C-Means clustering algorithm. Department of Computer Science, The University of Tennessee, Knoxville
20. Nagi, J., et al.: Max-pooling convolutional neural networks for vision-based hand gesture recognition. In: 2011 International Conference on Signal and Image Processing and Applications (ICSIPA), pp. 342–347 (2011)
21. Huynh, D.Q.: Metrics for 3D rotations: comparison and analysis. J. Math. Imaging Vis. **35**, 155–164 (2009)
22. Ong, S.C.W., Ranganath, S.: Automatic sign language analysis: a survey and the future beyond lexical meaning. IEEE Trans. Pattern Anal. Mach. Intell. **27**(6) (2005)

Current Situation and Problems of Hand-Transmitted Vibration Measurement and Monitoring Equipment and Benefits of Wearable Sensors for Everyday Monitoring

Setsuo Maeda[1]([⊠]), Leif Anderson[2]([⊠]), and Jacqui Mclaughlin[2]

[1] Faculty of Applied Sociology, Kindai University, 3-4-1 Kowakae,
Higashi-Osaka City, Osaka 577-8502, Japan
maeda@socio.kindai.ac.jp
[2] Reactec Ltd., 3 Cultins Road, Edinburgh EH11 4DF, UK
{LeifAnderson,jacquimclaughlin}@reactec.com

Abstract. Employers who expose their workforce to hazardous vibration from mechanized tools need to develop an understanding of the health risk their employees face as a consequence. ISO 5349-2 was developed as an international standard to define how vibration exposure should be measured to quantify the risk to the individual. The authors would contend that the standard does not facilitate the economic collection of data across a range of tool users on a routine basis. In this paper, the Current Situation of Monitoring equipment for Hand-Transmitted Vibration Measurement is summarized with a review of the equipment's functionality. The paper then examines evidence from a time in motion study of a group of skilled operators repeatedly performing a single task to highlight the broad variability in monitored exposure levels relative to that expected. The authors conclude with the benefits the wearable sensor offers as a practical everyday assessment relative to existing methodologies.

Keywords: Wearable sensor · Hand-transmitted vibration · HAVS ·
ISO 5349 · Regular monitoring · Exposure data analytics

1 Introduction

From the increasing number of exhibitions on wearable technologies including the Wearable EXPO in Tokyo, Japan on January 2019 and as published in the conference proceedings of AHFE2018 [1], wearable sensor technology developed for the purposes of health monitoring can be distinguished into two categories. The first category focuses on measuring parameters or vital signs from the human body as indicated on the right-hand side of Fig. 1. The second category, as illustrated on the left of Fig. 1, is focused on the detection of risks to the human body for the purpose of damage prevention to the body.

T. Ahram (Ed.): AHFE 2019, AISC 973, pp. 82–92, 2020.
https://doi.org/10.1007/978-3-030-20476-1_10

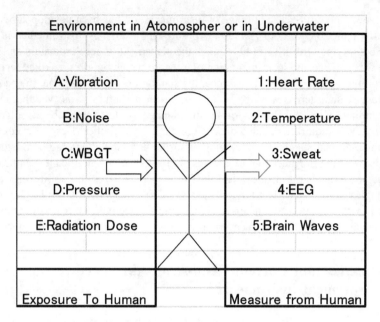

Fig. 1. Current wearable sensor technology.

The right-hand side of Fig. 1 concerns condition monitoring of the individual by sensing data measured from the human body, such as heart rate, body temperature, etc. From No.1 to No.5, the wearable equipment is measuring the effect on human body function in reaction to outside stimuli. Although these functional effects can be monitored using the wearable device, the underlying stimulus or physical agent is not monitored and therefore it is difficult to reduce or prevent adverse effects.

The left-hand side of Fig. 1 is focused on the prevention of risk to the human in the workplace by measuring the physical agents present in the environment such as A (Whole-Body Vibration and Hand-Arm Vibration measurement equipment: ISO 8041), B (Noise Dose measurement equipment: ISO 9612), C (Wet-bulb globe temperature measurement equipment: ISO 7243, 8996, 7726), D (Dive Computer: ISO 6425), and E (Radiological Protection: ISO 16637, 17025, 27048). All such wearable instruments have been measuring various physical quantities indirectly exposed to the human body in the work environment.

Exposure to vibration through the hand and arm system, known as Hand Arm Vibration (HAV), or through other areas of the body, known as Whole Body Vibration, are both known to cause irreversible damage. Exposure to HAV can cause neurological, vascular and musculoskeletal damage with the symptoms referred collectively as HAV Syndrome [2]. The established method for managing HAV exposure risk has been standardized in the form of ISO 5349 (BSI, 2001a), with employers being required to control exposure levels to predetermined limits of daily exposure within their respective territorial legislation. The standard also outlines the perceived probability of developing vascular damage from exposure to HAV based on an average daily exposure dose. Per ISO 5349, HAV risk is determined from the time duration of HAV exposure and a specific processing of the vibration magnitude during exposure. As will be expanded upon in Sect. 2, a range of instrumentation has been developed over time to quantify either one or both vibration exposure characteristics. Instrumentation aimed at fully complying with ISO 5349 requires the placement of accelerometers on the tool surface while in use and instrumentation of a nature which requires a skilled technician to interpret the gathered data. In essence, the standard does not facilitate the economic collection of data across a range of tool users on a routine basis, yet the accepted understanding of the likelihood of health damage is determined by average daily exposure levels requiring detailed data of regular daily activities. Despite the existence of the international standards concerning exposure assessment and regional legislation regarding working practices, reported cases of HAVS remain significant as indicated by disability benefit claims in the UK [3]. This paper explores the ability of wearable technologies to readily collect large data sets and examines whether utilization of data acquired in this fashion might provide for a more effective risk assessment and furthermore act as a preventative measure to this disease by ensuring timely intervention.

2   Comparison of Current Hand-Arm Vibration Measurement Instruments

In 2013 Pitts and Kaulbars [4] examined the limitations of measurement and instrumentation standards relative to the emergence of instrumentation aimed at easing the gathering of personal HAV exposure data. In doing so the definitions of Table 1 were

Table 1. Key elements of systems used for assessing vibration magnitude, exposure time and daily vibration exposure.

Name	Definition
Machine operation timer	Device for measuring the time for which a machine is operating (e.g. based on power take up or machine vibration)
Personal vibration exposure timer	Device for measuring the time for which either a hand is in contact with a vibrating machine or the body is in contact with the seat (e.g. using a switch to detect contact times with the vibrating surface)
Human vibration indicator	Instrument that measures a vibration value but whose measurements cannot conform with ISO 5349 and/or ISO 2631 - due to, for example, the transducer position being away from the hand grip or away from the centre of the seat.
Human vibration meter	Instrument conforming to ISO 8041 for measuring vibration magnitudes in accordance with ISO 5349 and/or ISO 2631
Personal human vibration exposure meter (PVEM)	Instrument for measuring daily vibration exposures, based on a measurement that directly determines the vibration exposure of a worker.[a]

[a]No assumptions are made at this stage as to whether a PVEM fully meets the requirements of ISO instrumentation or vibration measurement standards. Such an instrument might, for example, make a measurement on the hand of an operator that has been shown to be (within acceptable tolerances) equivalent to a measurement in accordance with ISO 5349-1.

developed to clarify instrumentation capabilities. ISO/TC108/WG33 as a consequence is considering the hand-arm vibration standard, such as ISO/NWIP 8041-2 standard. The examination of Pitts and Kaulbars [4] predated the emergence of wearable device to determine directly on the tool user the vibration risk transmitted to the user.

Pitts and Kaulbars recognized that prior to 2005, when the instrumentation standard was developed, most instruments on the market where aimed at researchers or vibration specialists. While the research market may need precision and reliability of measurement, employers aiming to protect their workforce have a need for practical, routine monitoring of vibration exposure which is not consistent with these instruments. Despite acknowledging the need for ease of data collection, recognition of the standard drove most of the instrumentation reviewed by Pitts and Kaulbars to arrange that sensors come into direct contact with the tool handle or to compromise and measure only vibration exposure time. Unfortunately, vibration magnitude has a greater influence on risk of injury relative to time as the dose is calculated from the square of the magnitude [5] and vibration magnitude is known to be largely variable depending on the condition of the tool, the material being worked on, the quality of tool accessories and the skill of the worker. An instrument realizing all the functionality of a PVEM was therefore considered the most useful. A PVEM was also considered the most challenging to develop a standard for, due to the practical challenges of the mounting of the sensors on the tool grip point. Ironically, Annex D of the ISO 5349-1 standard identifies the limitations of measuring hand transmitted vibration on the tool handle principally due to the tool operator's influence on the level of vibration transmitted to him while he is in contact with the tool. Within clause 4.3 of this standard, it is stated that the characterization of the vibration exposure is assessed from the acceleration of the surface in contact with the hand as the primary quantity. Therefore, ISO 5349-1 [5] assumes that the hand-transmitted vibration exposure magnitude is the tool handle vibration measurement, while the reality is that the hand-transmitted vibration is affected by the many factors listed in Annex D of ISO 5349-1 standard as shown in Fig. 2.

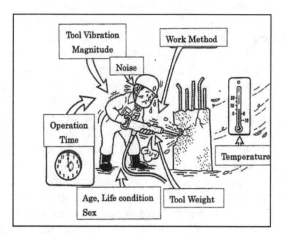

Fig. 2. Factors likely to influence the effects of human exposure the hand-transmitted vibration in the working conditions of Annex D of the ISO 5349-1 standard.

For many years, the factors outlined within Annex D of ISO 5349-1 have not been adequately captured when making an assessment of hand-transmitted vibration exposure for the purposes of prevention of HAVS in real world environments. A desire by employers to adhere strictly to the ISO 5349-1 standard may be contributing to inaccurate dose assessments and inferior outcomes for the worker. Wearable sensors mounted on the tool user offer the opportunity of determining the vibration transmitted to the user potentially offering a more credible answer to the need for practical everyday monitors.

3 New Consideration of Hand-Transmitted Vibration Measurement Wearable Sensor

Within this section the suitability of one wearable personal vibration exposure meter for use in the effective management and reduction of vibration exposure will be examined. Compliance with ISO 5349-1 & ISO 8041 is not claimed by the manufacturer and therefore this paper does not seek to demonstrate absolute equivalence, rather it merely seeks to demonstrate the relevance of data acquired from such a device in the effective management of risk. See Fig. 3.

The HAVwear system (Reactec Ltd) combines a number of emerging technologies to provide the user with real time vibration exposure data pertinent to their work activities via an LCD and audible warnings in addition to a suite of analytics tools to enable the design out of harmful vibration. A ruggedized wearable device for data acquisition is coupled with an IOT enabled docking station to upload vibration data to a cloud based analytics platform.

The wearable device mounts to an operators wrist by way of an adjustable nylon webbing strap and comprises a 3 – axis linear MEMS accelerometer sampling at 1.6 kHz for 0.66 s every 1.5 s. A frequency range from 3 Hz to 650 Hz is captured. Acceleration data from each axis is converted independently from time domain to frequency domain through a 1024 point Fourier analysis incorporating a Hanning window function to generate 512 discrete magnitude values for each axis. The corresponding frequency point magnitude values for each axis are then combined to create an overall magnitude value for each point on the spectrum. A proprietary transfer function is then applied to each value across the spectrum to calculate the instantaneous magnitude detected at the wrist. This transfer function compensates for attenuation of the biomechanical structures of the hand and is designed to provide parity with measurements taken on the tool in accordance with ISO 5349-1. Overall exposure is calculated by means of a rolling RMS for the duration of the trigger time. Parity with measurements taken in accordance with existing ISO standards is desirable for calculating exposure as all current research and knowledge on the pathogenesis and progression of the disease is based on exposure calculated to these standards.

The instrument's capabilities have been examined in two recent bodies of work. A UK independent research and consultancy services organization, the Institute of Occupational Medicine [6], carried out an extensive exercise to compare the vibration magnitude determined by the wearable sensor with that taken concurrently in full compliance to the standards over a broad range of tools. Maeda et al. [7] examined the strength of correlation of the wearable's vibration determination and the human response to vibration.

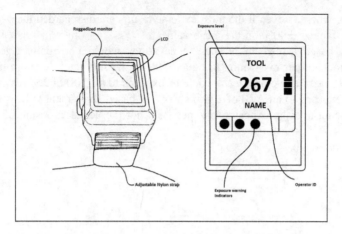

Fig. 3. Wearable Personal Vibration Exposure Meter (HVW-001, Reactec Ltd).

The true value of the wearable device comes from its ability to capture the variance in exposure and deviation from the anticipated risk. The traditional risk assessment approach of task based assessment relies on tool performance, substrate consistency, operator proficiency and vibrating tool work load allocation all remaining constant which in practice rarely happens. Additionally, it has been found that there is no typical working environment across many industries including manufacturing as tool use fluctuates from day to day and month to month. A risk assessment will be required to take this into consideration but will be open to error when relying on sampled data as the resulting average exposure estimation is based on a limited timeframe which can never hope to capture all conceivable fluctuations in activity and behavior present within a real-world environment. Continuous monitoring reduces the margin of error associated with this estimation [8].

To illustrate the variability from tasks a detailed risk assessment exercise was carried out on the work of 14 tool operators in multimen teams, each excavating the same size hole within the same grade of road surface. Each operator used the same tool type for which the duty holder had determined a vibration magnitude of 12 m/s^2 through the use of extensive live testing in accordance with ISO 5349. Note this was well in excess of the tool vibration data declared by the manufacturer of 4.2 m/s^2. The duty holder had determined that, was inappropriate for the type of road surface under test. A mix of site teams were used per excavation which consisted of between two and four man operator teams.

Figure 4 displays the calculated total exposure risk in A(8) m/s^2 for/at each excavation broken down by the contribution to each excavation by each operator. Each team accomplished the same task within different time durations. The colors within each excavation show the relative contribution from each team member of each excavation. The wide range of exposures clearly illustrate the limitations of a static risk assessment, whereby even if the most extreme results are disregarded the exposure per excavation ranged ±40% on the mean value.

A task based assesment of these excavations would potentially determine the average exposure per excavation to be an A(8) of 4.2 m/s^2 and therefore if the work was evenly shared a two man team would be exposed to an A(8) of 2.9 m/s^2 and a three man team exposed to an A(8) of 2.4 m/s^2. A task based assessment will typically only account for an average exposure risk per task, not the actual exposure of individual operators.

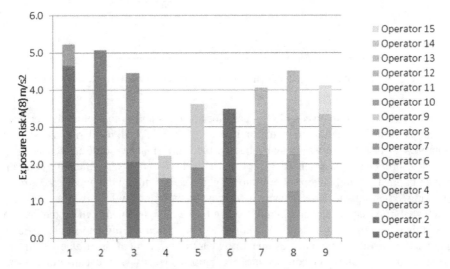

Fig. 4. Calculated total exposure in A(8) m/s^2 for each excavation

The assessment also included live tool vibration testing which took place during these excavations. Two methods were used to determine the vibration level.

1. An ISO5349-1 measurement using a reference instrument (HVM100, Larson Davis) operated by a skilled technician
2. Data collected using industrial wearable device (HVW-001, Reactec Ltd.) on a HAVwear worn by the tool operator.

The data from both is depicted in Fig. 5, which shows the numerical average of the RMS vibration level determined by the two devices. Strong correlation can be seen between the wearable sensor and the ISO5349 compliant reference instrument with an overall average for the reference instrument of 11.1 m/s2 and 11.9 m/s2 for the wearable device.

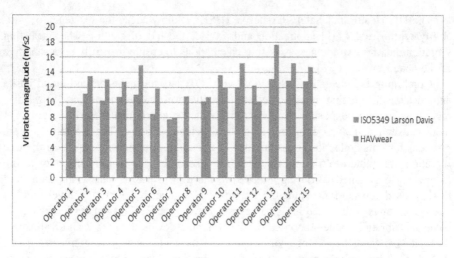

Fig. 5. The data from both companies, such as Larson Davis and HAVwear.

Figure 6 depicts the calculated maximum HAV exposure for each individual when excavating just one hole from the actual trigger time and real-time vibration data from the wearable device. Using a conventional task based assessment with a minimum two operator team the expected max risk would be an A(8) of 2.9 m/s². In reality, 5 out of the 14 operators exceeded this level. This data leads the investigators to conclude that job rotation is unlikely to be as expected without monitoring and furthermore each operator is unique in technique and physicality, which can lend itself to a high level of variability in exposure risk. Operator 5's exposure was close to the limit exposure level of 5 m/s² and 6 times that of some other operators tasked with the same duties. Given that there is no safe level and employers are required to reduce the exposure to as low as is reasonably practical it is clear in this case that the risk is being concentrated on certain individuals without their knowledge.

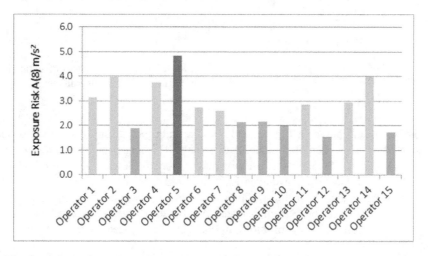

Fig. 6. Calculated maximum HAV risk exposure for each individual for one excavation

The UK Health and Safety Executive (HSE) acknowledges the existence of variability within real world application and advises the use of caution when selecting vibration data for extrapolated exposure calculations to ensure it is truly representative of the actual exposure [8].

Combining the wearable technology with IOT data transmission and cloud based analytics has for the first time allowed for the collection and review of truly "Big Data" regarding vibration exposure. Analysis of this data suggests that there is a significant delta between perceived exposure levels and actual exposure. The wearable system provides for data collection in two modes of operation. The first mode senses the true vibration transmitted to the tool operator and a second mode that calculates exposure from a pre-programed RFID tag attached to the tool in use. By analyzing data from heavily used tools we can plot the manufacturer's declared vibration level and the average assumed value used in risk assessments from the RFID tag against a histogram of actual vibration values recorded on the wearable device. Figures 7 and 8 illustrate the delta between the average vibration value used during risk assessments of one specific tool used and the histogram of actual vibration recorded by the wearable device over 2,850 h of trigger use of the tool. The X axis in Figs. 7 and 8 denotes the vibration magnitude in discrete ranges. The Y axis denotes the number of days in which individual operators experienced the level of vibration at each discrete range when using the tool.

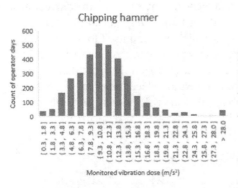

Fig. 7. Operator day instances of in-use monitored vibration magnitude – chipping hammer

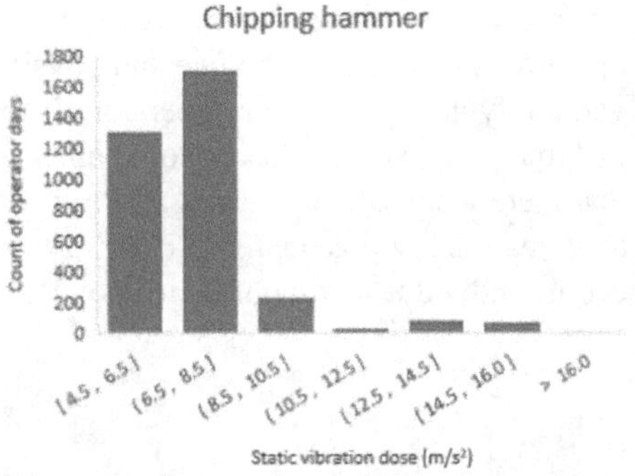

Fig. 8. Operator day instances of static vibration magnitude – chipping hammer

For the tool shown in Figs. 7 and 8, a separate tool testing exercise had been conducted on a live demolition site. A series of 30 X single minute duration trigger tests compliant to ISO5349 yielded an average vibration level of 11.9 m/s2 while a monitor worn during testing averaged 13.8 m/s2. Analysis of this large data set coupled with the field validation of the specific tool, lead the investigators to conclude that a significant delta must be present between the assumed static value used in the risk assessments of analysed organisations and the actual risk faced by many of their tool users.

Beyond ensuring exposure is kept within regulated limits in the short term, the ability to have real time continuous monitoring data which is traceable to tools, tasks and operators can aid significantly in the process of designing out exposure through identifying the source. Targeted procurement, process optimisation, improved tool maintenance and operator training are enabled through access to this data and can lead to significant reduction in exposure. The very nature of a wearable device that provides real time feedback to the operator can facilitate behavioral changes and a level of awareness that are not possible through periodic interventions.

In UK legislation, the Control of Vibration at Work Regulations of 2005 stipulates that simply keeping exposure below the action and limit values is not sufficient [9]. To ensure compliance, employers must reduce exposure levels to lowest practicable levels. The process of reducing exposure and also providing evidence of this is greatly aided with the advent of real time exposure monitoring and associated analytical tools. Figure 9 illustrates the progress made by a major global construction organisation in reducing exposure to hazardous hand arm vibrations through the use of real time monitory, and illustrates how leveraging the power of analytical data to drive decision making within the organisation can have significant and measurable impact. The Y axis of the graph denotes the exposure as a percentage of predefined exposure limit values for monitored operators utilizing the UK points based system.

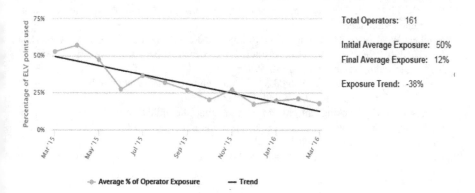

Fig. 9. Workforce average exposure within a major construction organisation following the introduction of wearable monitoring technology

4 Conclusions

The ISO 5349-1 AND the EU directive a relationship between health damage and the daily exposure to hand arm vibration. However, this standard and its emphasis of making measurements on the tool with instruments following ISO 8041-1 is recognized as suited to research environments, but not to regular employee risk monitoring. Given the wide range of emitted tool vibrations over a tools' normal course of use, it would seem logical that employers need access to practical everyday monitoring devices to prevent harm to their employees. The authors have presented data to highlight the variability that exists in every day tasks when examining closely a controlled assessment of work and the practical way in which a wearable devise can gather the necessary day that will affect an individual's direct health risk.

The authors believe there is an opportunity to more intelligently develop HAV exposure control measures with a more detailed insight to the drivers of HAV exposure risk than that developed from generic HAV risk assessments based on assumed static vibration dosages.

References

1. Advances in Human Factors in Wearable Technologies and Game Design, Proceedings of the AHFE 2018 International Conferences on Human Factors and Wearable Technologies, and Human Factors in Game Design and Virtual Environments. https://link.springer.com/book/10.1007/978-3-319-94619-1
2. Bovenzi, M.: Exposure-response relationship in the hand-arm vibration syndrome: an overview of current epidemiology research. Int. Arch. Occup. Environ. Health. **71**(8), 509–519 (1998)
3. UK Health and Safety Executive
4. Pitts, P.M., Kaulbars U.: Consideration of standardization requirements for "vibration dosemeters". VDI-Berichte Nr. 2190 (2013)
5. BS EN ISO 5349-1:2001 Measurement and evaluation of human exposure to hand-transmitted vibration
6. Mueller, W., Cowie, H., Groat, S., Graveling, R.: Report 606-02372-03, Review of Data Measured by Reactec HAVwear, Institute of Occupational Medicine, Research Avenue North, Riccarton, Edinburgh EH14 4AP, UK
7. Maeda, S., Taylor, M.D., Anderson, L.C., McLaughlin, J.: Determination of hand-transmitted vibration risk on the human. Int. J. Ind. Ergon. **70**, 28–37 (2019)
8. Anderson, L.C., Buckingham, M.-P.: Investigating the utility of wrist worn vibration monitoring in the effective management of exposure to dangerous vibrations within the work place. Reactec Ltd 290–135 (2017)
9. HSE L140: The Control of Vibration at Work Regulations 2005, UK Health and Safety Executive

Gait Monitoring System for Stroke Prediction of Aging Adults

Hongkyu Park[2(✉)], Seunghee Hong[1,2], Iqram Hussain[1,3],
Damee Kim[1,2], Young Seo[1], and Se Jin Park[1,2,3(✉)]

[1] Korea Research Institute of Standards and Science, Daejeon, South Korea
{hsh82622,iqram,dameeing,young2da,sjpark}@kriss.re.kr
[2] Electronics Telecommunication Research Institute, Daejeon, South Korea
hkpark@etri.re.kr
[3] University of Science & Technology, Daejeon, South Korea

Abstract. Health has become a major concern nowadays. People pass significant amount of time of daily life on walking, moving here and there and so on. Some health complexity happens during walking like heart problem, stroke etc. Stroke patient has unbalanced gait pattern compared to normal person. The Internet of Things (IoT) plays an important role in the development of connected people, which offers cloud connectivity, smartphone integration, safety, security, and healthcare services. Insole Foot Pressure sensor and accelerometer will be attached to the foot for gait speed, foot pressure and other gait pattern. Gait parameters of 68 Stroke patients and 208 Elderly healthy persons have been gathered in Chungnam National University Hospital Rehabilitation Center, Daejeon, South Korea. Gait parameters are foot pressure, gait acceleration etc. Dynafoot2 Insole sensor used for data acquisition. Subjects walked and perform activities like walking, sitting, standing, doing some regular activities during gait data acquisition. Area under curve (AUC) of performance curve for C4.5, SVM, Random Tree, Logistic Regression, LSVM, CART algorithms are 0.98, 0.976, 0.935, 0.909 and 0.906, 0.87 respectively. A gait monitoring system has been proposed for stroke onset prediction for stroke patient. IoT sensors are used to gather gait data and machine learning algorithms are used to classify gait pattern of stroke patient group and normal healthy group. In future, sensors such as EEG, EMG will be used to improve system reliability.

Keywords: Stroke · Gait · Health monitoring · Classification algorithms

1 Introduction

With the increased understanding of health and life; as well as advancement of medical technology, Health has become a major point of interest nowadays. For well-being, it is necessary to spend a significant amount of time of daily life on walking, moving here and there [1]. Some health abnormality happens during walking like heart problem, stroke and so on; ability of movement reduces due to health complexity such as stroke [2].

© Springer Nature Switzerland AG 2020
T. Ahram (Ed.): AHFE 2019, AISC 973, pp. 93–97, 2020.
https://doi.org/10.1007/978-3-030-20476-1_11

Stroke is one of deadly disease; especially for above the age of 60 years, and its proportion is rising [3, 4]. Many health abnormality happens after stroke. Stroke is the sudden collapse of brain cells due to lack of oxygen, caused by blockage of blood flow to the brain or breakdown of blood vessels [5]. The stroke symptoms are a weakness in the arm or leg or both on the same side, loss of balance, sudden headache, dizziness, coordination problems, vision problems, difficulty in speaking, and weakness in the face muscle [6]. Gait disorders is observed as one of the most common disabilities after stroke [7–9]. Stroke effected persons lose conscience and ability to contact emergency services or hospital. Without immediate detection and treatment of stroke, it is very difficult to prevent and recover complications [10].

For accessing gait pattern of an individual; acceleration, foot pressure are generally measured. In most of case, accelerometer, gyro sensor, insole pressure sensor, pedometer, GPS (Global Positioning System), footswitches are used to capture gait parameter [11]. Several Signiant parameters are extracted such as number of steps or walking bout, step time and stride time, step length and stride length, credence, GRP (Ground Reaction Force), speed and so on.

The Internet of Things (IoT) plays an important role in the development of connected people, which offers cloud connectivity, smartphone integration, safety, security, and healthcare services [12, 13]. Several researchers are working to develop IoT based health monitoring system for various purposes [14, 15]. Gait monitoring is one of most interesting application of wearable devices for clinical and daily regular activity monitoring. Gait monitoring is also widely used in healthcare, sports.

This paper focused on briefly explaining classification of Gait patterns of Stroke Patients and Elderly Adults using machine learning (ML) algorithms such as C4.5, SVM, Random Tree, Logistic Regression, LSVM, CART. Here we also proposed real-time gait monitoring system especially for elderly persons in order to successfully detect stroke onset.

2 Methodology

2.1 Gait Monitoring System

Proposed Gait monitoring system is consists of insole foot Pressure sensor and accelerometer which will be attached to foot as shoe insole for gathering gait speed, foot pressure and other gait signals. As shown in Fig. 1, an insole foot pressure and accelerometer is designed and prototyped and tested. Gait pattern such as gait speed, foot pressure etc. of normal person and stroke patient are significantly different from each other. Stroke patient has unbalanced gait pattern compared to normal person [7]. IoT devices and Machine learning technique are able to detect stroke onset of elderly adults. Entire system will feed subject's physiological data to cloud engine to compare real-time data and already stored reference data in order to detect stroke.

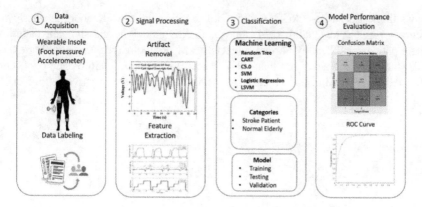

Fig. 1. Methodology of Gait classification.

2.2   Experimental Setup

Gait parameters of 63 Stroke patients and 208 Elderly healthy persons have been measured in Chungnam National University Hospital Rehabilitation Center, Daejeon, South Korea. Gait parameters are foot pressure, gait acceleration and Ground reaction force (GRP). Dynafoot2 Insole sensor used for data acquisition. Subjects walked and perform activities like waking, sitting, standing, doing some regular activities during gait data acquisition. Room temperature was maintained at 24 °C and relative humidity 40%.

2.3   Gait Feature Extraction

Matlab R2017a has been used to extract gait features from foot pressure, acceleration signal. Foot pressure and GRP (Ground Reaction force) are extracted from insole pressure signal and foot acceleration data has been extracted from accelerometer signal.

2.4   Gait Pattern Classification

IBM SPSS Modeler Software has been used to classify gait data of stroke patient and normal person. Logistic Regression, Support vector machine (SVM, LSVM) and CART, C4.5, Decision tree algorithms has been used to classify foot pressure, acceleration and GRP data of stroke patients and normal person. 50% Data has been used for training, 30% data for testing and 20% gait data for validation of classification models.

3   Results and Discussion

In order to evaluate classification accuracy, ROC (Receiver operating characteristic) curve and Gini coefficient are most effective tools [16]. Classification accuracy of different Machine Learning algorithms are listed as AUC (Area under curve) and Gini coefficient in Table 1. In the training phase of models, it is found that C5.0 model shows highest accuracy (AUC: 0.995, Gini: 0.993)) and CART model results lowest accuracy (AUC: 0.87, Gini: 0.74) among all ML models to classify gait pattern of stroke patients and healthy normal subjects. SVM, Random Tree, Logistic Regression, LSVM shows accuracy more than 90%; AUC: 0.978, Gini: 0.956; AUC: 0.94, Gini: 0.88; AUC: 0.91, Gini: 0.821 and AUC: 0.908, Gini: 0.816 respectively (Fig. 2).

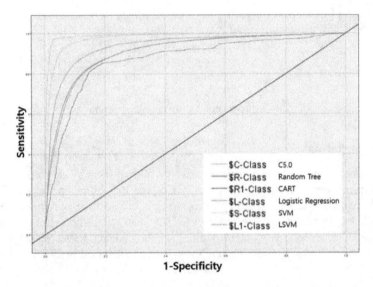

Fig. 2. Receiver operating characteristic (ROC) curve of six classifiers output.

In data testing and validation phase, classification algorithms shows similar trend as training phase. In pattern testing phase, C5.0 model shows highest accuracy (AUC: 0.983, Gini: 0.967)) and CART model results lowest accuracy (AUC: 0.866, Gini: 0.732) to classify gait pattern. In final validation step, C5.0 model again shows highest accuracy (AUC: 0.984, Gini: 0.968) and CART model results lowest accuracy (AUC: 0.869, Gini: 0.738) to classify gait pattern of stroke patients and healthy normal.

4   Conclusion

A gait monitoring system has been proposed for stroke onset prediction for stroke patient. IoT sensors are used to gather gait data and machine learning algorithms are used to classify gait pattern of stroke patient group and normal healthy group. In future, sensors such as EEG, EMG will be used to improve system reliability.

Acknowledgments. This work was supported by the National Research Council of Science & Technology (NST) grant by the Korea government (MSIP) (No. CRC-15-05-ETRI).

References

1. Lee, I.M., Buchner, D.M.: The importance of walking to public health (in eng). Med. Sci. Sports Exerc. **40**(7), S512–S518 (2008). Suppl
2. Clark, D.J., Ting, L.H., Zajac, F.E., Neptune, R.R., Kautz, S.A.: Merging of healthy motor modules predicts reduced locomotor performance and muscle coordination complexity post-stroke **103**(2), 844–857 (2010)
3. Lundstrom, E., Smits, A., Borg, J., Terent, A.: Four-fold increase in direct costs of stroke survivors with spasticity compared with stroke survivors without spasticity: the first year after the event (in eng). Stroke **41**(2), 319–324 (2010)
4. Norrving, B., Kissela, B.: The global burden of stroke and need for a continuum of care. Neurology **80**(3) Supplement 2, S5 (2013)
5. Lakhan, S.E., Kirchgessner, A., Hofer, M.: Inflammatory mechanisms in ischemic stroke: therapeutic approaches (in eng). J. Transl. Med. **7**, 97 (2009)
6. Faiz, K.W., Sundseth, A., Thommessen, B., Ronning, O.M.: Patient knowledge on stroke risk factors, symptoms and treatment options (in eng). Vasc Health Risk Manag. **14**, 37–40 (2018)
7. Balaban, B., Tok, F.: Gait disturbances in patients with stroke. PM&R **6**(7), 635–642 (2014)
8. Gor-García-Fogeda, M.D., Cano de la Cuerda, R., Carratalá Tejada, M., Alguacil-Diego, I.M., Molina-Rueda, F.: Observational gait assessments in people with neurological disorders: a systematic review. Arch. Phys. Med. Rehabil. **97**(1), 131–140 (2016)
9. Yavuzer, G., Kucukdeveci, A., Arasil, T., Elhan, A.: Rehabilitation of stroke patients: clinical profile and functional outcome (in eng). Am. J. Phys. Med. Rehabil. **80**(4), 250–255 (2001)
10. Ponikowski, P., et al.: 2016 ESC guidelines for the diagnosis and treatment of acute and chronic heart failure. **18**(8), 891–975 (2016)
11. Benson, L.C., Clermont, C.A., Bošnjak, E., Ferber, R.: The use of wearable devices for walking and running gait analysis outside of the lab: a systematic review. Gait & Posture **63**, 124–138 (2018)
12. Gubbi, J., Buyya, R., Marusic, S., Palaniswami, M.: Internet of things (IoT): a vision, architectural elements, and future directions. Future Gen. Comput. Syst. **29**(7), 1645–1660 (2013)
13. Hassanalieragh, M., et al.: Health monitoring and management using internet-of-things (IoT) sensing with cloud-based processing: opportunities and challenges. In: 2015 IEEE International Conference on Services Computing, pp. 285–292 (2015)
14. Park, S.J., Hong, S., Kim, D., Hussain, I., Seo, Y.: Intelligent In-Car Health Monitoring System for Elderly Drivers in Connected Car, Cham, pp. 40–44. Springer International Publishing (2019)
15. Park, S.J., et al.: Development of a real-time stroke detection system for elderly drivers using quad-chamber air cushion and IoT devices (2018). https://doi.org/10.4271/2018-01-0046
16. Fawcett, T.: An introduction to ROC analysis. Pattern Recogn. Lett. **27**(8), 861–874 (2006)

Facial Recognition on Cloud for Android Based Wearable Devices

Zeeshan Shaukat$^{(\boxtimes)}$, Chuangbai Xiao, M. Saqlain Aslam,
Qurat ul Ain Farooq, and Sara Aiman

Beijing University of Technology, Pingleyuan No 100, Chaoyang District,
Beijing, China
{zee,annie,sara}@emails.bjut.edu.cn,
cbxiao@bjut.edu.cn, se.saqlain@yahoo.com

Abstract. Facial recognition applications for Android Based Wearable Devices
(ABWD) Can benefit from cloud computing as they become easy to acquire and
widely available. There are several applications of facial recognition in terms of
assistance, guidance, security and so on. We can greatly reduce the processing
time by executing the facial recognition application on cloud, and clients will
not have to store the big data for the image verification on their local machine
(mobile phones, pc's etc.). Comparing to the cost of acquiring an equally strong
server machine, cloud computing increases the storage and processing power
with very less cost. In this research plan is to enhance the user experience of
augmented display on android based wearable devices, and for doing that, this
system is being proposed in which a person wearing Android based smart
glasses will send an image of an object to Hadoop (open-source software for
scalable, reliable, distributed computing) powered cloud server. Facial Recognition Application on cloud server will recognize the face from already present
database on server and then respond results to Android Based Wearable client
devices. Then android based wearable smart devices will display the detail result
in form of augmented display to the person wearing them. By transferring the
process of facial recognition and having the database on cloud server, multiple
clients no longer need to maintain their local databases and the device will
require less processing power which results in reduction of cost and processing
time.

Keywords: Keywords facial recognition · Android · Wearable devices ·
Cloud computing · Augmented reality

1 Introduction

Face recognition system is a computer application whose aim is to automatically
identify or affirm a person's face from a video frame from a live video source or a
picture [1]. One of the many ways to do this is by doing comparison of designated
facial features from the picture and a face record database. It is generally used in
security systems and there can be a comparison with other biometrics such as eye iris
recognition or fingerprint systems. There are many other uses of face recognition too,
but mainly all revolve around verification of the person. In simpler words, it can be said

© Springer Nature Switzerland AG 2020
T. Ahram (Ed.): AHFE 2019, AISC 973, pp. 98–110, 2020.
https://doi.org/10.1007/978-3-030-20476-1_12

as person of interest identification. Face recognition System plays a vital role in several applications such as person identification, face tracking, video surveillance, and human computer interaction. Algorithm for effective face and facial characteristic detection are required for applying to those tasks. In the present-day industry system of face recognition is proved to be highly effective. In today's technological world, the quick increase in usage of mobile devices and the explosive growth of mobile face recognition is one of significant application. At the same time there are many challenges mobile devices are facing in their resources such as low computing capacity, low battery life span, limited bandwidth and low storage [2]. Usually face recognition process is practiced by internal security and defense departments for the enforcement of law and order situation in the country at both public and private areas. Facial detection system performs an indispensable role in the variety of scenarios to recognize the target objects like extremist, kidnappers, offenders, malfeasants, missing and wanted people (Fig. 1).

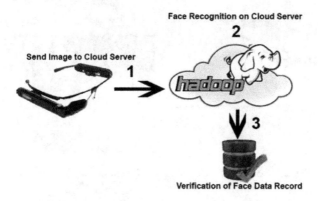

Fig. 1. Working of facial recognition system on cloud server

Currently, many biometrics techniques and strategies exist to verify any individual's identity by using specific and incomparable bio informatics features (e.g. face, iris and finger prints). However, the use of mentioned bio informatics features is very inconvenient for any on duty patrolling office because it is very time-consuming activity. Therefore, face detection of the targeted person has emerged as a dominating, promising and favorable technological process. Moreover, it is a real time and non-intrusive system that is capable of multiple object detection at the same time Hence, face recognition system is commonly deployed at public areas like banks, airports, patrol/gas stations, bus stops, shopping malls, roads, streets, subways, highways and seaports [3]. Surveillance Cameras are widely used for face-recognition in public places. But as it is observed, surveillance cameras provoke many glitches such as, installation of camera on roof or top corner's so that they remain unnoticeable and undetectable spots. In such scenario, top captured facial views are harder to recognize, another negative aspect is fixed location of camera. Criminals or intruders can trace out the location of fix camera, bypass it and plan to attack accordingly. Moreover, the easy obstruction of cameras provides outlaws more opportunities to evade or deactivate face recognition system completely. As a conclusion of above discussed deficiencies, there must be a cloud based computerized-eye wear (e.g. smart glasses) face detection system, in this method, a cloud service module (as server) and a computerized eye-wear (client) are considered as two basic elements. According to functionality, the officer employs a camera mounted eyewear client that detects the face and data to the cloud service. Cloud service matches the facial data with existing cloud database, identify the required person and return the result back to client. As a result, client warns the officer if desired person is identified and displays the related data on the screen of eyewear [4] (Fig. 2).

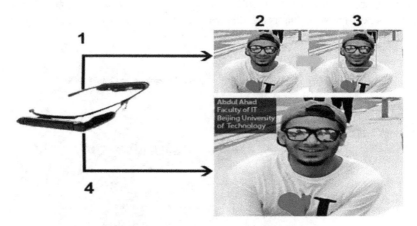

Fig. 2. Working of facial detection system on android client app

2   Literature Review

Steve Mann, the computerized eye-wear pioneer, has been designing and wearing computerized eyewear for spans, the gear increasing prominently in cleverness over time. In the development of wearable computer serial was exhibited by authors [5]. In the third generation, a see-trough augmented reality eyewear system named as STAR 1200XL was introduced its transparent widescreen video display empower and facilitates to observe the reality. Computerized information like pictures, script video and text are demonstrated on the screen. Smart Glasses is a type of wearable device with an ocular head-mounted display. The aim of developing smart glasses is to manufacturing a mass-market ubiquitous computer. Smart Glasses spectacles info in a format like smartphone i.e. hands-free format. People who wear smart glasses interconnect with the Internet using natural language speech commands.

2.1   Smart Glasses/Eyewear

Smart glasses are getting smarter and augmented reality specs are finally approaching prime time - and plenty of startups are getting into the space. It's not just about slapping a camera on your face, either. AR, fitness tracking and mixed reality are all powering the next generation of smart eyewear. From first-person videos and photos to turn-by-turn directions, health-sensing and facial recognition, the invasion of the smart glasses is very much alive. We don our future-specs to reveal both the best smart glasses on the market and the upcoming devices we believe have the potential to take connected specs mainstream in the next few years.

Smart Glass API(SGAPI): Smart Glass/Eyewear is a device which has Android Operating System. What we see via smart Glass is mainly a personalized Launcher or home screen application, it can be described as a timeline of cards about current and past events. It has a slightly different theme. This makes it really exciting for Android programmers to develop for smart glass, All the familiar Android framework APIs can be used for development. Though wearing smart glass feels completely different than using a mobile phone. There's a big alteration in designing applications. Not only the UI is different: An existing application cannot be ported to android based smart glass just like that. Use cases have to be intended specially for smart glass [7]. Some features of the application might not make sense on smart glass. Some other remarkable features may only be feasible on smart glass. It's nearly not possible to get a feeling for that without using smart glass for few days [8].

Presently it can be decided between two ways to do the development for smart glass: The Mirror API or Glass Development Kit (GDK).

Mirror API: The Mirror API has been the first API that has been introduced for Smart Eyewear's. It's a server- side API impact that the applications don't run on Glass itself but on your server and it's your server that intermingles with Glass. The Mirror API is prodigious for pushing cards to the timeline of Glass and sharing content from Glass with your server application [9]. Some examples of applications that could use the Mirror API are Twitter Client and Context-Aware Notifications.

Smart Glass Development Kit (SGDK): With the SGDK you can construct Android applications that can run on glass without any issue. Think of the SGDK as Android SDK with some extra APIs for Smart Glass. It's constituent to mention that the SGDK is at this time in an early promo state [10]. Presently some important parts are missing and the API is incomplete. When developing Glass, you have two options how your application should show up on Glass (Fig. 3).

The application shows up left of the Glass clock as a card in the timeline. We have two rendering options for these cards:

High Frequency Rendering: The background service renders directly on the live card's surface. We can draw anything and are not limited to Android valuations. Additionally, the card can be updated many times within a second.

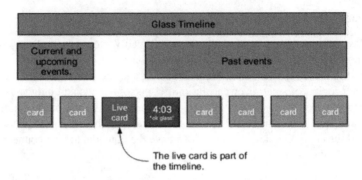

Fig. 3. Live card shows up in timeline of smart glass

Low Frequency Rendering: Remote Views are used to render cards. Ruminate of it as a Home screen widget on Android phones. For updating these views, a background service is used. We only update the views every now and then.

Immersion: An immersion is at the bottom of regular Android activity. For activity to look like a timeline card no need to allocate a theme to activity or use the Device Default theme as base for customization. Even though touch pad of Smart Glass can be used almost like a d-pad: Most input- related Android widgets can be avoided. They don't make much sagacity on Glass because of not using a touch screen. As a substitute try to use gestures with the *GestureDetector* class or voice input. Use the Card class and its *toView()* method for creating a view that looks like regular Glass card (Fig. 4).

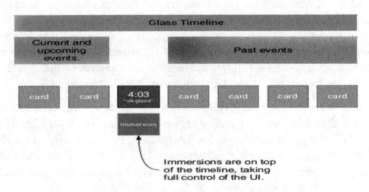

Fig. 4. An immersion is not part of timeline but replacing it.

2.2   Face Recognition

Face recognition is done by comparing digital image captured by camera or from video frame with the existing image in database. There are multiple algorithms for face recognition based on special feature of faces which are taken out from features of target's face. Dr. Bonsor used shape and size of nose, eyes, jaw, cheeks and their relative position in the algorithm [11]. One of popular face recognition approaches practice Principal Component Analysis via eigen faces [12], another pattern is Elastic Bunch Graph Matching that uses Fisher face algorithm [13]. Similarly, face feature is extracted by Local Binary Pattern Histogram (LBPH) [6]. In addition, some professional enterprises have developed their own face recognition systems such as "Lambda", "Google's Picasa", "Apple's iPhoto". There is a three-dimensional (3D) face recognition trend with improved accuracies because its data is pose variant and consist of more facial surface information [14] A comprehensive pose invariant automatic system was presented by, Prof. Loannis [15]. According to design, registration of 3D facial scan was done by using composite alignment technique. As compared to 2D systems, 3D face recognition system achieves higher accuracy. But 3D system is much expensive alongside computational delay due to which has less practical implementation. Furthermore, 3D face recognition requires subject's cooperation to capture data from 3 dimensions that delays the process. The author [16] used OpenCV for face detection due to open source and highly supportive behavior for Eigen faces, Fisher faces and Local Binary Patterns Histograms (LBPH). Eigen faces [17, 18] demonstrates illumination-robust property reliable under multiple illuminations. Eigen faces approaches also support to update face recognizer due to which it lessens the running delays and memory space [19].

2.3 Cloud Computing

Cloud computing has been envisioned as the next generation computing model for its major advantages in on demand self-service, pervasive network access, locality independent resource pooling and transference of risk [20]. Cloud Computing is the latest developments of computing models after distributed computing, parallel processing and grid computing [21]. In recent years, rapid growth and remarkable success of cloud computing has gained the attention though out the globe. It comprises on a pool of configurable, customizable and visualized computer resources including hardware and software over the network. It transfers amenities to host and perform immediate analysis on the datasets [22]. Map Reduce is considered as a famous programming models in the cloud technology [23]. This model is practiced on cluster of commodity servers for large scale of information in distributed processing. Another improved cluster file system was introduced by Ananthanarayanan et al. [24] for Map Reduce applications. Instead of using traditional cluster file management, they arranged files in a consecutive set of blocks on the same disk called Meta block. In addition, analysis on bulky datasets can be also be done by Apache Pig [25] platform which was designed by using Map Reduce on the top of Hadoop. The name Map Reduce comes from the two kinds operations in functional programming language: Map and Reduce. In functional programming language, function has no side-effect, which means that programs written by functional programming language can be more optimized in parallel programming. In functional programming language, Map and Reduce take functions as parameters, which are fully used for Map Reducing [26].

2.4   Apache Hadoop

For manipulating the large data sets in distributed computing, an open source Java based programming framework Hadoop is used [27]. Apache Software Foundation has funded this project. By using Hadoop, it is being tried to run the applications with hundreds of nodes with thousands of tera bytes. It consists of distributed file system that supports rapid data transfer rates among nodes. it further allows the system to keep functioning with no interruption in case of a node failure. This methodology goes under the risk of catastrophic system failure, even if a most important number of nodes turn into inoperative. The Apache Hadoop framework consists of the succeeding modules [28].

Hadoop Common: Covers utilities and libraries required by other Hadoop sections.

Hadoop Distributed File System: It's a scattered filesystem that provisions data on the commodity machineries, giving hefty extent of combined bandwidth across cluster.

Hadoop YARN: It is a resource-controlling podium that has key role to cope compute resources in clusters and using them for planning of users' applications.

Hadoop Map-Reduce: It is a software model for massive scale data processing
All the components in Hadoop are intended with an essential supposition that hardware failures (of individual machines, or racks of machines) are common and so should be automatically controlled in software by the framework. Apache Hadoop's MapReduce and Hadoop Distributed File System are originally derived respectively from Google's Map-Reduce and Google File System (GFS) papers (Fig. 5).

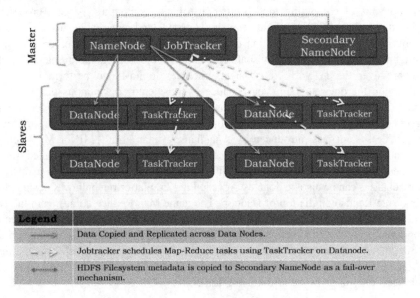

Fig. 5. Overview of apache hadoop

3   System Architecture and Requirements

System is designed such that it does not need much time to load pages. Develop each Interface page to be fast-loading. All images, graphics and multimedia are optimized to the appropriate size and quality to ensure that they load within a few seconds. The design of system is user friendly that any type of user shall learn to operate the system after a short training. The interface by which user interact is easy to navigate throughout the system. Its reliable in its best capacity to detect face and recognize the person in it. System availability is the time when the application must be available for use. Application provide a secure access to the cloud data and 24/7 availability to allow the users to access the system anytime they want. The system is able to recover or heal up when an error occurs. If it does not heal then it should at least display an error message.

3.1   System Hardware Requirements

Following are the systems hardware requirements for the system of cloud based facial recognition for android based smart glasses (Table 1).

Table 1. Hardware requirements

Component	Web server	Cloud server
Processor	Intel Core i5 5th Gen	Intel Core i7 5th Gen
Hard Drive	1 Terabyte	1 Terabyte
RAM	8 GB 1600 MHz	16 GB 1600 MHz
Network capability	Wireless and LAN both	Wireless and LAN both

3.2   System Software Requirements

Succeeding are the software requirements to run the system of cloud-based facial recognition for android based glass (Table 2).

Table 2. Software requirements with version

Application	Web server	Cloud server	Smart glasses
Android OS	✗	✗	✓ 4.0+
SSH	✓	✓	✗
PHP	✓	✗	✗
Java JDK	✗	✓ 8+	✗
Apache	✓	✗	✗
OpenCV	✗	✗	✓
Maven	✗	✓	✗
Hadoop	✗	✓ 2.6	✗
CURL	✓	✗	✗

3.3   System Architecture

System architecture is the conceptual model that defines the configuration, behavior, and more views of a system. An architecture portrayal is a formal portray and representation of a system, structured in a way that supports rational about the structures and behaviors of the system. Face recognition refers to a very broad range of applications including secure access control, crowd surveillance, forensic facial reconstruction and police identification. All systems belong to the category of static face recognition, and their performance is bound by the hardware specification of machine on which they run. The purpose of this study is to remove the hardware bound from any face recognition method, by doing the processing using cloud computing. The cloud-based facial recognition system for android based smart glass has three main modules. Following figure gives an overview of system architecture (Fig. 6).

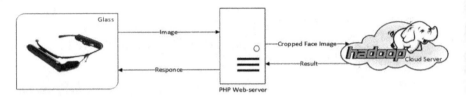

Fig. 6. An overview of system architecture

4   Implementation of the System

In this section we will discuss state of the art implementation details supported by procedural approach of the system which are function-oriented. This system is a networked system, so it requires multiple platforms to run. As mentioned in name, this system is mainly for android based glass. For web server Linux (Ubuntu) has been used, as it is more convenient to use and configure as web servers. The third element of this system is cloud server. For that too Linux is used, because it is Hadoop based cloud server.

4.1   Implementation of Face Detection and Displaying Result on Glass

Smart glass application appears in the timeline of Android based glass after installation. User can use it by taping on it. After the application starts running, it will open a camera with active face detection feature. On detecting a face, user should take the picture, and then wait for the result. After that smart glass is listing for further instructions. Home screen for smart glass is just like any other android mobile device. By taping on this screen smart glass will take user to the installed application list. After taping on this screen user will be taken to the screen of available applications. There user can swipe forward or backwards to bring the desired application in main screen. In this case, the desired application is called "Smart Glass". By tapping on Smart Glass, application will run, and user will then experience the application interface. On Opening the application, smart glass will open up camera, with active face detection feature. When the person wearing smart glass will look towards any other person, the application will try to detect his/her face. When a face is detected, smart glass will mark the face with a square box as shown in Fig. 2. It means now it is okay to take the picture to add it on live Camera feed.

4.2 Program Code for Face Detection:

```
public Mat onCameraFrame(CvCameraViewFrame inputFrame) {
    mRgba = inputFrame.rgba();
    mGray = inputFrame.gray();
    myRgba = mRgba;
    Imgproc.equalizeHist(mGray, mGray);
    if (mAbsoluteFaceSize == 0) {
        int height = mGray.rows();
    if (Math.round(height * mRelativeFaceSize) > 0) {
mAbsoluteFaceSize = Math.round(height * mRela-
tiveFaceSize);}}   MatOfRect faces = new MatOfRect();
if (mJavaDetector != null)
mJavaDetector.detectMultiScale(mGray, faces, 1.1, 2, 2,
        new Size(mAbsoluteFaceSize, mAbsoluteFaceSize),
    new Size());
        Rect[] facesArray = faces.toArray();
            for (int i = 0; i < facesArray.length; i++)
                Core.rectangle(mRgba, facesAr-
    ray[i].tl(), facesArray[i].br(), FACE_RECT_COLOR, 1);
                return mRgba;
}
```

After picture has been taken the user will be displayed a "Please Wait" message. Behind this message application will send this image to server, and server will response in the form of person information. Smart glass application will take a little while for getting response from server, and mean while a "Please Wait" message will remain on user screen. After getting the response from the server, smart glass application will display the result on the display as shown in Fig. 2.

4.3   Program Code for Displaying Result

```
public void onCreate(Bundle savedInstanceState) {
   super.onCreate(savedInstanceState);
try {String response = im-
gUpload.uploadImageToServer(ImgPath);
            String[] responceArray = response.split("##");
            String successTag = "correct";
            String wrongTag = "WRONG";
   if(responceArray[1].toLowerCase().contains(successTag.
   toLowerCase())){Log.i(TAG, "IMAGE UPLOADED");
         mCard.setText(responceArray[0]);}
   if(responceArray[1].toLowerCase().contains(wrongTag.to
   LowerCase())){mCard.setText(responceArray[0]);}else{
   mCard.setText("Recognition Failed" + response);}
   } catch (Exception e) {e.printStackTrace();}}
```

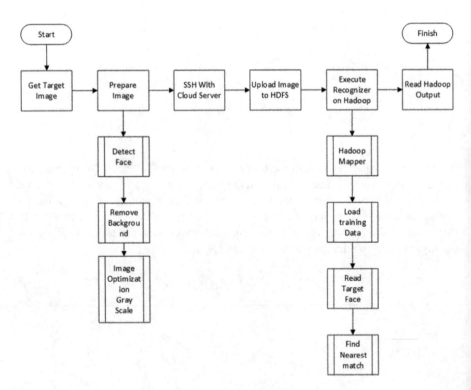

Fig. 7. Flowchart of recognition process

4.4   Implementation of Recognition Process on Cloud

Recognition component is the core component of the system. It will run on cloud computing using Hadoop. Therefore, all the resources required by this operation will be managed by Hadoop. The application will be performed as a background service. Which means, user will not be aware of what is going on. User will be shown a please wait message, and in background the system will perform all the required tasks, then the system will display the result to the user who is using smart glasses on top of the image he sent for recognition (Fig. 7).

5   Conclusion

The implementation of the smart glass project has clearly demonstrated the concepts underlying the system to be entirely feasible. This system is very helpful for running heavy applications on mobile device, by storing the big data and performing the heavy processing on cloud server. This system has many applications in real life, for example finding a person of interest (thieve or a criminal) using android smart glasses with a database on cloud server. Law enforcement agencies can make use of this system to pinpoint suspects very efficiently by saving almost 99.5% of time. There can be many other useful applications for this system such as helping disabled people who are unable to remember people, helping a tourist to meet new people and helping the job interviewer to view candidate's previous records etc.

References

1. Lenc, L., Král, P.: Automatic face recognition system based on the SIFT features. Comput. Electr. Eng. **46**(Supplement C), 256–272 (2015)
2. Aminzadeh, N., Sanaei, Z., Ab Hamid, S.H.: Mobile storage augmentation in mobile cloud computing: taxonomy, approaches, and open issues. Simul. Model. Pract. Theor. **50** (Supplement C), 96–108 (2015)
3. Wang, X., et al.: Person-of-interest detection system using cloud-supported computerized-eyewear. In: 2013 IEEE International Conference on Technologies for Homeland Security (HST) (2013)
4. Chaudhry, S., Chandra, R.: Face detection and recognition in an unconstrained environment for mobile visual assistive system. Appl. Soft Comput. **53**(Supplement C), 168–180 (2017)
5. Mann, S., Mann, S.: My Augmediated Life. IEEE Spectrum (2013)
6. Wikipedia. Google Glass. (2018, January 5). In: Wikipedia. 2018; Available from https://en.wikipedia.org/wiki/Google_Glass
7. Rahman, S.A., et al.: Unintrusive eating recognition using Google Glass. In: Pervasive Computing Technologies for Healthcare (PervasiveHealth), 2015 9th International Conference on. IEEE (2015)
8. Lv, Z., et al.: Hand-free motion interaction on google glass. In: SIGGRAPH Asia 2014 Mobile Graphics and Interactive Applications. ACM (2014)
9. Tang, J.: The Mirror API, in Beginning Google Glass Development. Springer, pp. 297–336 (2014)

10. Ha, K., et al.: Towards wearable cognitive assistance. In: Proceedings of the 12th annual international conference on Mobile systems, applications, and services. ACM (2014)

11. Bonsor, K., Johnson, R.: How facial recognition systems work. HowStuffWorks. Com Np (2001)

12. Turk, M., Pentland, A.: Eigenfaces for recognition. J. Cognit. Neurosci. 3(1), 71–86 (1991)

13. Lee, H.-J., Lee, W.-S., Chung, J.-H.: Face recognition using Fisherface algorithm and elastic graph matching. In: Proceedings of 2001 International Conference on Image Processing, 2001. IEEE (2001)

14. Abate, A.F., et al.: 2D and 3D face recognition: a survey. Pattern Recogn. Lett. 28(14), 1885–1906 (2007)

15. Kakadiaris, I.A., et al.: Three-dimensional face recognition in the presence of facial expressions: an annotated deformable model approach. IEEE Trans. Pattern Anal. Mach. Intell. 29(4), 640–649 (2007)

16. Baggio, D.L.: Mastering OpenCV with practical computer vision projects. 2012: Packt Publishing Ltd

17. Ojala, T., Pietikainen, M., Maenpaa, T.: Multiresolution gray-scale and rotation invariant texture classification with local binary patterns. IEEE Trans. Pattern Anal. Mach. Intell. 24(7), 971–987 (2002)

18. Zhang, B., et al.: Local derivative pattern versus local binary pattern: face recognition with high-order local pattern descriptor. IEEE Trans. Image Process. 19(2), 533–544 (2010)

19. Karakashev, D.Z., Tan, H.Z.: Exploring How Haptics Contributes to Immersion in Virtual Reality (2016)

20. Juan Fang, Z.S., Ali, S., Zulfiqar, A.A.: Cloud computing: virtual web hosting on infrastructure as a service (IAAS). In: 13th International Conference on Mobile Ad-hoc and Sensor Networks, MSN. Springer (2017)

21. Mollah, M.B., Islam, K.R., Islam, S.S.: Next generation of computing through cloud computing technology. In: Electrical & Computer Engineering (CCECE), 2012 25th IEEE Canadian Conference on. IEEE (2012)

22. Wen, Y., et al.: Forensics-as-a-service (FAAS): computer forensic workflow management and processing using cloud. In: The Fifth International Conferences on Pervasive Patterns and Applications (2013)

23. Dean, J., Ghemawat, S.: MapReduce: simplified data processing on large clusters. Commun. ACM 51(1), 107–113 (2008)

24. Ananthanarayanan, R., et al.: Cloud analytics: do we really need to reinvent the storage stack? In: HotCloud (2009)

25. Fuad, A., Erwin, A., Ipung, H.P.: Processing performance on apache pig, apache hive and MySQL cluster. In: Information, Communication Technology and System (ICTS), 2014 International Conference on. IEEE (2014)

26. Xu, G., Xu, F., Ma, H.: Deploying and researching Hadoop in virtual machines. In: Automation and Logistics (ICAL), 2012 IEEE International Conference on. IEEE (2012)

27. Joshi, S.B.: Apache hadoop performance-tuning methodologies and best practices. In: Proceedings of the 3rd ACM/SPEC International Conference on Performance Engineering. ACM (2012)

28. Shaukat, Z., Fang, J., Azeem, M., Akhtar, F., Ali, S.: Cloud based face recognition for google glass. In: Proceedings of the 2018 International Conference on Computing and Artificial Intelligence (ICCAI 2018). Association for Computing Machinery, pp. 104–111 (2018)

Eye Tracking in the Diagnosis of Aggressive Behaviors and Emotions: A Systematic Review of the Literature

Noemi Gabriela Gómez-Ochoa[1](✉), Patricia Ortega-Chasi[1,2],
Omar Alvarado-Cando[1,3], Martha Cobos-Cali[1,4],
and Sonia Artega-Sarmiento[1,5]

[1] Grupo de Neurociencias, Universidad del Azuay, Cuenca, Ecuador
ggomezochoa@es.uazuay.edu.ec,
{portega,oalvarado,mcobos,marteaga}@uazuay.edu.ec
[2] Escuela de Ciencias de la Computación, Universidad del Azuay,
Cuenca, Ecuador
[3] Escuela de Ingeniería Electrónica, Universidad del Azuay, Cuenca, Ecuador
[4] Escuela de Psicología Clínica, Universidad del Azuay, Cuenca, Ecuador
[5] Unidad de Idiomas, Universidad del Azuay, Cuenca, Ecuador

Abstract. Eye Tracking it is a sensor technology that allows - through a device – to exactly know where a person's gaze is focused. In the field of Neuropsychology, it is important to know studies with this tool to start new research. Therefore, this study explores and synthesizes studies on the diagnosis of aggressive behaviors and emotions by using the Eye tracking technology. This systematic review of the literature was carried out under the PRISMA criteria. The studies included in the review consist of articles from empirical studies, peer-reviewed published between 2010 and 2018. The databases included in the study are: Pubmed, Scopus, PsycInfo, Scielo, DOAJ. We obtained 53 preliminary articles. 16 studies were selected according to the inclusion and exclusion criteria of the investigation.

Keywords: Aggressive behaviors · Gaze fixation · Gaze tracking · Emotions

1 Introduction

Eye Tracking is a technique applied through a device called a tracker, which allows researchers to monitor and record the way person looks at a particular stimulus, specifically the areas in which he fixes his attention, the time he invests and in what order this activity is carried out [1]. This physiological indicator measures the response of the subjects through the movement of the eyeballs. The analysis of eye movements is not a neuroscience technique itself, but it is a type of biometric measurement that can help to understand the unconscious of the study subjects, even the data relative to blinking, speed of movement and dilation of the pupil to infer the emotional involvement with what is being observed [2].

© Springer Nature Switzerland AG 2020
T. Ahram (Ed.): AHFE 2019, AISC 973, pp. 111–121, 2020.
https://doi.org/10.1007/978-3-030-20476-1_13

Aggressive behaviors and emotions have usually been diagnosed through traditional processes, which required broad time for the application and measure. The type of responses obtained through these processes are subjective because of the social moral burden in the issue of aggressiveness.

The existing literature on aggressiveness refers to conceptualization, studies of possible etiologies, diagnosis at different stages of the development of the human being and with a diversity of tools, comorbidity with psychopathologies, among others.

The present study aims to systematically review the literature regarding the use of eye-tracking technology in the diagnosis of aggressive behaviors and emotions. This review focuses mainly on analyzing the studies regarding the population participating in these studies and their conditions, the methodology used, the areas of study and the results. This document is organized the following way: the methodology used to carry out this review is described in the following section. A description of the search criteria, criteria for inclusion and exclusion of studies is included, and the data extraction process is described. Section three presents the results of the analysis of the selected studies in relation to the research questions and discusses the findings. Finally, in section four the conclusions are included.

2   Methodology

The systematic literature review was developed under the PRISMA model (Preferred Reporting Items for Systematic Reviews and Meta-Analyzes statement) criteria [3, 8]. The research questions that guided this literature review include:

1. What areas of study have the diagnostic investigations about aggressive behaviors and emotions that use eye-tracking technology been addressed?
2. What methodological approaches have been used to diagnose processes of aggressive behaviors and emotions using eye-tracking technology?
3. What tools have been used to diagnose processes of aggressive behaviors and emotions that use eye-tracking technology?

2.1   Literature Search

The main search terms used to find studies in the field were selected based on the research questions. Synonyms and slang were also included as alternative terms to incorporate additional studies related to the results. The search chains were built according to these terms in two languages: English and Spanish. Additionally, in order to limit the results of the searches, the Boolean operators AND and OR, and quotation marks were used to indicate specific phrases. The search terms were applied to the metadata, specifically to the title, summary and keywords where these options were available. For such cases, a manual review process of the articles was carried out to verify that the title, summary or keywords contain the search terms. The keywords used for the search are shown in Table 1.

Table 1. Searching keywords.

Searching keywords
"Aggressive behavior" and "eye tracking"
"Emotions" and "eye tracking"
"Emotions" and "eye tracking" and "aggressive behavior"
Conductas agresivas y emociones
Emociones y seguimiento de mirada

Once the keywords were defined, the researchers looked for relevant articles. This search was conducted between October 11 and November 8, 2018 in the following academic databases: Pubmed, Scopus, PsycInfo, Scielo and DOAJ.

2.2   Search Criteria

To carry out this research, the following criteria were included:

- *Type of text:* empirical and peer reviewed studies due to the rigor required in the field of health, the update and application of appropriate methods and the incidence or impact on patients.
- *Language:* articles written in English and Spanish.
- Year of publication: the publication year needed to be within the last 8 years, depending on the advances of technology and their use in the health, education and psychology fields.

After the search stage, the researchers read the summary of each article to verify whether they meet the inclusion and exclusion criteria or not. The inclusion and exclusion criteria applied are detailed below:

- *Inclusion Criteria:* Only studies dealing with eye-tracking technology applied to the diagnosis of aggressive behaviors and diagnostic studies of emotions were included.
- *Exclusion criteria:* Studies conducted with eye tracking technology irrelevant to health and/or education were excluded.

2.3   Data Extraction Strategy

The selected articles were explored to answer the research questions. The data extraction form designed for this purpose included an article identifier, the author, the title of the study, year of publication, DOI, summary, and fields to evaluate the research questions. Once the data extraction was completed, the results were analyzed and synthesized. The following diagram shows the process followed for the identification of the studies to be included in this literature review (Fig. 1).

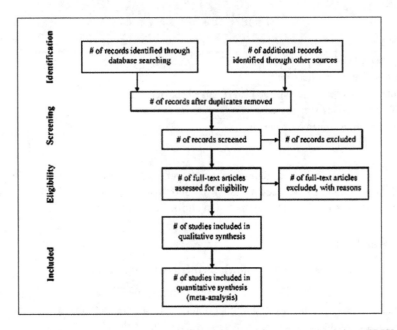

Fig. 1. Flow of information through the different phases of a systematic review (PRISMA).

Initially, only the title and summary were reviewed to determine the relevance of the article. However, because the abstracts did not include enough information to determine if it was a study on aggressive behaviors and emotions with eye tracking technology, it was necessary to review the methodology and results. This clarified the inclusion and exclusion criteria.

3   Results

The focus of this study is to examine the literature to investigate the scope in the use of eye-tracking technology applied to the diagnosis of aggression and emotions. This results section was organized the following way: first, the results of this review are presented, next, the subsequent sections focuses on answering and discussing the findings of following research questions:

(a) What areas of study have been addressed to diagnose aggressive behaviors and emotions by means of the eye-tracking technology?
(b) What methodological approaches are used in diagnostic processes of aggressive behaviors and emotions through eye-tracking technology?
(c) What tools have been used to diagnose processes of aggressive behaviors and emotions through the use eye-tracking technology?

3.1   Number of Studies

The search process yielded fifty-three studies from all five academic databases queried in this process. Duplicate studies were eliminated, as well as studies that did not comply with the inclusion and exclusion criteria. A total of sixteen studies were obtained. The following diagram shows the results in each phase of the process (Fig. 2).

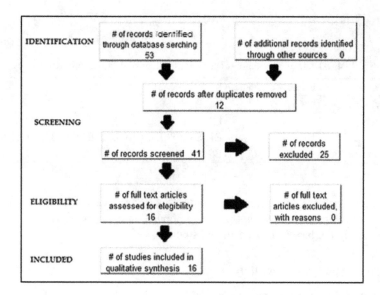

Fig. 2. Number of papers identified in each phase of the searching process.

3.2   Scope of the Study

The existing research is mostly oriented to the study of aggressiveness and eye tracking from the psychological and psychiatric fields. However, there is also the approach from Neurology and Genetics combined with Psychology and Psychiatry, and a study with a neuropsychological nuance. These results are shown in Table 2. In Tables 3 and 4 it is presented the summary of the resulting studies regarding the clinical conditions of participants, and details of population gender that participated in the different studies respectively.

Table 2. Detail of the fields of study focused in which the investigations

Field of study	# of studies	Study code
Psychology	8	E1, E6, E7, E11, E13, E14, E15, E16
Psychiatry	2	E2, E4
Genetics/Neuroscience	1	E3
Psychology/Neuroscience	3	E9, E10, E12
Psychiatry/Neuroscience	2	E5, E8

Table 3. Detail of the clinical conditions of participants

Sample conditions	# of studies	Study code
People with diagnosed with autism, depression, borderline, behavioral disorders, etc.	9	E1, E2, E4, E5, E6, E8, E9, E13, E14
People without clinical diagnostic	6	E3, E10, E11, E12, E15, E16

Table 4. Detail of population gender that participated in the different studies.

Gender	Number of studies	Study code
Women	3	E4, E15
Men	1	E2
Mixed	10	E1, E3, E8, E9, E10, E11, E12, E13, E14, E16
Not specified	1	E6

- A study analyzes the advantages of tools such as eye tracking in aggressive behavior. However, it does not present a population under study and the results of this section are based on other fifteen studies.

3.3   Tools Applied in the Study

The different studies have a convergent point regarding the diagnosis of emotions and aggressive behaviors when using eye- tracking technology. Therefore, the researchers assure that all the studies considered in this literature review have applied this technology. Two of the studies have used this tool exclusively, as is the case of [4, 5].

The other studies combine the use of eye tracking with other tools, such as: electroencephalogram [6], electromyography [7], functional magnetic resonance [4, 6] and depth sensor imaging technology [9]. These studies have been used at a medical level. Regarding the psychological field, complementary tools like questionnaires and scales of depression, alexithymia, needs, affects, stress, aggression, emotional processing and social performance have been used. The studies of: [10–18] are an example of this type.

The images used with eye tracking to detect emotions, are mostly oriented to detect faces expressing emotions. A study included an analysis of pupil posture and dilation [9]. Other two studies focused on emotionally charged faces and daily scenarios: [10, 15] (Fig. 3).

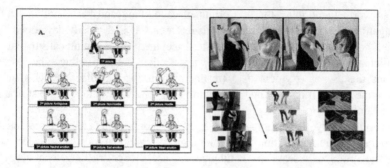

Fig. 3. Examples of images used in [10, 15, 6] studies.

Finally, it is necessary to indicate that two studies highlight the importance of this technology in the psychological field. One of them, through an analysis of relevant issues and the processes developed so far, [19] consider eye tracking as an adequate tool to analyze gazes in the relationship of culture-perception regarding emotions. In the other study, [14] analyzes and highlight the use of eye tracking as a means of coding the processing of Social Information in children in real time.

3.4 Eye Tracking Technology Data

The studies carried out and addressed in this literature review use the following variables of eye tracking technology to develop their studies:

- First saccade
- Gaze fixation
- Gaze latency
- Gaze tracking

These variables were used according to the objectives of each study. The results were accordingly analyzed and used to reach the corresponding conclusions.

Researchers indicate that they have considered this tool for their studies because of the benefits it brings to the type of responses in the physiological order, the precision of the evaluation as to where the participant's gaze is directed [9, 10, 14, 20].

3.5 Study Objectives

The main objective of this study was to determine whether there was or not a correlation between the orientation of the gaze (fixation, tracking, latency) towards hostile stimuli and aggressiveness. The studies of [9, 10, 13, 17], corroborate their study approaches in this sense.

It is worth mentioning that the origin of each study and the conditions in each investigation are variables important to this research. Thus, the study developed by [4], in which he uses eye tracking technology to detect the type of stimuli calls the attention of participants` gaze. The participants of this study were borderline. The researcher mentions the use of oxytocin as a treatment form for this disorder. This helps to decrease the levels of anxiety and orientation to hostile information in the surrounding environment.

In another study, the participation of a gene in modulation is analyzed to determine the accuracy in the recognition and exploration of emotional facial expression. In this study, the group of researchers combined the use of eye tracking technology with neurological imaging studies to concluded that the 5-HTTLPR gene influences social perception, modulating the general surveillance of social signals [12].

Other studies took into consideration the amygdala in the development of morals, the relationship between information processing and aggressive tendencies, emotion and body posture, social exclusion and aggression, chronic depression-pain, and perception of emotions and culture. All these studies were based on eye tracking technology with valuable contributions that can be counted on both the results of correlation and the gap left for future research.

4   Conclusions

Although there is literature regarding the diagnosis of aggressive behaviors, emotions and related situations, the majority of studies focus on adulthood, which demonstrate a gap in the research. Then, it would help to confirm that aggressiveness is formed at an early age as well as moral. Therefore, these investigations would allow knowing certain characteristics or patterns that can be addressed early.

Of the sixteen studies, nine expose the participation of people with specific clinical diagnostic criteria, which represents an axis to follow in the preparation of future studies in which aggression is analyzed from other types of etiologies, in which the underlying aggressiveness of a painting is manifested and analyzed in a more natural way.

In relation to the type of tools used to detect emotions, when using images of faces with expression of emotions, it is necessary to consider the use of real scenes, the exposure to real scenarios. In this way, participants can respond naturally to a specific situation. This would make the evaluation more accurate in terms of gaze and how participants move their eyes.

Finally, it is necessary to express the value of eye tracking technology in the diagnostic of aggressive behaviors to approach other studies from the neuropsychological field, in order to generate tools or possible areas to improve in situations with this type of manifestations.

Appendix: Selected Studies

See Table 5.

Table 5. Selected studies

Id	Title	Author(s)	Year
E1	Eye Gaze Patterns Associated with Aggressive Tendencies in Adolescence	Laue C, Griffey M, Lin PI, Wallace K, Van der Schoot M, Horn P, Pedapati E, Barzman D	2018
E2	Emotional face recognition in male adolescents with autism spectrum disorder or disruptive behavior disorder: an eye-tracking study	Bours CCAH, Bakker-Huvenaars MJ, Tramper J, Bielczyk N, Scheepers F, Nijhof KS, Baanders AN, Lambregts-Rommelse NNJ, Medendorp P, Glennon JC, Buitelaar JK	2018
E3	5-HTTLPR modulates the recognition accuracy and exploration of emotional facial expressions	Boll, S. Gamer, M	2014
E4	Interpersonal Threat Sensitivity in Borderline Personality Disorder: An Eye-Tracking Study	Bertsch K, Krauch M, Stopfer K, Haeussler K, Herpertz SC, Gamer M	2010
E5	Oxytocin and reduction of social threat hypersensitivity in women with borderline personality disorder	Bertsch K, Gamer M, Schmidt B, Schmidinger I, Walther S, Kästel T, Schnell K, Büchel C, Domes G, Herpertz SC	2013
E6	In the eye of the beholder: eye-tracking assessment of social information processing in aggressive behavior	Horsley TA, de Castro BO, Van der Schoot M	2010
E7	Social information processing, emotions, and aggression: conceptual and methodological contributions of the special section articles	Arsenio WF	2010
E8	Predicting Aggressive Tendencies by Visual Attention Bias Associated with Hostile Emotions	Lin PI, Hsieh CD, Juan CH, Hossain MM, Erickson CA, Lee YH, Su MC	2016
E9	Early detection of intentional harm in the human amygdala	Hesse E, Mikulan E, Decety J, Sigman M, Garcia Mdel C, Silva W, Ciraolo C, Vaucheret E, Baglivo F, Huepe D, Lopez V, Manes F, Bekinschtein TA, Ibanez A	2015

(*continued*)

Table 5. (*continued*)

Id	Title	Author(s)	Year
E10	Perception of face and body expressions using electromyography, pupillometry and gaze measures	Kret, M., Stekelenburg, J., Roelofs, K., De Gelder Beatrice	2013
E11	Social exclusion leads to attentional bias to emotional social information: Evidence from eye movement	Chen, Z., Du, J., Xiang, M., Zhang, Y., Zhang, S	2017
E12	Emotions' impact on viewing behavior under natural conditions	Kaspar, K., Hloucal, T., Kriz, J., Canzler, S., Gameiro, R., Krapp, VB., König, P	2013
E13	Processing of Emotional Faces in Patients with Chronic Pain Disorder: An Eye-Tracking Study	Giel, K., Paganini, S., Schank, I., Enck, P., Zipfel, S., Junne, F	2018
E14	Recognition of Facial Expressions in Individuals with Elevated Levels of Depressive Symptoms: An Eye-Movement Study	Wu, L., Pu, J., Allen, J. J. B., & Pauli, P.:	2012
E15	Emotion perception acorss cultures: the role of cognitive mechanisms	Engelmann, J. B., & Pogosyan, M	2013
E16	Novel paradigms to measure variability of behavior in early childhood: Posture, gaze, and pupil dilation	Hepach, R., Vaish, A., & Tomasello, M.:	2015

References

1. Iniestra, E., y Mancilla, E.: La técnica de eye tracking aplicada en algunas láminas del test de Relaciones Objetales. Educación y Humanidades, 210–216 (2017)
2. Quiñones, M.: La Neurociencia como oportunidad estratégica. Cuadernos Latinoamericanos de Administración **6**(11), 97–110 (2010)
3. Moher, D., Liberati, A., Tetzlaff, J., Altman, D.G.: Preferred reporting items for systematic reviews and meta-analyses: the prisma statement. PLoS Med. **6**(7), e1000097 (2009)
4. Bertsch, K., Gamer, M., Schmidt, B., Schmidinger, I., Walther, S., Kästel, T., Herpertz, S.C.: Oxytocin and reduction of social threat hypersensitivity in women with borderline personality disorder. Am. J. Psychiatr. **170**(10), 1169–1177 (2013)
5. Kaspar, K., Hloucal, T.-M., Kriz, J., Canzler, S., Gameiro, R.R., Krapp, V., König, P.: Emotions' impact on viewing behavior under natural conditions. PLoS ONE **8**(1), e52737 (2013)
6. Hesse, E., Mikulan, E., Decety, J., Sigman, M., Garcia, M. del C., Silva, W., Ibanez, A.: Early detection of intentional harm in the human amygdala. Brain **139**(1), 54–61 (2015)
7. Kret, M.E., Stekelenburg, J.J., Roelofs, K., de Gelder, B.: Perception of face and body expressions using electromyography, pupillometry and gaze measures. Front. Psychol. **4** (2013)

8. Liberati, A., Altman, D., Tetzlaff, J., Mulrow, C., Gotzsche, P., Loannidis, J., Clarke, M., Devereaux, P., Kleijnen, J., Moher, D.: The PRISMA statement for reporting systematic reviews and meta-analyses of studies that evaluate health care interventions: explanation and elaboration. PLoS Med. **6**(7), e1000100. https://doi.org/10.1371/journal.pmed.1000100

9. Hepach, R., Vaish, A., Tomasello, M.: Novel paradigms to measure variability of behaviorin early childhood: posture, gaze, and pupil dilation. Front. Psychol. **6** (2015)

10. Laue, C., Griffey, M., Lin, P.-I., Wallace, K., Van der Schoot, M., Horn, P., Barzman, D.: Eye gaze patterns associated with aggressive tendencies in adolescence. Psychiatr. Q. **89**(3), 747–756 (2018)

11. Bours, C.C.A.H., Bakker-Huvenaars, M.J., Tramper, J., Bielczyk, N., Scheepers, F., Nijhof, K.S., et al.: Emotional face recognition in male adolescents with autism spectrum disorder or disruptive behavior disorder: an eye-tracking study. Eur. Child Adolesc. Psychiatr. (2018)

12. Boll, S., Gamer, M.: 5-HTTLPR modulates the recognition accuracy and exploration of emotional facial expressions. Front. Behav. Neurosci. **8** (2014)

13. Horsley, T.A., de Castro, B.O., Van der Schoot, M.: In the eye of the beholder: eye-tracking assessment of social information processing in aggressive behavior. J. Abnorm. Child Psychol. **38**(5), 587–599 (2009)

14. Arsenio, W.F.: Social information processing, emotions, and aggression: conceptual and methodological contributions of the special section articles. J. Abnorm. Child Psychol. **38** (5), 627–632 (2010)

15. Lin, P.-I., Hsieh, C.-D., Juan, C.-H., Hossain, M.M., Erickson, C.A., Lee, Y.-H., Su, M.-C.: Predicting aggressive tendencies by visual attention bias associated with hostile emotions. PLoS ONE **11**(2), e0149487 (2016)

16. Chen, Z., Du, J., Xiang, M., Zhang, Y., Zhang, S.: Social exclusion leads to attentional bias to emotional social information: evidence from eye movement. PLoS ONE **12**(10), e0186313 (2017)

17. Giel, K.E., Paganini, S., Schank, I., Enck, P., Zipfel, S., Junne, F.: Processing of emotional faces in patients with chronic pain disorder: an eye-tracking study. Front. Psychiatr. **9** (2018)

18. Wu, L., Pu, J., Allen, J.J. B., Pauli, P.: Recognition of Facial Expressions in Individuals with Elevated Levels of Depressive Symptoms: An Eye-Movement Study

19. Engelmann, J.B., Pogosyan, M.: Emotion perception across cultures: the role of cognitive mechanisms. Front. Psychol. **4** (2013)

20. Bertsch, K., Krauch, M., Stopfer, K., Haeussler, K., Herpertz, S.C., Gamer, M.: Interpersonal threat sensitivity in borderline personality disorder: an eye-tracking study. J. Personal. Disord. **31**(5), 647–670 (2017)

Research on Smart Wearable Device Design Based on Semiotic Approach to Product Architecture Design

Huaxiang Yuan$^{(\boxtimes)}$, Lu Han, and Hongzhi Pan

Huazhong University of Science and Technology, Wuhan, Hubei, China
990252480@qq.com

Abstract. It aims to improve the wearable device interaction design for old people who are easy to get lost. Typical rehabilitation behavior and related objects of the users are recorded by non-participant observation and interview. Based on the SAPAD frame, the relationship between behavior-object-signification is analyzed. It has provided some solutions to optimize the APP interaction design.

Keywords: SAPAD · Interface design · Interface evaluation

1 Introduction

Great concerns have arisen about the lives of older people due to the increasing number of aging populations. And the lost elderly becomes the biggest concern because of their memory loss. In order to decrease the probability of this dangerous event, people use smart wearable devices to located old people. However, this kind of product pays more attention on old people and pay little attention on mobile users who ultimately makes the buying decisions. Furthermore, this kind of smart wearable devices can greatly help users to deal the anxiety of old people safety. In this case, the smart wearable device interaction demands more suitable framework to improve itself in different situation.

There have been a great number of studies in framework for HMI improving. Zheng Liu proposed the classic software framework of Model-View-Controller (MVC) can be used in the civil aircraft cockpit. The aim of this software framework is decoupling the logic dependency and system model [7]. Wanyu Zhang and Xun Yuan tried to improve APP interface design localization in difference aspect [2], Xun Yuan studied how to improve WeChat interface design with the local style. Wanyu zhang combining "Mobile Internet" and "Productive Protection of Qiang people's silver jewelry" into the APP interface design [1] from the perspective of user experience and cognitive psychology. Hairong Long designed the APP interface and verified it's feasibility by the framework including user factors, environmental factors and emotional factors. Both of these two researches improved the APP interface design with user's mental model [3]. As for evaluation of HMI before improving, there are many researches about evaluation of HMI in different situations. Kuowei su investigated interactive virtual reality navigation system by testing 15 people with Delphi method

© Springer Nature Switzerland AG 2020
T. Ahram (Ed.): AHFE 2019, AISC 973, pp. 122–129, 2020.
https://doi.org/10.1007/978-3-030-20476-1_14

and Heuristic Evaluation. All of these researches have not taken the semantic of user's behavior into account to provide a perspective of interpret users behavior.

The framework used in this research is SAPAD framework which was proposed by Fei Hu and Keiichi Sato in 2011. It was used in product design, service design and interaction design. It aims to reconstruct the functional service modules of community and build the service system for the elderly rehabilitation to realize the design innovation [8]. Researchers mapped relationship between the three dimensions which are behavior, object and signification. From this frame, the relationship between behavior-object-signification is analyzed. It could provide new approach and proposal for some problem in different aspects. However, there is no conclusive research to talk about smart wearable device interface design, especially about the interaction design in specific scenario.

The rest of the paper is organized as follow: Sect. 2 introduces the SAPAD frame . Sect. 3 describes the process of building user model by observing user's behavior. Section 4 analyzes the relationship between behavior-object-signification when people use the APP based on SAPAD frame. Section 5 give solutions for improving this interaction design. Section 6 discusses the gaps and challenges of SAPAD frame for interface design. Finally, in Sect. 7, the conclusion is summarized.

2 SAPAD Frame

SAPAD (Semiotic Approach to Product Architecture Design) is the Product construction under the full name of Semiotic Approach, which was developed by professor Hu Fei in cooperation with professor Keiichi Sato of the Design school of Illinois institute of technology in the United States during his study visit to the United States in 2011. This method forms three dimensions of behavior-meaning-product between products and users by introducing the interpretation of user behavior meaning in semiotics, and each dimension is divided into several levels accordingly. By analyzing the corresponding relationship between user's behavior and object and behavior and meaning when using the product, the mapping relationship between object and meaning can be obtained, and the design opportunity can be explored to improve the product (Fig. 1).

3 Building User Model

It is designed to better monitor elderly people whose memories are fading, mainly by their children, volunteers and CARE worker. For SAPAD, non-participatory observation is adopted to analyze the user's task flow by recording the user's operation steps.

Through user interviews, some important significant behavioral variables are selected to form a number of variable axes, and then the relative position of the interviewees on the variable axis is divided. Generally speaking, the mutual location relationship between users in a certain range is more important than the exact location, which plays a role in subdividing the user group. In this paper, object and behavior

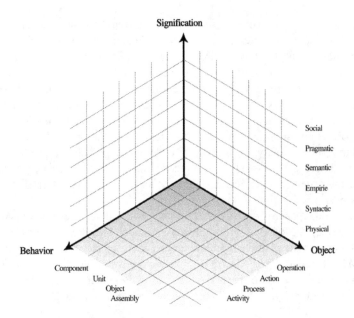

Fig. 1. Semiotic approach to product architecture design

variables are corresponded through in-depth interviews with 5 users. Each user was interviewed separately in a relaxed environment, and the usage of several users was recorded by means of chat and pre-prepared interview scripts. It is possible to observe the similarities between the user's needs for the elderly information management, positioning mode and behavioral operation. If the user has certain commonality in listening, sharing and operation, it indicates that they have similar group behavior pattern.

Record the operation activities related to APP for 5 users with non-participatory observation. They are then interviewed in depth to get a full picture of how users behave when using the APP. The main behaviors include clicking the APP, linking the devices, reading the health information, getting the location information.

4 Analyzes the Relationship Between Behavior-Object-Signification

Based on SAPAD (Semiotic Approach of Product Architecture Design) framework, the user-centered Product Design strategy was established by analyzing the mapping relationship between behavior, meaning and Product (Fig. 1).

5   Analysis of User Behavior and Key Items

After recording the operation behaviors of users in three usage scenarios, the corresponding behaviors and objects in each scenario are studied. The object can be any element related to user operation in APP interface, such as chart, text, virtual button or whole page. Users may use multiple interface elements in each step of operation, which are all related to behaviors. Among them, key elements are directly related to operations and essential interface elements (Table 1).

Table 1. Analysis of key components

Active	Process	Operation	Correlate	Key item
Binding	Master account binding	Open the Bluetooth	APP icon, Bluetooth	Bluetooth
		Find the device	Search devices, choose the device	Choose the device
		Scan the code	Camera, code, Scan the code	Code, scan the code
		Input the information	Registry	Registry
	Vice account binding	Send request	Send	Send
		Master account agrees	Receive the request, agree	Agree
Find the elder	Locate	Locate	Location	Location
	Follow the location	Open Navigation	Navigation, map, voice guide	Map
		Guide over	Navigation, voice guide	Navigation
Know the number of steps	Open the step panel	Open the step panel	Step panel, Statistics Panel	Statistics Panel
Know the heart rate	Open the heart rate panel	Open the heart rate panel	Heart rate panel, Statistics Panel	Statistics Panel
Know the sleep quality	Open the sleep panel	Open the sleep panel	Sleep quality panel, Statistics Panel	Statistics Panel
Schedule reminder	Choose the event	Edit the events	Keyboard, project bar	Project bar
		Choose the event	Project bar, Y	Project bar
	Setup time	Setup time	Time bar, clock	Time bar
		Choose the way of remind	Event list, time list, selection bar	Selection bar

This part completes the analysis of behavior-object, and finds out the interface elements in APP corresponding to user behavior through observation. The relationship between user behavior and key objects in each scene has been clarified. And then, it is necessary to further explain the meaning of user behavior from the perspective of human.

5.1   Significance Analysis Based on User Behavior Observation

This section is a behavior-signification analysis, focusing on explaining the meaning of the user's behavior. Firstly, the signification of user behavior is qualitatively analyzed from the physical level, the semantic level, the syntactic level, the empirical level, the pragmatic level and the social level through video. In order to avoid the difference between the content of qualitative analysis and the actual thoughts of participants, the user was interviewed again after the qualitative analysis, and the content of qualitative analysis was checked and revised with the participants. Finally, the analysis of signification is completed (Table 2).

Table 2. Meaning construction based on observed behavior

Operation	Physical	Empiric	Syntactic	Semantic	Social
Log in	Click the APP	Click the APP	Enter the APP	Want to get elder's information	
Binding	Search and choose the device	Link the mobile phone and device	Synchronize data		
know the number of steps	Open the step panel	Get the information of physical activity	Analyze the fitness data	Care of the elder's health and get security	
Know the heart rate	Open the heart rate panel	Get the information of physiological safety	Analyze the life data	Care of the elder's health and get security	
Know the sleep quality	Open the sleep panel	Get the information of sleep	Analyze the sleep data	Care of the elder's health and get security	
Schedule reminder	Setup time	Remind the check time	Receive the reminders and take the action	Make sure the important things are done	
Locate	Follow the guide	Find the elder	Get the elder's location and follow it	Ensure safety of elder	

5.2   Cluster the Significant of User Behavior

In the SAPAD framework, the behavior-signification relationship is corresponding related objects. Therefore, the association construction of core signification may reassemble the objects, point out the gap of the interaction, and improve the existing design. Since the signification of the physical level and the syntactic level reflects the objective logical relationship between the interactive elements of the APP and the interactive elements, the result of clustering can only reflect the original interaction design of the APP. Therefore, the focus here is to cluster the empirical level related to interactive optimization.

By cluster analyzing, we have gained the core significance cluster, and constructed the core significance relationship (Table 3). The strong or weak relevance of meaning can be divided into 4 levels: 0, 1, 2, and 3. "0" represents no correlation, "1" represents weak correlation, "2" represents strong correlation, and "3" represents core correlation. The meaning cluster can be clearly seen from the operation results.

Table 3. Significant cluster of empirical level

	Click the APP	Link the mobile phone and the device	Get the information of physical activity	Get the information of physiological safety	Get the information of sleep	Remind the check time	Find the elder
Click the APP	3	1	0	0	0	0	0
Link the mobile phone and the device	1	3	0	0	0	0	0
Get the information of physical activity	0	0	3	3	3	2	1
Get the information of physiological safety	0	0	3	3	3	2	0
Get the information of sleep	0	0	3	3	3	2	0
Remind the check time	0	0	2	2	2	3	3
Find the elder	0	0	1	0	0	3	3

The empirical level emphasizes users' skills and life experience, and the clustering analysis results in 2 meaning clusters, which are getting information of elders' health in daily life and ensuring elders will not get lost.

6 Solution

Optimize the information architecture:

The goal of information architecture optimization is to reduce the complexity of information acquisition and shorten the distance between users and information. As for start page, there are two scenarios of using the APP which are daily scenarios and emergency scenarios. Users prefer to get the information on health in old age in daily scenarios but get bored of this kind of information when the elders get lost. It better to move the relevant information from start page to health information page.

Optimize product functions:

The goal of product function optimization is to find out the difficulties or inconveniences when users use the APP and solve them or propose better ways to replace the original functions.

Optimization of visual details of product:

The goal of visual detail optimization is to use the visual method to make the information display more clearly and do the appropriate beautification. Through users' operation and feedback, it can be found that the first part that can be improved is to emphasize some contents and functions that users pay more attention to. For example, the font size of some fonts should be enlarged and optimized within the current visual design style of the whole APP.

7 Gaps of the Research

There are some gaps and disadvantages through the research. Firstly the research of SAPAD method select users randomly when conducting user research, which is not typical, and it is not very helpful for the iterative update of APP products. Secondly, SAPAD does not take into account the user's usage scenarios when analyzing the user behavior, while the APP user usually has multiple usage scenarios. Therefore, it is necessary to distinguish the user's usage scenarios.

8 Conclusion

In this paper, the effects of SAPAD frame were investigated. This framework not only focuses the user needs but also excavates the users' demand from the behavior to signification in order to get the core requirement. The APP of elder wearable device are analyzed with this framework in empirical level. The result of analysis shows that users require more efficient interaction when they locate elders urgently and they call for humanize interaction when they learn the health information about the elder. However, the study ignore users preform differently in different scenarios, although it is discussed in solutions part.

References

1. Wanyu, Z.: APP interface design of Qiang people's silver jewelry based on user experience. In: Proceedings of the 3rd International Conference on Contemporary Education, Social Sciences and Humanities (ICCESSH 2018) (Advances in Social Science, Education and Humanities Research, Vol. 233)
2. Yuan, X.: Introduction to We Chat interface design localization. In: Proceedings of The 4th International Conference on Education, Language, Art and Inter-cultural Communication (ICELAIC 2017) (Advances in Social Science, Education and Humanities Research, Vol. 142)
3. Long, H.: Self-guided tour APP interface design based on user experience. In: Proceedings of the 2016 3rd International Conference on Education, Language, Art and Inter-cultural Communication (ICELAIC 2016)
4. Yan, L.: User experience for landscape teaching mobile device software interface design. In: Proceedings of the International Conference on Mechatronics Engineering and Information Technology (ICMEIT 2016)
5. Hu, F.: Human-centred product architecture from UPPA to SAPAD. Science and Engineering Research Center. In: Proceedings of 2015 International Conference on Sustainable Energy and Environmental Engineering (SEEE 2015)
6. Su, K.-W.: The interface design and usability evaluation of interactive virtual reality navigation system. In: Proceedings of 2013 International Conference on Mechanical Engineering and Materials (ICMEM 2013)
7. Liu, Z.: A study of cockpit HMI simulation design based on the concept of MVC design pattern. In: Proceedings of 2018 3rd International Conference on Modelling, Simulation and Applied Mathematics (MSAM 2018)
8. Fei, H.U., Kun, Z., Zhang-Sheng, L.: Service design of community rehabilitation for the elderly based on SAPAD framework. Packag. Eng. (2018)

Wearable Smart Emotion-Aware Gloves: Develops a New Way People Interact with Others

Ping Shan[1], Shijian Luo[1,2(✉)], Chengyi Shen[1], Zhitong Cui[1],
Ze Bian[1], Huan Lin[1], Wenyin Zou[1], and Yufei Zhang[1]

[1] Department of Industrial Design, Zhejiang University,
Hangzhou 310027, China
403191990@qq.com, 2370343775@qq.com, 359243864@qq.com,
stellazou@qq.com, 634906847@qq.com, sjluo@126.com,
shenchengyi@zju.edu.cn, linhuan_design@163.com
[2] Design Industrial Innovation Center, China Academy of Art,
Hangzhou 310024, China

Abstract. The paper introduces a wearable parent-child emotional interaction device based on physiological detection. We locate product features based on the user's wishes and usage scenarios. Then we conducted a design study on the prototype of the product system, and summarized the comprehensive influencing factors and potential contradictions and divergence points that the target users usually have in the process of using the product. Finally, we built an intelligent prototype that implements the code writing and hardware construction of each module. In the process, we performed two evaluation tests on the prototype.

Keywords: Interactive design · Smart devices · Parent-child

1 Introduction

Parent-child relationship, as the first relationship that people can reach after birth, is of great significance to people's lives. There is a connection between parent-child relationship, family learning environment and school preparation [1]. Parental psychological flexibility is closely related to children's mental health [2], and is also related to children's behavior problems, children's anxiety, etc. [3, 4]. How to establish a good parent-child relationship to promote the growth of young children is an eternal hot issue. In the early childhood, if parents and children lack emotional communication, children cannot get a direct emotional experience from them. This will lead to alienation of parent-child relationships and a decrease in intimacy and trust between parents and children.

However, in the increasingly fierce social competition, people's life rhythm is accelerating, and how to deal with parent-child relationship efficiently becomes a problem. Busy parents often have problems with parent-child relationships [5]. Although early intervention in child growth is very important [6], children of younger

© Springer Nature Switzerland AG 2020
T. Ahram (Ed.): AHFE 2019, AISC 973, pp. 130–144, 2020.
https://doi.org/10.1007/978-3-030-20476-1_15

age often fail to accurately express and feel emotions because of limitations in their ability to express themselves. In the long run, it is easy for parents and children to have a dilemma in communication, which makes the parent-child relationship alienated. In a good parent-child relationship, it is very important to properly understand each other's feelings. Emotional survey methods such as empathy design models have gradually developed and can be used in the development of children's products in the future [7].

Behind the heavy social pressure of modern people and the phenomenon of Grandparents parenting grandchildren [8–11], people's parent-child relationship has also undergone many changes. How to deal with parent-child relationship efficiently becomes a problem. First of all, considering the busy work of parents, they need to take more narrow gaps to take care of their children during this limited time, which is obviously a quick and direct way. Then consider the recent common Grandparents parenting grandchildren phenomenon, for parents who go out to work, less home, their communication with the child is obviously a deep and efficient way. Other members besides parents are also increasingly important in the growth of children [12]. Parents who lack experience and guidance often need special supplements if they want to communicate with their children. Younger age children also need a medium that makes it easier for them to express and feel their emotions.

From this background, we did relevant market research and target user research. A design idea of a new wearable intelligent emotion-sensing device is proposed. The fabric gloves were selected as the carrier to carry out the modular design of the product, which was carried out from three parts: design research, product design and program test. We used desktop research, user interviews and other methods to sort out the design features of parent-child devices in the market. A group of inexperienced urban young parents were surveyed and their needs were collected and analyzed through in-depth interviews. After comprehensive analysis of its advantages and disadvantages, we can draw our design opportunities.

There are not many smart products currently focusing on parent-child emotional interaction in the market. Most of them are relatively simple non-interactive non-intelligent products. This is a market application direction worth exploring. The device we studied provided it with a port of exploration.

2   Method

2.1   User Interviews with Target Groups

In order to better understand the needs of parent-child interaction products and users, we conducted a single interview with the target users. The main target users are parents of children aged 4–8. In this interview, we invited a total of 4 people. The basic information is shown in the Table 1. There are already some parent-child interaction products in the family. We arranged a 40 min in-depth interview for each of them, recording them in a comfortable environment with recordings and pens. The content and purpose of the interview can be seen in the Table 2.

Through interviews, we learned that although the participants had ordinary emotional comfort products at home, the utilization rate was not high.

Table 1. Basic information maps of the interviewed participants

Code	Identity	Gender	Age	Child age	Background description
A	Married and have 1 child	Male	33	6	Engineer, bachelor degree Fragmented parenting knowledge learning (most of the article class)
B	Married and have 2 children	Female	37	5	Clerk, college degree Fragmentation learning parenting knowledge (WeChat friends circle forwarding)
C	Married and have 2 children	Male	38	4	Mechanical repair, technical secondary school Busy work, no time to pay attention
D	Married and have 1 child	Female	35	7	Clerk, college degree Fragmentation learning parenting knowledge (WeChat friends circle forwarding)

Table 2. Contents and purposes of the interview outline

Interview stage	Time	Content	Purpose
First stage	5 min	Self introduction; the rules described; warm field	Allow participants to relax and gain trust, paving the way for the smooth progress of the following interviews
Second stage	10 min	The situation of the participants; the traditional parent-child interaction products and the intelligent parent-child interaction products that the participants usually contact	Get the basic personal information of the participants, and explore their interests in the product, the motivation to purchase and the high frequency factors considered
Third phase	15 min	Situation of common parent-child interaction products and intelligent parent-child interaction products in the participant's home	Learn about the parent-child interaction products currently in use and find our design opportunities
Fourth stage	10 min	Understand the relevant content of the emotional management knowledge background of the participants; the daily activities of the participants at home and their future emotional management plans	Understand the relevant content of participants' parenting knowledge and emotional management knowledge background; understand the daily activities of the participants and their future parenting plans

Through interviews, we learned that although the participants already have some parent-child interaction products at home, the utilization rate of these products is not high. Mainly because these types of products interact in an overly simple way, or require users to pay too much for learning.

Users are often attracted to the promotion of these products to purchase such products. But after they buy such products for a while, parents often can't insist on using simple interactive products, and children are more likely to like new ones. Therefore, both of them will immediately lose interest in entertainment. Such products are prone to long-term uselessness due to the user's loss of interest in the product, resulting in a situation in which the utilization rate is not high and the living space is occupied.

Even if users don't give up on such products, they are often inconvenient to use. Because according to the interview results, users generally lack parenting knowledge and emotional management knowledge, products cannot provide advice for them. Due to the above various reasons, these products often receive some negative evaluations, such as easy accumulation of dust and occupation of living space.

Through interviews, we learned some key points:

1. Enhance the interactive links of parent-child products, which is conducive to attracting users to use for a long time.
2. Users want to improve the level of intelligence of parent-child products. Most parent-child interactive products are not intelligent enough, requiring users to have certain parenting knowledge and emotional management knowledge, and the cost of learning to use the products reasonably is high.
3. Users want to increase the utilization rate of parent-child products.
4. Users want these products to be small, easy to store and clean.
5. The factors that users often consider are: puzzle, safety, price, utilization, use effect, aesthetics, land occupation and other factors.

2.2 Design Testing and Verification Based on Emotional Signal Acquisition and Analysis

2.2.1 Test Purposes

According to the questionnaire filled out by the participants, the participants who need to improve the parent-child relationship are selected. We mimic the preferences of specific sensory stimuli when parent-child interactions in the home environment determine the main functions of our product prototypes. And test the expression of specific sensory stimuli suitable for parent-child interaction. For example: music, stories, jokes, family members' voices, etc.

2.2.2 Participants Selection

On the basis of understanding the characteristics of parent-child relationship, the questionnaire was compiled and supplemented by the Eysenck Personality Questionnaire to screen out the people with poor parent-child relationship or who did not care about this aspect. 12 participants were selected as typical samples.

2.2.3   Test Environment

The laboratory is arranged in an environment similar to the living room, with dim light and room temperature controlled at a suitable temperature of 26 °C. During the experiment, the experimenter and the participant use the voice plug-in to communicate with each other, and use the front camera of the computer to record the test video (Fig. 1).

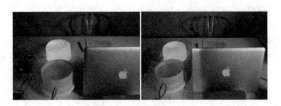

Fig. 1. Lab environments

2.2.4   Test Content

Based on previous research, we decided to use the GSR (Galvanic Skin Response) as the primary measurement data. Specific sounds and LED lights are used as the primary mood-inducing material, supplemented by olfactory stimuli (e.g., fruity aromas) to induce the emotions of the participants. After the experiment, the oral report method and questionnaire method were used to evaluate the feedback effect.

2.2.5   Test Results

The participants generally showed good stimuli feedback under the auditory and visual channels. Changes in sound and lighting will affect the emotional performance of the participants to some extent. Users are often able to accept sound suggestions, light stimuli, and show consistency with certain emotions. However, in terms of smell, individuals vary widely and it is difficult to find a certain pattern. So, we choose the sound, the light interacts with the user.

In terms of expression, changes in the color of the lights and the sound of story-telling are considered to be good tips and suggestions.

2.3   User Requirements Summary

Target users: Urban young parents who lack parenting experience earn $15,000 to $20,000 a year.

1. User insight
   a. Users expect to improve the quality of parent-child interaction, but often lack parenting knowledge.
   b. The fast-paced modern life leads to insufficient allocation of parent-child interaction time and lack of systematic learning energy.
   c. The principle of pursuing convenience and efficiency, the consumer psychology brought by public opinion.

2. Commitment to profit
   a. Reasonable advice to provide parent-child interaction advice for young urban parents who lack parenting experience.
   b. Efficient interaction saves time for young urban parents who want to improve parent-child relationships but are busy with work.
   c. Reduce costs and reduce the physical and mental exhaustion caused by poor parent-child interactions for young urban parents.
3. Support points
   a. GSR monitoring module.
   b. Music selection player module and light prompt module.

3 The Experimental Description

3.1 Hardware System Concept Design

The design of this equipment is mainly divided into three parts: product function and modeling design, intelligent hardware and circuit design, experimental design and evaluation.

Based on the results of the survey, we used a hand-worn product solution to study a wearable glove device that senses emotions. By designing an interactive way that matches the natural behavior and psychological experience of the parent and child, the device allows the parents to feel each other's emotions and improve the efficiency of communication and contact between parents and children. The device consists of a sensor module, a feedback module and a processing module, which are integrated into a wearable glove. The user's emotional data is collected by the sensor module embedded in the palm of the glove, thereby inferring the user's emotional data; the user is provided with easy-to-detect feedback and contextual suggestions through the feedback module; the data is interpreted and processed by the processing module, and the control module is controlled Operation.

Realize 3 product features:

1. Real-time monitoring of GSR data, making judgments and issuing instructions. The feedback method is adjusted according to the valence and arousal of the emotion. Through preliminary research and experimental verification, the most intuitive and convenient monitoring method was selected. Follow the humanized improvement trend of monitoring methods and read and record in a natural way. Simplify the operation process and save users the cost of learning. Improve the intelligence and automation level of the monitoring module and optimize the monitoring and recording process.
   a. Monitor user GSR values in real time and respond quickly to read data;
   b. Based on past research results, make reasonable judgments on physiological monitoring data and classify emotions;
   c. Issue corresponding instruction in accordance with the degree of emotion;
2. According to the received instructions, adjust the light of different colors for feedback to make the emotions visible. The instruction transmits the command to

the feedback module through the processor, turns on the target response mode, and visually prompts the suggested solution that will be provided later. Using the principle of color psychology, it is suitable for indoor entertainment scenes, and naturally provides visual representation of emotions in daily life.

   a. After receiving the emotion grading instruction, the corresponding light color can be controlled according to the preset type to visually display the emotional change, and the user can observe the state of mind of the contact object.

   b. Set the color and time period of the light in different modes.

   c. Touch to turn off the light feedback.

3. Perform different audible feedback based on the instructions received to provide emotional comfort suggestions. The instructions are passed through the processor and passed to the feedback module to implement a targeted response mode, providing a solution for monitoring and judging results, using voice prompts. After receiving the classification instructions, select the appropriate recommendations based on the emotional degree.

   a. After receiving the emotion grading instruction, the corresponding sound suggestions (music, voice, etc.) can be played according to the preset type to provide an auxiliary solution for the emotional reaction.

   b. The SIM card stores the sound library. After receiving the command, you can adjust the corresponding track.

   c. Touch to turn off the sound feedback.

3.1.1 Hardware Prototype System Principle

The intelligent hardware part of the monitoring module and the main body will be designed to be integrated to simplify the tedious operation of the user. In the structure of the device prototype, the monitoring module and the feedback module are separate, and data transmission is performed through the processing module. The main body of the monitoring module is an electrode piece and a signal processing template, and the feedback module comprises a ring-shaped LED full-color lamp and a power amplifier board and a speaker. The device prototype can be worn at the end of the user's limb, such as the palm of the hand, thereby reducing the impact on the user's activity experience. Considering that the electrode sheet has a certain degree of vulnerability, it is designed as a separate, integrally detachable component. The layout also helps to replace physiological indicators monitors of different characteristics according to the user's monitoring preferences, and the overall design has certain environmental adaptability. The sound and light feedback module is hidden in the product, and the position of the voice feedback module is slightly higher, thereby improving the convenience of sound quality and voice input and output. The position of the sensor is arranged at the outermost level, and the information collected for different physiological parameters has a certain degree of adaptation.

If we analyze the structure of the prototype from the outside to the inside, it can be divided into three levels: the inner layer is an intelligent hardware board, which is used to store the single-chip microcomputer and the power amplifier board; the inner layer external unit is equipped with a power supply part such as a battery; the outer layer is monitoring and feedback unit. All sensors, voice devices, and bulb components need to be exposed for good results, so all related components are placed separately.

Through previous research, we began to develop prototypes with functional original features based on our design concepts. The basic working principle of the hardware system is: First, the monitoring module performs physiological signal acquisition. Then send the information to the motherboard, let the motherboard make judgment decisions, such as emotional awakening, valence, and so on. After that, the hardware of the feedback module receives the instruction from the motherboard and sends it to the corresponding feedback module. Thereby changing the mode of the feedback module, performing a voice interaction mode or a visual interaction mode.

The key function points are:

1. The information transmitted by the monitoring module to the hardware is mainly the human physiological signal data collected by each component of the monitoring module. Maintain real-time feedback to prepare for the next calculation.
2. Based on the results of the previous experiment, select the sound track with the apprehension suggestion and store it in the TF card in advance. The sensor of the GSR module detects the change of the GSR reading. When the MCU (Microcontroller Unit) of the motherboard judges that the value is greater than the reference value, the speech feedback module plays the corresponding sound track, and the sound is diffused through the sound board and the speaker.
3. At the same time, the sensor of the GSR module detects changes in physiological readings, and when the MCU of the motherboard determines that the value is greater than the reference value, the bulbs are transformed in different colors and durations.
4. Communicate the monitoring module and the feedback module through the motherboard.

3.1.2   Motherboard and Sensor Selection

1. Motherboard

The core of the functions in this system is the open source motherboard microcontroller. Arduino's Arduino UNO, the motherboard size is only 53 × 19 × 12 mm, easy to operate, very suitable for the development of the project. Although the Arduino Nano is smaller and fits well with current wearable prototyping needs, we used a more stable Arduino UNO board for stability during the actual build process of the prototype.

2. Bionic sensor

We chose the Grove - GSR module to collect the user's physiological data. This module measures skin electrical response to measure skin conductivity. Strong emotions can stimulate the sympathetic nervous system, causing sweat glands to secrete more sweat, which affects skin conductivity. Simply connecting the two electrodes to the appropriate skin surface reveals strong mood swings. This fun piece of equipment is great for creating emotionally relevant projects that are great for our projects.

3. Broadcast module

We chose Waveshare's Music Shield module as an audio player to provide music interaction. It is based on the Arduino standard interface design and is compatible with UNO and Leonardo development boards. The board has a 74VHC125 level conversion chip that is compatible with 3.3 V and 5 V microcontroller interfaces. The module has a TF card slot, and can access the TF card to play audio files therein. The module is based on the VS1053B audio decoder chip and supports common audio file formats such as MP3/AAC/WMA/WAV/MIDI. Moreover, there is a MIC recording function on the board, and 2S and MIDI interfaces are introduced in this area, which is advantageous for function expansion.

4. Lighting module

We chose the intelligent full color RGB light ring development board as a light feedback to provide light interaction. The module is powered by a 5 V power supply. Eight bright smart LED NeoPixels are arranged on a circular PCB with a 32 mm outer diameter. These ring-shaped RGB lamps can be easily cascaded through the output pins to the other input pins (DI → DO). Use a single-chip pin to control as many LEDs as possible through cascading. Each LED integrates a driver chip to make a single LED smart and addressable. Each LED has a constant current of 18 mA, so the color of the LED is very consistent even when the voltage changes. In the actual prototyping, the LED RGB 140C05 module was used for soldering stability (Table 3).

Table 3. Monitoring module materials

Name	Model	Number
Arduino Uno board	R3	1
Arduino data transmission line	/	1
GSR module	Grove - GSR	1
GSR electrode	/	2
Music shield module	VS1053	1
speaker	/	2
Amplifier board	PAM8403	1
Intelligent full color RGB light ring development board	8 digits WS2812 5050 RGB LED	1
Touch switch	TTP223B	1
Battery	/	1
Battery Holder	/	1
DuPont line	/	Several

In order to implement the product module function, the following materials are required (Fig. 2):

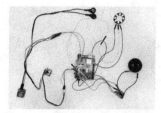

Fig. 2. Circuit design of the prototype

3.1.3   Hardware Prototype Design and Implementation

The overall product intelligent hardware prototype is divided into five areas: sensing area, switch area, control area, sound area and lighting area. The sensing area monitors the user's emotional physiological signals and transmits them back to the motherboard; in the control switch area, the work of the prototype of the device is controlled to be turned on and off by the touch switch; in the main area, the MCU on the motherboard acts as a central processing unit for coordinating the users, identify emotional grading commands, and control the operation of the entire system and other hardware devices; in the voice area, the MP3 Player Shield receives the playback command, and the sound amplifier board expands the volume through the speaker. In the lighting area, the WS2812 chip in the intelligent full color RGB light ring development board controls the color change of the light. Obtaining hardware control instructions of the control area to feed back or obtain monitoring module information for feedback, for example, the sound area has voice input and output devices related to the voice service. The modular system is powered by battery or USB.

The prototype's monitoring function enables the sending and receiving of commands via the GSR electrode, microcontroller and Bluetooth module. It monitors the user's GSR value in real time and responds quickly to read data. After the physiological information is collected, it will be quickly transferred to the main area. The MCU in the main area processes the data and analyzes it to assess the emotional arousal and potency of the contacted user, classify the rating, and finally output the control command. The instructions are simple and the operation is easy to control. The hardware prototype of the monitoring function involves three areas: the control area, the sensing area, and the main area (See Fig. 3).

Control area: This area turns the prototype on or off with a touch capacitive switch.

Sensing area: This area integrates two GSR electrodes and a Grove-GSR module board. The GSR electrode detects the GSR signal and transmits it to the Grove-GSR module board for processing and conversion. Thereby measuring the physiological data of the collected user.

Main area: This area is the Aduino UNO controller for the recording program. It needs to implement the functions of accepting GSR data, processing the data for classification, and transmitting the classified instructions to the feedback module.

The feedback function of the prototype realizes the interaction of sound and light through the music module and the LED module. Due to the limitations of the wearable application scenario, the size of the monitoring module and the feedback module is limited. Therefore, the size of the board should be reasonably reduced, and the appearance should be properly installed, and the space should be carefully arranged. The hardware prototype used to implement the feedback function is divided into four areas: the main board area, the voice area, and the light area (see Fig. 3).

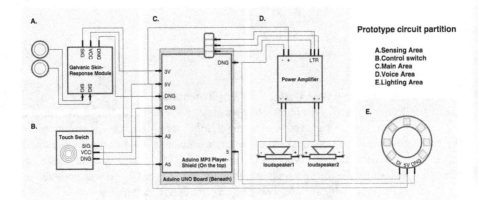

Fig. 3. Circuit design of the prototype

Motherboard area: This area is the Aduino Uno controller and the program has been programmed. It mainly includes four functions: identifying and classifying the received GSR signal; giving a color change instruction of the control lamp according to the determined classification; and giving an instruction to control the playback of the voice track according to the determined classification.

Lighting area: The motherboard controls the change of the color of the light, prompting the user's emotional changes in the user currently touched. This area is mainly composed of LED bulbs and WS2812 chips.

Voice area: The Music Shield board accepts the classification signal from the motherboard output and plays the corresponding music track in the TF card. A speaker that boosts the sound quality through the power amplifier emits a sound that is representative of the suggestion (Fig. 4).

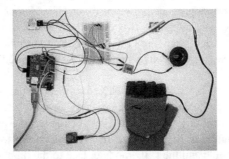

Fig. 4. Prototype construction

3.2   Product Design System Testing

3.2.1   Test Purposes

We conducted two experiential evaluation experiments in total, one at the time the basic implementation of the functional unit, and one at the time the initial model was built. In order to adapt to the user's demand for the product, it is necessary to quickly adjust the product function, and explore the real needs of the user in the iteration. This will more realistically verify the user experience of the product.

The first experiment tested the main functions of the product: finding the comfort suggestions that the user thought would be reasonable in various situations, and creating an audio library with the best effect audio type; finding the color and duration of the emotional cue that the user deems appropriate.

The second experiment tested the user experience of the product: correcting the bad user experience details found during the prototype experience and improving the unreasonable setup of the product prototype.

3.2.2   Participants Selection and Environment

On the basis of grasping and understanding the characteristics of urban young parents, the questionnaire was compiled, supplemented by the Eysenck Personality Questionnaire, and the unreasonable participants were screened out, and 5 typical participants were selected.

The laboratory is placed in an environment similar to a living room where the light is regulated to a comfortable level and the room temperature is controlled at a suitable temperature of 26 °C.

3.2.3   Test Content

To test the experience in wear mode, we hid the monitoring electrodes of the previously constructed prototype into the glove and produced a prototype for the test evaluation. Evaluation tests were conducted using observation methods, oral reporting methods, and questionnaire survey methods.

Learning Adaptation: Use observation methods combined with video observation to record the duration, fluency and difficulty users encounter when learning products.

Describe the experience directly: Use a loud thinking approach to visually describe the user's experience with the product.

Suggestions for improvement: Use the questionnaire model to obtain user improvement suggestions (Fig. 5).

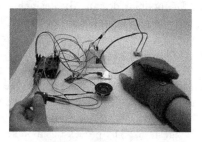

Fig. 5. Product prototypes for testing

3.2.4 Test Results

Participants were curious about the way the prototype detected emotions and provided suggestions. The prototype was a novelty product for them and the operating experience was brand new. Most of the time, the participants got good product prototype feedback and showed a positive emotional response. In the process, their emotions are elevated toward a positive emotional dimension. They proposed:

1. Pay attention to modularizing the product to make it more changeable and adapt to the needs of different ages.
2. Be interesting while maintaining its usefulness and increasing utilization.
3. To optimize the shape and lightweight circuit.

Most of the participants can quickly learn how to use the prototype. Some people think that the motherboard and speakers are too large and affect the activity experience. Others have suggested improving the color and shape of the light modules, hiding more exposed wires in the prototype, and associating them with handheld mobile applications to make their recommendations richer. This makes it easier to operate and smarter.

A few of the participants encountered some difficulties in the experience process. One person was dissatisfied with the construction of the prototype of the product, because his large-scale activities were hindered by the circuit connection, which made him feel very annoyed and difficult to achieve the desired experience. After the experience ended, he also expressed concerns about product safety.

4 Discussions

In this paper, we propose a design idea for a new wearable intelligent emotion-sensing device. We have done relevant market research and target user research, from design research, product design, and program testing. We used desktop research, user interviews and other methods to sort out the design features of parent-child devices in the market. A group of inexperienced urban young parents were surveyed and their needs were collected and analyzed through in-depth interviews. After comprehensive analysis of its advantages and disadvantages, we can draw our design opportunities.

Then we defined the needs of the target users and the functional design of the products, improved the design ideas of the products, and developed the system design of the products. Based on the results of the survey, we used a hand-worn product solution to study a wearable glove device that senses emotions, and finally chose a fabric glove as a carrier for modular design. By designing an interactive way that matches the natural behavior and psychological experience of the parent and child, the device allows the parents to feel each other's emotions and improve the efficiency of communication and contact between parents and children.

The device is mainly composed of a sensing module, a feedback module and a processing module, and is integrated on a wearable glove. The user's emotional data is collected by the sensor module embedded in the palm of the glove, thereby inferring the user's emotional data; The feedback device is used to give the user easy-to-detect feedback and contextual suggestions; The processing module interprets and processes the data and controls the operation of the device.

We use the product demo to conduct user evaluation tests, and the test results are good, which verifies the feasibility of the design. Most users say the product is helpful and active. There are not many smart products currently focusing on parent-child emotional interaction in the market. Most of them are relatively simple non-interactive non-intelligent products. This is a market application direction worth exploring. The device we studied provided it with a port of exploration.

Parents and children can be our users in the usage scenario. All they need to do is wear emotionally stimulated gloves and then shake hands or touch the forehead to touch the surface of the other's skin. It takes only a few seconds for a short test time to know each other's emotions and to get a solution for this situation. These programs are usually presented by voice, stories, jokes, songs, etc., giving parents some tips, bringing joy to the children and driving the atmosphere of parent-child interaction.

With the development of open source hardware platforms such as Arduino, we can use a combination of various sensors, Bluetooth modules and microcontrollers to transmit user sentiment judgments and user status information to users' mobile applications in real time. Improve the intelligence of the product and make it more human.

Acknowledgments. This research was supported by the National Natural Science Foundation of China (No. 51675476 and No. 51175458).

References

1. Parker, Boak, F.L.: Parent-child relationship, home learning environment, and school readiness. J. School Psychol. Rev. **28**(3), 413 (1999)
2. Brassell, A.A., Rosenberg, E., Parent, J., et al.: Parent S psychological flexibility: associations with parenting and child psychosocial wellbeing. J. Context Behav. Sci. **5**(2), 111–120 (2016)
3. Cheron, D.M., Ehrenreich, J.T., Pincus, D.B.: Assessment of parental experiential avoidance in a clinical sample of children with anxiety disorders. J. Child Psychiatr. Hum. Dev. **40**(3), 383–403 (2009)
4. Simon, E., Verboon, P.: Psychological inflexibility and child anxiety. J. Child Fam. Stud. **25**(12), 3565 (2016)
5. O'Brien, K., Mosco, J.: Positive parent–child relationships. In: Roffey, S. (Eds.) Positive Relationships. Springer, Dordrecht (2012)
6. Eyberg, S.M., Bussing, R.: Parent–child interaction therapy for preschool children with conduct problems. In: Murrihy, R., Kidman, A., Ollendick, T. (Eds.) Clinical Handbook of Assessing and Treating Conduct Problems in Youth. Springer. New York, NY (2011)
7. Yu, H.C., Chou, C.J., Luh, D.B., et al.: User-centered empathy design: a prototype of school-age children learning aids system. J. Ambient. Intell. Human Comput. (2018)
8. Brophy, S.: Grandparents raising grandchildren. In: Singh, A., Devine, M. (Eds.) Rural Transformation and Newfoundland and Labrador Diaspora. Transgressions (Cultural Studies and Education). Sense Publishers, Rotterdam (2013)
9. Hayslip, B., Kaminski, P.L.: Grandparents raising their grandchildren: a review of the literature and suggestions for practice. J. The Gerontologist. **45**(2), 262–269 (2005)

10. Pan, Y.J., Sun, L., Dong, S.S., Tu, Y.: Intergenerational conflicts and transmission of values in raising 0-2-year-old chinese babies. In: Li, L., Quiñones, G., Ridgway, A. (Eds.) Studying babies and toddlers. In: International Perspectives on Early Childhood Education and Development, Vol. 20. Springer, Singapore (2017)
11. Fuller-Thomson, E., Minkler, M., Driver, D.: A profile of grandparents raising grandchildren in the united states. J. Gerontologist. **37**(3), 406–411 (1997)
12. Vandeweghe, L., Moens, E., Braet, C. et al.: Perceived effective and feasible strategies to promote healthy eating in young children: focus groups with parents, family child care providers and daycare assistants. J. BMC Public Health. **16**, 1045 (2016)

The Research of Visual Characteristics for the Deaf Wearing AR Glasses

Zhelin Li[1,2(✉)], Xiaohui Mo[1], Chengchuan Shi[3], Shanxiao Jiang[1], and Lijun Jiang[1,2]

[1] School of Design, South China University of Technology, Guangzhou, China
{Zhelinli,ljjiang}@scut.edu.cn,
id_moxiaohui@163.com, mykakaqq@163.com
[2] Guangdong Human-Computer Interaction Design Engineering Technology Research Center, Guangzhou, China
[3] Guangzhou Voibook Technology Co., Ltd, Guangzhou, China
ccshi@voibook.com

Abstract. In order to design AR (Augmented Reality) glasses for the deaf, this paper studies the virtual display interface of AR glasses. It is found that 89.8% of the deaf have willingness to use AR glasses based on questionnaires of 1900 deaf people. An experiment with 10 deaf subjects and 10 hearing subjects was done to calculate times of completing specific tasks. The results show that: (1) 80% of the deaf subjects have 3 to 5 zones in the interface with significant differences on cognitive efficiency, compared with the 6 to 7 zones from 80% of the hearing subjects. (2) The heat maps of the cognitive efficiency distribution shows that there is an obvious difference on the distribution patterns of regional sensitivity between the two groups. (3) The significant difference on cognitive efficiency of the deaf shown in different display zones is less than that of hearing people.

Keywords: AR glasses · Visual interface · Cognitive efficiency · The deaf

1 Introduction

According to the World Health Organization, there are around 466 million deaf people (hearing loss greater than 40 dB in the better hearing ear in adults and a hearing loss greater than 30 dB in the better hearing ear in children) worldwide, and the number is estimated to reach more than 900 million by 2050 [1]. The deaf have inconvenience to communicate with the outside world, and mainly communicate with sign language, lip language, writing and etc. However, a certain communication mode only shows effectiveness in a special situation [2]. At present, the solution to restore hearing of the deaf is not ideal. For example, due to high prices and unsatisfactory speech clarity [3], the penetration rate of hearing aids in the global deaf population is less than 10% [1].

With the development of artificial intelligence technology, as AR (Augmented Reality) glasses is combined with voice technology, the sound can be converted into visual information on the glasses and will not obstruct user to see the real world [4]. It is conceivable that the communication will be improved efficiently for the deaf by

© Springer Nature Switzerland AG 2020
T. Ahram (Ed.): AHFE 2019, AISC 973, pp. 145–153, 2020.
https://doi.org/10.1007/978-3-030-20476-1_16

using those AR glasses. Sony has worked with American theater chain Regal Entertainment to provide a new kind of glasses technology that can display closed captions for the deaf [5]. Dhruv Jain et al. developed an AR glasses software which experimented on 24 deaf people to explore their interaction preferences for using AR glasses to locate the spatial position of sounds [6].

Human perceive the world through visual, auditory, tactile and other sensory systems. Modern cognitive neuroscience has proven that the loss of hearing can have an impact on vision [7, 8]. Based on the formal studies, it is assumed that the deaf and hearing people wearing AR glasses have differences on visual characteristics. However, there was few research on the types of sound information that are demanded for deaf in different situation, and there was also few research on the visual cognitive efficiency of virtual display interface of AR glasses for deaf. This paper carried out the following research: Firstly, through the questionnaire survey, we studied what sounds are important to the deaf in different situation; then, an experimental research on visual cognitive efficiency of virtual display interface was conducted.

2 Survey for the Deaf

A questionnaire survey was conducted to explore the willingness of the deaf wearing AR glasses to obtain sound information. According to the results of 1900 questionnaires, 89.8% of the deaf expressed their willingness to use that glasses, and they hoped to use it in work, home and other situations, as shown in Table 1.

Table 1. The result about where the deaf wearing AR glasses to obtain sound information

Situations	Work	Home	School	Outside	Meeting	Other
Percentage	84.9%	56.9%	34.5%	57.3%	56.3%	10.6%

Human activities can be divided into four types: subsistence, maintenance, recreation, and travel [9, 10]. The main types of sounds in these activities are human voice, physical sound, natural sound, etc. Questionnaires were issued on internet, and the deaf in China (hearing thresholds >40 dB, aged 18 to 45 years old) were invited to fill out. 110 valid responses were received. It was found that more than 55% of the participants want to obtain the voices from real people in each activity.

3 Experimental Research on Visual Cognitive Efficiency

3.1 Subjects

There are 20 subjects (normal vision; average age, 24 ± 6) in this experiment. 10 hearing impaired people were in the deaf group (numbered H1 ~ H10, pre-linguistic deaf, hearing thresholds ≥ 80 dB), and 10 normal people were in the hearing group (number N1~N10, hearing thresholds ≤ 25 dB).

3.2   Experimental Equipment and Environment

The AR glasses in this experiment is shown in Fig. 1a. The right frame of AR glasses is equipped with a control panel. Participants can slide up and down, slide back and forth or click on it to control the interface content. The virtual display interface absorbing 50% of ambient light has a resolution of 800px*480px and a size of 32 inches (69 cm wide, 39 cm high). The focus position is 240 cm in front of the eyes. The environment is shown in Fig. 1b. The field illumination is good. A gray non-transparent curtain (250 cm wide, 150 cm high) is used as the background. Participant is arranged at the eye position 240 cm (± 5 cm) from the curtain.

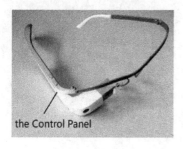   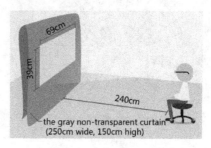

(a) AR glasses   (b) Lab environment

Fig. 1. Lab Equipment & Environment

3.3   Experimental Programming

In the experimental interface, according to the common geometry in the human-machine interface [11], the circle with symmetric features particularly was selected. The observing icon (hereinafter referred to as the three-circle icon) is composed of three circles randomly displaying the numbers 1, 2 or 3(size 32px [12]). The center of those circles are respectively distributed at the vertex of the equilateral triangle (the side length = 76px), as shown in Fig. 2. So that, the visual jumping distances are the same when participant watches each number. The circular background color is green (#99cc33). The number color is white (#ffffff) [11, 12]. The actual effect on the AR glasses is shown in Fig. 3.

According to the right visual field distribution map proposed by Theodore C. Ruch and John F. Fulton [13], the right visual field through the glasses interface is shown in Fig. 4, and the rectangular area represents the interface zone, as shown in Figs. 1b and Fig. 2a.

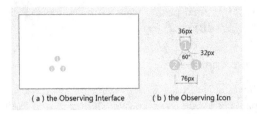

Fig. 2. The observing icon (three-circle icon)

Fig. 3. The actual effect

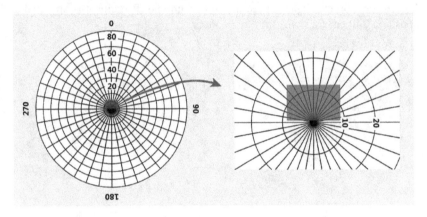

Fig. 4. The right visual field through the glasses interface

According to the optimal field of view proposed by ergonomics [14], the interface is divided into 4 zones (A, B, C and D) by visual field latitude lines of 2.5°, 5°, and 10°, as shown in Fig. 5a. Then, the zone B, C and D are evenly divided into 3 zones separately by longitude lines of 60° and 330°, as shown in Fig. 5a. What is more, the zone A is too small to be divided. Therefore, there are 10 zones in total which is shown in Fig. 5b.

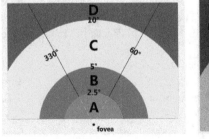
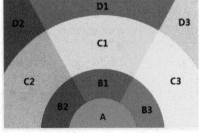

(a) The divided visual interface

(b) Zoning and naming the areas

Fig. 5. Define the display interface area

In the experiment, a three-circle icon is displayed in only one zone in each observing interface. The three-circular icon with 10 different coordinate positions are displayed in each zone, a total of 100 observing interfaces which are divided into 5 groups.

3.4   Experimental Procedure

Each subject is required to finish a single task in this experiment. An application program(App) was written by Android Studio to collect data. Experimental steps are listed as follows.

(1) App presents the boot page. After reading the experiment description, the subject needs to click on the control panel;
(2) App presents a 3 s countdown page to to unify the position of the subject's visual focus;
(3) App presents the observing interface. The subject is required to confirmed three numbers shown on the interface and click the control panel;
(4) App presents the question interface to provide two addition results for the option. The subject selects the correct answer on the control panel by the front or back slide operation. According to the observing interface groups, each group needs to continuously follow the steps b, c, and d.
(5) At the end of each group, the subject receives a reminder of taking rest presented on the interface, and needs to take off the glasses to take a rest for 4 min. After that, the subject should wear the glasses to continue next group.

During the experiment, the App records the subject's stay time in the observing interface and the selection in the question page.

4   Result

4.1   Significance Analysis

The total response time of the subject in the observing interface is $T = T1+T2+T3$. In this paper, we defined that T1 is the time to cognize the three-circle icon, which is related to the position and numerical value of the icon; T2 is the time to calculation the sum of the three numbers, which is related to the counting ability of the subject; T3 is the time to operate the control panel of AR glasses, which is related to the operational ability of the subject. For a specific subject, the operational ability of the same action is consistent, and the change of the numerical value 1, 2, and 3 has no effect on the numerical recognition time and calculation time. Therefore, the difference in T is reflected in the influence of the display position of the three-circle icon on T1. In other words, the icon display position has an influence on the user's cognitive efficiency.

By excluding the wrong answer data and using the Grubbs guidelines to eliminate extreme data, the valid data was retained. Variance Analysis and Multiple Comparison (LSD) were used by applying SPSS 23.0 in the valid data. The zones which of the statistically significant are consistently will be classified into one category. In the same category, each zone shows significant differences (P < 0.05) to the same areas which are not in this category. And then, sorting the categories into different level according to the magnitude of the average response time T'. The level of the category is higher, the lower the response time and the higher the cognitive efficiency of those zones in this category.

The statistical results of all the subjects were summarized to obtain a table (Table 2) which shows the cognitive efficiency characteristics of each subject.

Table 2. The classification of cognitive efficiency in interface zones

Subjects	Zone A	Zone B1	Zone B2	Zone B3	Zone C1	Zone C2	Zone C3	Zone D1	Zone D2	Zone D3	Total levels	PD (%)
H1	2	4	4	4	3	3	3	3	1	3	*4	63.6
H2	5	7	3	6	5	5	5	1	2	4	7	#44.4
H3	3	4	4	3	3	3	1	2	4	4	*4	#50.0
H4	1	3	4	3	3	3	3	3	2	3	*4	80.0
H5	4	5	3	4	3	5	2	1	4	2	*5	#44.4
H6	3	2	2	2	2	3	3	1	2	1	*3	#0.0
H7	2	2	2	2	3	2	3	1	2	2	*3	#0.0
H8	1	4	3	4	3	3	3	3	2	3	*4	75.0
H9	4	2	5	1	4	7	3	6	2	4	7	#35.7
H10	2	2	3	2	2	2	2	2	1	2	*3	#0.0
–	–	–	–	–	–	–	–	–	–	–	–	–
N1	5	3	3	4	5	5	3	1	2	6	6	61.1
N2	4	7	7	7	6	7	5	1	2	3	7	90.9
N3	1	5	2	4	5	5	4	1	3	6	6	80.0
N4	3	6	2	3	5	3	3	1	2	3	6	70.6
N5	3	2	2	3	3	3	3	1	4	3	*4	81.8
N6	6	7	7	4	6	1	5	2	3	7	7	81.8
N7	4	6	5	5	5	5	5	1	3	2	6	73.3
N8	3	7	5	7	7	6	6	1	4	2	7	72.7
N9	1	6	5	5	6	4	5	2	3	5	6	84.2
N10	4	4	4	4	3	4	4	1	2	4	*4	#50.0

The numbers from area A to D3 represent the cognitive efficiency level. For example, for the subject H10, the first-level cognitive efficiency zone is D2; the second-level zones are A, B1, B3, C1, C2, C3, D1 and D3; the third-level zone is B2. * represents the total level ≤ 5; # represents the PD ≤ 60%.

From Table 2, it can be found that: (1) 80% of the deaf subjects (H1, H3 ∼ H8 and H10) have 3 to 5 zone categories in the interface with significant differences on cognitive efficiency, compared with the 6 to 7 zone categories from 80% of the hearing subjects (N1 ∼ N4, N6 ∼ N9). (2) The statistical number of extreme significant differences (n(P < 0.01)) divided by the statistical difference of significant differences (n(P < 0.05)), so that the proportion of data(PD) with extreme significant differences can be known. It can be seen that there are 3 subjects in the deaf group who have a value of more than 60%, and 3 subjects who have a value of 0%. In the hearing group. However there are 9 subjects who have reached more than 60%, and one subject who has a value of 50%. It indicates that the significant difference in cognitive efficiency of the deaf in different display zones is smaller than that of hearing people.

4.2 Heat Maps Analysis

The average response time T' to each zone of the deaf group and the hearing group were summarized and normalized. Variance Analysis and Multiple Comparison(LSD) were used by applying SPSS 23.0. The average regional response time ordering of the deaf group was T'D1>T'D2>T'A>T'D3>T'B3>T'C3>T'C1>T'B1>T'B2>T'C2, and the hearing group was T'D1>T'A>T'D2>T'C2>T'B1>T'D3>T'C3> T'B3>T'B2>T'C1. By comparing the P values, the zones with consistently significant differences will be classified into one category. And then, according to the average regional response time ordering, sorting categories into different level. The level of the category is higher, the lower the response time and the higher the cognitive efficiency of those zones in this category. Visualized the results and drew a heat maps (Fig. 6) showing the cognitive efficiency of the interface.

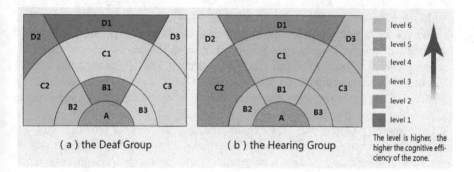

(a) the Deaf Group   (b) the Hearing Group

level 6
level 5
level 4
level 3
level 2
level 1

The level is higher, the higher the cognitive efficiency of the zone.

Fig. 6. The regional heat maps showing the cognitive efficiency of the visual interface

Comparing the regional heat maps, it can be seen that the distribution pattern of regional sensitivity is inconsistent between the deaf group and the hearing group. The deaf group shows a trend of lower cognitive efficiency from the left to the right, from the middle to the lower and upper sides; the hearing group shows a trend of lower cognitive efficiency from the middle and the right to the left, from the middle to the lower and upper sides.

5 Discussion and Conclusion

In previous studies, it was suggested that the deaf is willing to be able to hear various types of voices and environmental sounds. Tara Matthews et al. have interviewed some deaf and learned that they were expecting to hear various types of sounds in schools, offices, streets, etc. [15, 16]. However, they only have a small investigation sample size including 24 people, and there is a lack of comparative research on the demand for the sound types in different activities. In this research, a questionnaire survey is sent out to gather 110 valid responses, and the difference among the demand of different kind of sound types in several activities is also considered in this research.

For the difference in visual characteristics between the deaf and the hearing people, in the present research, the researchers have leaded to conclusions with different methods. Stevens et al. found that the visual field of the deaf is 10 cm^2 larger than hearing person by using automated peripheral kinetic and foveal static perimetry [17]. Loke et al. found that peripheral vision of the deaf were more sensitive by designing an experiment in which the deaf students and hearing students were required to respond to signals presented at the gaze point or out of the gaze point area respectively [18]. Previous studies have mainly used large-screen display devices for experiments. However, the research on the difference in visual characteristics wearing AR glasses is temporarily blank.

This research studied the willingness and visual characteristics of the deaf by the investigation and experiment. The survey found that the deaf have a high willingness to use the AR glasses to obtain the sound content by convert the sound information into visual information,and there is more than 55% of the demand for voices in the subsistence, maintenance, recreation, and travel activities. Experimental study has shown that: In the 32-inches virtual display interface of AR glasses, 80% of the deaf subjects have 3 to 5 zones in the interface with significant differences on cognitive efficiency, compared with the 6 to 7 zones from 80% of the hearing subjects; The heat maps of the cognitive efficiency distribution from the deaf and the hearing subjects shows that the distribution pattern of regional sensitivity is inconsistent between the two groups; The significant difference in cognitive efficiency of deaf people in different display zones is less than that of hearing people.

The research results of this paper could provide references for providers who are determined to improve the life quality for deaf, and it's also helpful for the AR glasses application developers. In the next step, researches on visual style and interaction mode of sound visual information displayed on the AR glasses will be carried out, and the number of samples will increase to get more accurate conclusions.

Acknowledgments. This research is supported by the Fundamental Research Funds for Central Universities (2017ZX013).

References

1. World Health Organization. https://www.who.int/zh/news-room/fact-sheets/detail/deafness-and-hearing-loss
2. Mi, K.: A review of the research on communication mode of hearing impaired. Modern Spec. Educ. **08**, 72–76 (2016)
3. McCormack, A., Fortnum, H.: Why do people fitted with hearing aids not wear them? Int. J. Audiol. **52**(5), 360–368 (2013)
4. Schweizer, H.: Smart glasses: technology and applications Student report Ubiquitous computing seminar FS (2014)

5. Sony. Sony Access Glasses. http://goo.gl/0DKFoQ
6. Jain, D., Findlater, L., et al.: Head-mounted display visualizations to support sound awareness for the deaf and hard of hearing. The CHI, 241–250 (2015)
7. Levine, A., Billawa, S., Bridge, L., et al.: FMRI correlates of visual motion processing in hearing and deaf adults. J. Vis. 14(10), 297 (2014)
8. Bavelier, D., Dye, M.W.C., Hauser, P.C.: Do deaf individuals see better? Trends Cognit. Sci. 10(11), 512–518 (2006)
9. Lu, X., Pas, E.I.: Socio-demographics. activity participation and travel behavior. Department of Civil and Environmental Engineering, pp. 3–4
10. Timmermans, H., et al.: Analyzing space-time behaviour: new approaches to old problems. Progress Hum. Geogr. 26(2), 175–190 (2002)
11. Shao, J.: Research on the information encoding method of helmet mounted display system interface based on visual perception theory. Southeast Univ. 77, 113–118 (2016)
12. Google Inc. https://developers.google.com/glass/design/style
13. Ruch Theodore, C., Fulton, J.F.: Medical physiology and biophysics (1960)
14. Ding, Y.: Ergonomics. Beijing Institute of Technology Press (04), 39–41 (2011)
15. Matthews, T., Carter, S., et al.: Scribe4Me: Evaluating a Mobile Sound Transcription Tool for the Deaf. Technical Report No. UCB/EECS 9, 159–176 (2006)
16. Matthews, T., Fong, J., et al.: Evaluating non-speech sound visualizations for the deaf. Behav. Inf. Technol. 25(2), 333–351 (2007)
17. Stevens, C.: Neuroplasticity as a double-edged sword: deaf enhancements and dyslexic deficits in motion processing. J. Cognit. Neurosci. 18(5), 701–714 (2006)
18. Loke, W.H., Song, S.: Central and peripheral visual processing in hearing and non hearing individuals. Bull. Psychon. Soc. 29(5), 437–440 (1991)

Ergonomic Evaluation of Pilot Helmet and Neck Injury

Xiangyu Ge$^{(\boxtimes)}$, Qianxiang Zhou, and Zhongqi Liu

Beihang University School of Biological Science and Medical,
Beijing 100191, China
09121@buaa.edu.cn

Abstract. In addition to protecting pilot's head, flying helmet is also a mounting platform for head-mounted display, tracking and sighting systems, night vision devices, oxygen masks, and wireless communication devices. Although these helmets can greatly improve the combat effectiveness, these systems may increase the support load of head and neck, and the irrational design of helmet ergonomics will lead to the shift of Centre of Gravity (CG) of helmet and the increase of joint torque of neck, which increases the risk of neck injury to pilot. In response to this problem, this paper developed a system based on the three-point method for measuring CG of helmet, which is used to measure CG and Moment of Inertia (MI) of helmet, and to measure the physical parameters of two flying helmets. Measurement results are follows: mass M1 = 1.143 kg, M2 = 1.020 kg, CG C1 = (0.002, 0.542 and 7.630 cm), C2 = (0.314, 0.117 and 2.446 cm).MI J1 = (0.072, 0.089, 0.016 kg•m2), J2 = (0.041, 0.056, 0.015 kg•m2).The error analysis results show that the measurement error of system is less than 2 mm, which indicates that the system has high calculation accuracy and simple testing steps. In addition, the neck muscle fatigue test of wearing a helmet was carried out, and the fatigue degree of each muscle was analyzed quantitatively by using muscle fatigue model. This paper can provide some method support for man-machine ergonomics design of helmet, analysis of helmet mass parameter and evaluation of pilot's neck injury.

Keywords: Flight helmet · Ergonomics evaluation · Neck injury ·
Centre of gravity · Moment of Inertia · sEMG · Muscle fatigue

1 Introduction

More and more Head-Mounted Devices (HMDs), such as NVGs, oxygen masks, radio communication systems, visual tracking and targeting systems, and night vision devices, are installed on flying helmets [1] to enhance pilot situational awareness and improve weapon operations. While these HMDS do improve combat effectiveness, but it will result in an increase in the total mass of helmet system and a shift in Centre of Gravity (CG), which in turn increases the pressure on pilot's neck muscles and skeletal system. This kind of pressure is more likely to induce pilot's neck injury [2–4].

Pilot helmet not only plays a role in protecting pilot safety [5, 6], but also a flight and air combat information display terminal. It is the versatility of pilot's helmet that puts new demands on the ergonomic design of helmet. Helmet CG is one of the most

© Springer Nature Switzerland AG 2020
T. Ahram (Ed.): AHFE 2019, AISC 973, pp. 154–166, 2020.
https://doi.org/10.1007/978-3-030-20476-1_17

basic and important parameters of pilot helmet kinematics and dynamics research. It is an important indicator related to pilot health, safety and the full play of human - machine performance. The application environment of the carrier-based pilot helmet is quite different from that of other flight helmets, especially during take-off and landing phases, on aircraft carrier deck. Many carrier-borne flight helmets have problems of heavy mass and CG offset. The lighter the helmet is, the less inertia force head will suffer, and the better the comfort will be. When mass is constant, helmet deviates from the center of head, which may cause uneven distribution of muscle strength in head and neck, causing muscle fatigue or injury, and reducing comfort [7–9]. When pilot is subjected to a vertical load of 8–10 g, the actual mass of the helmet on the neck can reach 20–5 kg. During the take-off and landing phases, the horizontal impact acceleration forced on pilot's head and neck can be as high as 4–5 g in the horizontal direction due to the "whiplash" effect. When landing on the deck, calculation of the rotational moment and MI of the helmet, as well as analysis of neck damage caused by the number of takeoff and landing, becomes more difficult. Therefore, the mass characteristic measurement and ergonomics analysis of pilots' helmet need to be carried out urgently.

According to the structural characteristics of the helmet and the flight environment, the comprehensive ergonomics evaluation of the carrier pilot helmet was carried out. Aiming at the analysis and calculation of the helmet's CG and MI, the flight helmet centre of gravity measurement system was developed to obtain the position of the helmet's CG and MI. And a pilot neck muscle injury evaluation model based on sEMG was established to investigate the effects of helmet mass change and CG offset on the pilot's neck injury

2 Methods

In order to evaluate the neck damage caused by the increase in the mass and CG drift of helmet, a helmet physical properties measurement system was designed and developed. Due to the difference in mass and shape of the measured object and the requirements for measurement accuracy, there are big differences in the design and measurement methods of measuring equipment, each with its own characteristics and pertinence, but all adopt the basic principles of mechanics. We propose a three-point inversion measurement method that the CG of helmet can be calculated by measuring the distribution of the force on the three sensors based on the principle of moment balance. The object to be measured is placed on the measuring pallet which is supported by three force sensors. CG can be acquired by calculating the position of each sensor relative to the reference center. The CG measurement system consists of a spatial coordinate measuring system and a weighing system. The distribution of the three sensors is shown in Fig. 1.

Where, L_d = distance between sensor #1 and sensor #3, L = distance between sensor #2 and sensor #3.

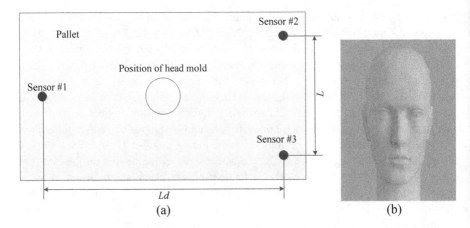

Fig. 1. (a) The distribution of sensors and (b) A Chinese male head mold

2.1   Measuring Coordinate System

The three-dimensional measuring coordinate system was established based on the sensors distribution. The line connecting sensor #2 and sensor #3 is the y-axis, and its' midpoint is the coordinate origin O. X-axis is the line connecting the sensor #1 and the origin O. Z-axis is a line perpendicular to the intersection of X and Y-axis as shown in Fig. 3. The head mold was fixed on to an orthogonal holding fixture so that it could be oriented along three orthogonal planes: X–Y, X–Z, and Y–Z.

Where, m_1 is helmet mass, its CG is (x_1, y_1, z_1); m is the mass of pallet and the auxiliary box with uniform density, which CG is (x_c, y_c, z_c); F_1, F_2 and F_3 represent the total force on the three sensors.

2.2   Principle of Measurement

In theory, the mass of the helmet is equal to the sum of the three sensors minus the mass of the pallet, auxiliary box and head mold. The calculation is as shown in Eq. 1.

$$m_1 = \frac{(F_1 + F_2 + F_3) - mg}{g} \tag{1}$$

Firstly, fix the head mold (as shown in Fig. 2) in the auxiliary box, and then place combining on the measuring platform smoothly. Using the measurement system to obtain the distribution of the head mold. Based on the principle of zero torque, the two-dimensional CG value of the head mold (xh, yh) can be calculated as Eq. 2.

$$x = \frac{F_1 L_d - mgX_c}{m_1 g}, y = \frac{(F_2 - F_3) \cdot L}{2m_1 g} \tag{2}$$

Next, rotate the combining 90° around the Y-axis. As a result, X-axis is converted to Z'-axis, Z-axis to X'-axis. The CG in Z-axis is calculated as Eq. 3.

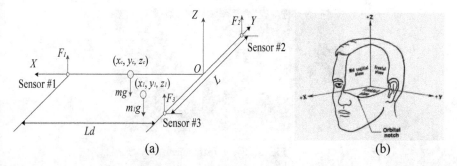

Fig. 2. (a) The three-dimensional measuring coordinate system (b) Head coordinate system [1]

$$z = \frac{F'_1 L_d - mgX_c}{m_1 g} \tag{3}$$

Mount the helmet on the head mold and fix it in the auxiliary box, and calculate the CG of the helmet according to the above method.

2.3   Moment of Inertia

MI is the inertia of a rigid body rotating around an axis. It can be understood as the inertia of an object for rotational motion, which is used to establish the relationship between angular momentum, angular velocity, moment and angular acceleration. For a particle, the MI can be expressed as Eq. 4.

$$I = mr^2 \tag{4}$$

Where, I is MI; m is mass; r is the vertical distance from the particle to the axis of rotation.

According to Eq. 4, the MI of the helmet around the X, Y and Z-axis can be calculated as Eq. 5.

$$I_x = m\left(y^2 + z^2\right), I_y = m\left(x^2 + z^2\right), I_z = m\left(x^2 + y^2\right) \tag{5}$$

2.4   Structural Design

The hardware components of the measurement system mainly includes pressure sensors, signal amplifiers, inspection instruments, RS-485 to USB conversion plugs, power supplies and computers. System structure design is shown in Fig. 3.

Fig. 3. System structure

2.5 Test Results

In order to verify the accuracy and validity of the system, the CG and MI of two helmets are measured by the system. The helmets for testing is shown in Fig. 4 and the test results are shown in Table 1.

Fig. 4. The flying helmets

Table 1. The test results.

Helmet number	Mass (kg)	CG (cm)	MI (kg • m2)
1	1.143	(0.002, 0.542, 7.630)	(0.072, 0.089, 0.016)
2	1.020	(0.314, 0.117, 2.446)	(0.041, 0.056, 0.015)

The error analysis results show that the error of measurement mass is less than 2.5% and the error of CG calculation is less than 2 mm.

3 Evaluation of Neck Injury

sEMG is widely used to evaluate muscle activity. Researchers can attach electrodes to the surface of the skin overlying a muscle and measure the amount of electricity produced by muscle fibers contraction [10]. Studies have shown that local fatigue characteristics can be reflected by sEMG signal after muscle exertion [11–13]. When the external load is equal, the amplitude change of sEMG can be understood as local muscle fatigue [14, 15]. In this paper, sEMG is used to investigate the effects of helmet mass and CG migration on neck muscles.

Since the CG of head and the CG of helmet are located in front of the atlantoaxial joint in the sagittal plane, the head mass and the helmet produce a flexion moment. To maintain the head neutral balance, semispinatus capitis, levator scapulae, scalp gripper and cervical gripper that produce greater muscle strength will be activated [16]. The sternocleidomastoid muscle is located between the posterior mastoid, the sternum and clavicle. Its function is to turn the head to the opposite side, helping to maintain head stability when head keeps still; and works with the trapezius muscle to bend the head and neck [17, 18]. Therefore, sEMG of scalp, sternocleidomastoid and trapezius muscles were collected to analyze the fatigue and injury of the neck muscles.

3.1   Neck Injury Assessment Model

Dragomiretskiy et al. proposed a new signal decomposition method, Variational Mode Decomposition (VMD) [19] that is much more robust to sampling and noise. Its overall framework is a variational problem, which minimizes the sum of the bandwidths of each modal estimation. Assuming each modal has a finite bandwidth of different center frequencies, the modals and their center frequencies are constantly updated with alternating direction multipliers. The modes are gradually demodulated to the corresponding baseband, and finally the modalities and the corresponding center frequencies are extracted together. VMD transforms the signal into non-recursive and variational mode decomposition modes, and has a solid theoretical foundation. Its essence is multiple adaptive Wiener filter groups, showing better noise robustness. By controlling reasonable convergence conditions, the sampling effect of VMD is much smaller than the Empirical Mode Decomposition (EMD) and Local Mean Decomposition (LMD). In terms of modal separation, VMD can successfully separate two pure harmonic signals with similar frequencies. When analyzing physiological time signals, Costa et al. proposed Multi-Scale Entropy (MSE) based on sample entropy to measure the dynamic complexity in physiological systems on a series of time scales [20]. Moreover, this method is widely used in physiological signal analysis and rotational machinery fault diagnosis [21–24].

Based on the advantages of the above signal processing methods, VMD and MSE were used to extract the feature vector of sEMG signals. The feature vector can be used as the input of Fisher discriminant to realize the analysis of the neck muscle fatigue and injury induced by the helmet mass change and CG offset. The specific analysis process is as follows.

3.2   Adaptive Variational Mode Decomposition

Analysis processing of VMD can refer to Dragomiretskiy's paper [19]. Here we highlight the implementation of the adaptive process. In VMD processing, K represents the number of modals that are decomposed. When K value is given, the residual signal after VMD processing may still contain useful information, or the decomposed modal component contains a large amount of noise, which increases the computational complexity. To solve this defect, the correlation coefficient r of the sequence can be used to determine the correlation between the residual signal and the original signal, thereby implementing adaptive VMD. First, initialize K value and calculate r value. Then, recalculate K value according to the following conditions:

$$\begin{cases} K = K+1, \ r > 0.09 \\ K = K-1, \ r < 0.05 \\ K = K, \ 0.05 \leq r \leq 0.09 \end{cases} \quad (6)$$

3.3   Injury Classification Based on MSE

After VMD processing, sEMG is decomposed into K modal components, each of which corresponds to a central frequency. As shown in Eq. 7.

$$sEMG = \sum_k u_k(t) \quad (7)$$

Next, MSE of the above modal functions is calculated and used to construct the sEMG feature vector F. For more details on the calculation of MSE, please refer to Costa's papers.

$$F = \{MsEn_1, MsEn_2, \ldots, MsEn_k\} \quad (8)$$

And, the feature vector F is used as the input of Fisher discriminant to realize the analysis of the neck muscle fatigue and injury induced by the helmet mass change and CG offset.

4   Experimental Results

In order to verify the validity of the model which we proposed, 10 volunteers were recruited to wear helmets with different mass and CG. The sEMG of scalp, sternoclei-domastoid and trapezius muscles were synchronously collected using a sEMG acquisition device. The physical parameters of helmets for testing are shown in Table 2.

Table 2. Mass and CG coordinates of helmets and head

Headgear	Helmet #1	Helmet #2	Head	Head + Helmet #1	Head + Helmet #2
Mass (kg)	1.143	2.143	4.130	5.273	6.273
CGx (cm)	−0.002	0.001	0.567	0.444	0.292
CGy (cm)	0.542	0.289	0.000	0.117	0.176
CGz (cm)	7.630	13.514	0.377	1.949	5.900

sEMG without helmet was collected for 5 min as the baseline of signal analysis; after 10 min rest, sEMG wearing helmet 1 was collected, and collection duration was 5 min; after 10 min rest, sEMG wearing helmet 2 was then collected for a duration of 5 min. During the collection process, volunteers need to look straight ahead and keep their heads steady.

VMD analysis of sEMG of scalp gripper muscle is shown in Fig. 5.

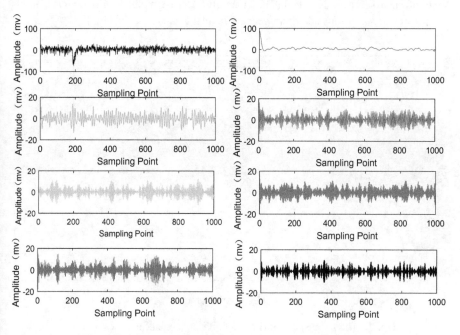

Fig. 5. VMD modal components of the scalp muscle's sEMG

As shown in Fig. 5, after VMD analysis, the original signal was decomposed into seven modal components. Since the signal length is too long, only the data of the first 1000 sampling points is displayed here. Each modal component corresponds to a center frequency, as shown in Fig. 6.

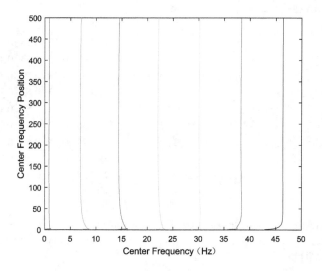

Fig. 6. Center frequency of the modal component of scalp muscle

Each modal component is divided into n segments in units of 1000 points, and MSE of each segment is calculated separately. The segmented MSE of each modal component of scalp muscle when wearing helmet 2 is shown in Fig. 7.

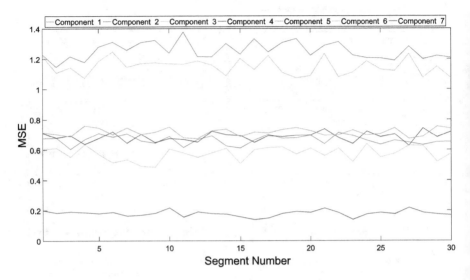

Fig. 7. The segmented MSE of each modal component of scalp muscle

The total MSE of seven modal components of sEMG of scalp muscle was calculated. The results are shown in Table 3.

Table 3. The total MSE of scalp muscle

Parameter		MSE						
		1	2	3	4	5	6	7
Baseline	Scalp	0.57	1.04	1.22	0.67	0.72	0.56	0.16
	Trapezius	0.64	1.22	1.25	0.65	0.72	0.53	0.15
	Sternocleidomastoid	0.70	1.26	1.20	0.62	0.73	0.50	0.14
Helmet 1	Scalp	0.66	1.22	1.19	0.63	0.72	0.52	0.15
	Trapezius	0.67	1.23	1.26	0.66	0.73	0.54	0.16
	Sternocleidomastoid	0.67	1.23	1.26	0.66	0.73	0.54	0.16
Helmet 2	Scalp	0.50	1.05	1.24	0.74	0.71	0.58	0.18
	Trapezius	0.37	1.01	1.06	0.64	0.70	0.56	0.17
	Sternocleidomastoid	0.84	1.29	1.09	0.54	0.70	0.46	0.13

According to the characteristics of sEMG of muscle injury, Fisher's discriminant analysis of feature vectors was used to classify neck muscle injury. The analysis results are shown in Table 4.

Table 4. Fisher discriminant analysis results of scalp muscle injury

Case number	Actual group	Highest group					Second highest group			Discriminant Scores	
		Predicted group	P(D > d | G = g) P	df	P(G = g | D = d)	Squared Mahalanobis distance to centroid	Group	P(G = g | D = d)	Squared Mahalanobis distance to centroid	Function 1	Function 2
Original 1	1	1	0.948	2	0.870	0.106	2	0.130	3.902	-2.118	-0.506
2	1	1	0.948	2	0.898	0.106	2	0.102	4.459	-2.435	0.063
3	2	1[a]	0.693	2	0.593	0.733	2	0.407	1.482	-1.557	0.241
4	2	2	0.477	2	0.726	1.482	3	0.266	3.49	0.865	0.49
5	3	3	0.494	2	0.996	1.412	2	0.004	12.221	2.712	-1.329
6	3	3	0.494	2	0.975	1.412	2	0.025	8.744	2.533	1.041
7	Ungrouped	3	0.000	2	1.000	66.504	2	0.000	120.638	10.545	1.788
8	Ungrouped	2	0.110	2	0.505	4.41	1	0.495	4.453	-1.839	1.843
9	Ungrouped	3	0.002	2	1.000	12.665	2	0.000	42.59	6.174	0.082
10	Ungrouped	1	0.026	2	0.712	7.299	2	0.286	9.12	-0.974	-2.588
11	Ungrouped	3	0.026	2	1.000	7.271	2	0.000	32.174	5.315	0.004
12	Ungrouped	3	0.081	2	1.000	5.032	2	0.000	21.425	3.501	-2.208
13	Ungrouped	1	0.000	2	1.000	26.593	2	0.000	51.472	-7.262	-1.541
14	Ungrouped	1	0.000	2	1.000	45.777	2	0.000	76.973	-8.472	-2.942
15	Ungrouped	3	0.000	2	1.000	905.459	2	0.000	1085.952	32.326	4.672

[a] Misclassified case

In Table 4, group 1 to group 6 are training data, group 7 to group 15 are data to be classified. Prediction Groups indicates classification result. It can be concluded that Fisher discriminant can perform fatigue damage grouping on the feature vector composed of MSE. Actual Group 1 indicates no fatigue damage, Group 2 is mild injury, and Group 3 represents moderate injury.

5 Discussion

Since only 10 volunteers' sEMG were measured in this study, however, these data may not be regarded as a basis for analyzing the classification of neck injury caused by changes in helmet mass and CG. In further study, a large number of experiments should be performed to collect a certain amount of data to support injury classification. In the experiment design, we should have controlled a single factor to facilitate the analysis of whether the injury is induced by the change of helmet mass or centroid. In fact, only two types of helmets are available due to experimental conditions, which increases the difficulty of analyzing experimental data. In the injury level classification, only the analysis of non-, mild and moderate fatigue injuries were implemented. The training data may not be representative of the characteristics of sEMG in muscle injury. Therefore, in the subsequent data processing, MSE eigenvectors for various muscle injuries should be identified and enriched in future. In the experiment, only sEMG in which volunteers remained a static state were collected for analysis, and the data acquisition duration was short. In the special case of high-speed maneuvering, overload, long-haul flight and landing, the injury to neck caused by helmet mass and CG is also worthy of further study.

When VMD is performed on sEMG, the longer the signal is, the longer the decomposition takes time, which makes the decomposition efficiency slower. Therefore, the optimization of VMD algorithm is worthy of further discussion.

6 Conclusion

Aiming at the problem of measuring CG and MI of helmet, a measurement system with higher precision was developed based on multi-point method. In order to further study the neck injury caused by the increase of helmet mass and CG shift, we proposed a method to extract the feature vectors of sEMG based on VMD and MSE model, and Fisher discriminant is applied to analyze the feature vector composed of MSE and to realize the classification evaluation of neck injury.

We designed and performed an experiment to acquire the sEMG of sternocleido-mastoid, trapezius and scalp muscle. And the collected sEMG was analyzed by the method we proposed. The results show that the model proposed in this paper can reasonably classify neck injury.

References

1. Gaur, S.J., Joshi, V.V., Aravindakshan, B., Aravind, A.S.: Determination of helmet CG and evaluation of neck injury potentials using 'knox box criteria' and neck torque limits. Ind. Aerosp. Med **57**(1), 37–44 (2013)
2. Manoogian, S.J., Kennedy, E.A., Duma, S.M.: A literature review of musculoskeletal injuries to the human neck and the effect of head supported mass worn by soldiers. USAARL contract report no. CR-2006-01 (2005)
3. Merkle, A.C., Kleinbeiger, M., Mannel, O.U.: The Effect of Head Supported Mass on the Risk of Neck Injury in Army Personnel. John Hopkins APL Technical digest, 26 (2005)
4. Bevan, M. G., Ward, E., Luong, Q.: Neck torque study induced by head-borne visual augmentation systems (VAS) in ground-based applications. NSTD-09-1007 (2010)
5. Leclair, B., O"Connor, P.J., Podrucky, S., Lievers, W.B.: Measuring the mass and center of gravity of helmet systems for underground workers. Int. J. Ind. Ergon. **64**, 23–30 (2018)
6. Bogerd, C.P.: A review on ergonomics of headgear: thermal effects. Int. J. Ind. Ergon. **45**, 1–12 (2015)
7. Mathys, R., Ferguson, S.J.: Simulation of the effects of different pilot helmets on neck loading during air combat. J. Biomech. **45**(14), 2362–2367 (2012)
8. Mh, V.D.O., Steinman, Y., Sluiter, J.K., et al.: The effect of an optimised helmet fit on neck load and neck pain during military helicopter flights. Appl. Ergon. **43**(5), 958–964 (2012)
9. Ibrahim, E.: The Effects of Neck Posture and Head Load on the Cervical Spine and Upper Extremities. Master's thesis. McMaster University, Hamilton, ON (2015)
10. Ankrum, D.R.: Questions to ask when interpreting surface electromyography (sEMG) results. Hum. Factors Ergon. Soc. Ann. Meet. Proc. **44**(30), 5530–5533 (2000)
11. Mannion, A.F, Dolan, P.: Electromyographic median frequency changes during isometric contraction of the back extensors to fatigue. Spine **19**(11), 1223–1229 (1994)
12. Masuda, K., Masuda, T., Sadoyama, T., et al.: Changes in surface EMG parameters during static and dynamic fatiguing contractions. J. Electromyogr. Kinesiol. **9**(1), 39–46 (1999)
13. Falla, D., Rainoldi, A., Merletti, R., et al.: Myoelectric manifestations of sternocleidomastoid and anterior scalene muscle fatigue in chronic neck pain patients. Clin. Neurophysiol. **114**(3), 488–495 (2003)
14. Stapley, P.J., Beretta, M.V., Dalla Toffola, E., et al.: Neck muscle fatigue and postural control in patients with whiplash injury. Clin. Neurophysiol. **117**(3), 610–622 (2006)
15. Bronfort, G., Evans, R., Nelson, B., et al.: A randomized clinical trial of exercise and spinal manipulation for patients with chronic neck pain. Spine **26**(7), 788–797 (2001)
16. Xiaohong, J., Junbing, M., Rencheng, W., et al.: Effect of helmet mass and mass center on neck muscle strength in military pilots. J. Med. Biomech. **27**(4), 416–420 (2012)
17. Yang, L.F., Kang, B.: Study on human neck muscles' comfort of different height levels based on sEMG method. In: Proceedings of the 6th International Asia Conference on Industrial Engineering and Management Innovation. Atlantis Press, Paris, pp. 563–574 (2016)
18. Chowdhury, S.K., Nimbarte, A.D., Jaridi, M.A.: Discrete wavelet transform analysis of surface electromyography for the fatigue assessment of neck and shoulder muscles. J. Electromyogr. Kinesiol. **23**(5), 995–1003 (2013)
19. Dragomiretskiy, K., Zosso, D.: Variational mode decomposition. IEEE Trans. Signal Process. **62**(3), 531–544 (2014)
20. Costa, M., Goldberger, A.L., Peng, C.K.: Multiscale entropy analysis of complex physiologic time series. Phys. Rev. Lett. **89**(6), 705–708 (2002)

21. Costa, M., Healey, J.A.: Multiscale entropy analysis of complex heart rate dynamics: discrimination of age and heart failure effects. Comput. Cardiol. IEEE (2003)
22. Takahashi, T., Cho, R.Y., Mizuno, T., et al.: Antipsychotics reverse abnormal EEG complexity in drug-naive schizophrenia: a multiscale entropy analysis. Neuroimage **51**(1), 173–182 (2010)
23. Escudero, J., Abásolo, D., Hornero, R., et al.: Analysis of electroencephalograms in Alzheimer's disease patients with multiscale entropy. Physiol. Measurement **27**(11), 1091 (2006)
24. Catarino, A., Churches, O., Baron-Cohen, S., et al.: Atypical EEG complexity in autism spectrum conditions: a multiscale entropy analysis. Clin. Neurophysiol. **122**(12), 2375–2383 (2011)

Human Factors in Game Design

Devilish Dilemma Games: Narrative Game Engine and Accompanying Authoring Tool

Johan de Heer[1(✉)], Thomas de Groot[2], Rafal Hrynkiewicz[2], Tije Oortwijn[2], and Paul Porskamp[1]

[1] Thales Research & Technology, Gebouw N, Haaksbergerstraat 67, 7554 NB Hengelo, The Netherlands
{Johan.deHeer, Paul.Porskamp}@nl.thalesgroup.com
[2] TXchange Game Based Learning Solutions, Gebouw N, Haaksbergerstraat 67, 7554 NB Hengelo, The Netherlands
{Thomas.deGroot, Rafal.Hrynkiewicz, Tije.Oortwijn}@txchange.nl

Abstract. In this paper we will focus on a narrative game engine and accompanying authoring tool that allows for easy development of game scenarios, including in-game and summative feedback in terms of key performance indicators and decision-making styles. Using our model driven game engine, we co-developed numerous games together with several modelling and simulation experts from various domains. The resulting single player turn-taking games are offered as a game service via on-line accessible learning management systems, and currently used in a variety of leadership training programs making trainees aware of their natural decision-making behaviors.

Keywords: Game based learning · Leadership skill development · Dilemma training

1 Introduction

Professional 21st century leadership skills and competencies are a sine-qua-non in modern day management; "hard skills get you the job but soft skills bring you to the top" is a popular saying. However, during preparation trainings and exercises we tend to focus on clearly defined scenarios, and how to act accordingly based on protocols. But how do you deal with ill-defined situations characterized by ambiguities and uncertainties? How is your judgment in these complex and sometimes time critical situations; and, how does your decision-making style unfold over the course of time when confronted with devilish dilemmas where there is no a priori right or wrong decision? At the end of the day, it is about your responsibility and accountability. Becoming aware of your (implicit) decision-making style proves to be useful: how do I collect relevant information, do I consult others, am I biased towards certain issues or people, et cetera.

In this paper we will focus on a narrative game engine and accompanying authoring tool that allows for easy development of games, including in-game and summative feedback in terms of decision-making styles. Using our model driven game engine, we

© Springer Nature Switzerland AG 2020
T. Ahram (Ed.): AHFE 2019, AISC 973, pp. 169–178, 2020.
https://doi.org/10.1007/978-3-030-20476-1_18

co-developed numerous game stories together with several modelling and simulation experts from various domains. The resulting single player games are offered as a game service via an on-line learning management system, and currently used in a variety of leadership training programs making trainees aware of their natural decision-making behaviors. For example, the Mayor game is an educational and didactical instrument that is used to train Dutch Mayors with respect to governmental dilemmas during crisis management operations. Another example is our Ethics game that focuses on your moral compass, which becomes transparent when confronted with a series of business dilemmas. Mind matters is all about various hard and soft tactics to influence others. The cybersecurity game brings awareness regarding reputational risks and how that frame of reference drives your judgment and decision-making.

Further, we will demonstrate the value of game analytics based on game log data providing additional insights in the continuous learning journey that can be represented on several levels of abstraction viz. at individual and organizational levels. Further, it becomes interesting to use these games (even within organizations) as a methodological research tool to collect game play data for hypothesis testing and further statistical analysis. The game logs can also be used for predictive analytics utilizing machine learning techniques that can be applied in adaptive gaming environments aligning the game 'level' with the 'level' of the player.

The outline of the paper is as follows. First, we describe the game engine and authoring tool. Secondly, we present some games for various applications. Third, we focus on analyzing game logs. We end with some lessons learned and take always.

2 DILIMMA Game Engine and Authoring Tool

Conceptually, DILEMMA is based on elucidating biases in decision-making and grounded in the works amongst others of Kahneman [1]. In his subliminal book 'Thinking Fast and Slow' he explains in various ways the characterization of how and why two distinct system functionalities during various choice conditions operate. Metaphorically, he describes the two systems in terms of animal behaviors, namely rabbits and turtles. Rabbits act fast and turtles act slow. Slow thinking decreases decision biases that occur during fast thinking. He argues that our fast thinking system is always on, however, sometimes is overruled by our slow thinking system if we intentionally and deliberately 'turn it on'. Choice behavior is one thing, understanding and elucidating the dynamics of human reasoning requires studying Dynamic Decision-making (DDM) taking into account sequences of decisions to reach a goal, interdependence of decisions on previous decisions, dynamics of a changing environment, and that decisions are made in real time. DILLEMA is considered offering microworld games that enables study of both choice behavior and DDM [2].

The underlying game model of DILEMMA is best described with Shell's [3] game design framework where a game consists of the balanced interplay between four game components: mechanics, story, technology and aesthetics (see Fig. 1). Mechanics are the procedures and rules of the game, describing the goal, and how players can try to achieve it. Story is the sequence of events that unfolds in the game, which may be linear and pre-scripted m or may be branching and emergent. Aesthetics is how the game looks and feels. Technology could be any material and interactions that make the game possible – from papers and pencils to high use of technological means. Note that the level of fidelity of the individual game components could vary from far away to very close to reality. The trade-off is mostly driven by development time, resources and budget. A high-fidelity flight simulator resembles the real world as close as possible making the transfer from the interacting learning experience with the simulator to daily practices as easiest and fast as possible. On the other hand, if one or more game components are resembling the real world to a lesser extent the transferability becomes more challenging. The design choice for DILEMMA was to optimize on the Story component of the game-based learning solution as compared to the other three components [4]. The result is a low-cost solution focusing on the value of storytelling. Storytelling is considered a fundamental interpersonal communication and learning tool. Capturing and analyzing narrative structures based on player interactions with the game reveals interesting behavioral aspects for further analyses. We developed several game solutions using the DILEMMA game engine (see next paragraph).

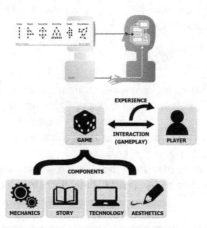

Fig. 1. DILEMMA consists of the interplay of 4 game components, player interaction to induce an experience that is revealed by examining game narratives.

The main functionalities of the DILEMMA game engine are an event driven system with feedback and logging functions. The DILEMMA game engine allows for the development of single player (branching) narrative games. Still, it requires serious game design skills to transfer real world phenomena and learning objectives to game formats, which elucidate decision-making behavior and enable experiential learning. Therefore, we developed and standardized a simple but effective game format, including an accompanying authoring tool allowing non-game designers developing their own game scenarios.

The game format is based on a decision room environment. During game play several dilemmas pop up that require a decision. Non-player characters may provide information in the form of messages that could be relevant to the decision at hand. The player could also request information from the non-player characters or may even ask for advice regarding the required decisional choices. All player actions are logged and can be used for further analyses. Decision choices result in scores on pre-defined (student) model dimensions that could be used as in-game and/or post-game feedback. An individual summative feedback file is generated after game play.

The authoring tool supports the scenario developer in scripting a DILEMMA game in a structured way, including describing the overall context of the scenario, write the dilemmas, select some available characters and backgrounds for each dilemma, define the non-player characters, their messages and advices, define the model dimensions and scoring scheme. The tool furthers allows to put all elements in a narrative structure. The resulting game can be tested on the fly and published as a SCORM package in a Learning Management System. Players/trainees can log into the LMS to select their course, including the game.

3   DILEMMA Games

With DILEMMA around one hundred game scenarios have been build. For illustration purposes we focus on some of them that have been developed for different applications purposes.

Mayor Game (Fig. 2).

The Mayor game [5] is implemented as a training intervention tool to train Dutch mayors in time-critical decision-making during crisis management operations. Currently, within the LMS around 15 game scenarios are available varying from flooding situations, dealing with toxic calamities to family dramas, and what to do when neo-Nazis are organizing a BBQ in the backyard of your hometown. All scenarios are defined by a subject matter expert from the Dutch Association of Mayors and resemble realistic crisis situations. For each scenario about eight dilemmas are defined. Thus, the

Fig. 2. Mayor Game.

total set of different dilemmas in the mayor game environment is about 120. Players are assessed in terms of three leadership dimensions. These leadership dimensions make sense to the Dutch context but in general indicate (a) an external orientation towards the community/municipality (People person), (b) the organizer of formal decision-making processes (Figurehead), and (c) an internal orientation in making sure that internal organization procedures are followed (Administrator). Games are embedded in a training program offered by the Institute for Safety. The mayor game is available since 2012 and generates over a 1000 game plays per year. Number of trainees during a training is about 8 people – most of the times a local crisis management response team participates. A trained trainer briefs the trainees. Each individual plays one or more scenarios. Results on the three leadership dimensions, the time it took to solve the dilemmas, the information generated by the non-player characters that was considered relevant, et cetera becomes available as summative feedback for each player. In addition, based on how the trainee played the game, a newspaper article is automatically generated that portrays how he/she will appear in tomorrows' newspaper. In the debriefing phase individual game results are group wise discussed, and pros, cons, reasoning are shared. Debriefing is moderated by the trainer.

Architecting Game (Fig. 3).

The Architecting game [6] was developed for a specific investigation purpose to elucidate how system architects think and act during bid phase in terms of decision-making styles. Focus was to examine if decision-making was driven by a. the voice of the customer, b. pushing the introduction of new technologies to the solution, or c. focus on selling the existing product portfolio. Game scenario and dilemmas were defined by several subject matter experts and pointed to various tradeoffs during bid phase. Messages and advises from the non-player characters provided several perspectives on the dilemma at hand. A sales manager, capture team lead, system engineer,

Fig. 3. Architecting Game.

subcontractor, product portfolio manager, pricing coordinator and proposal writer were present in the virtual decision room. The game was played during an architecting convention by 40 system architects across several business units from a one industrial company. Simple descriptive statistics and analysis revealed the differences in decision-making styles across business lines, level of experience, et cetera. From an organizational perspective a valuable insightful finding. The game turned into a training game for this company supporting competence and skill development for system architects.

Cyber Security Game (Fig. 4).

The Cyber Security game [7] is available on-line (cis.txchange.nl). During 2015–2018 the game was played 1500 times. The web-based game is available in two languages: English and French. A subject matter expert on cyber security resilience defined the scenario, dilemmas and scoring on playing styles [8]. The game targets managers that are responsible for cyber resiliency in their organization, and players receive feedback in terms of their risk taken and avoidance behaviors. In this case player actions generated scores on the dimension of reputational risk and operational risk behaviors. The story line resembles a crisis situation that starts slowly based on some weak signals but gradually moves into a full-blown cyber-attack crisis situation that finally resolves. During the episodes various dilemmas pop up that require a decision. Information and advices came from non-player characters that represent the CEO, operations, communication, legal affairs, IT, business continuity. It was found that for a first analyses based on 377 game logs gathered in the period 2015–2017 reputation risk taking behavior was 42% versus reputation risk avoidance 58%, operation risk taking behavior was 29% versus operation risk avoidance 71%. Figures are in line with the negativity bias in a plethora of situations related to risk-averse behaviors.

Fig. 4. Cyber Game.

Business Ethics Game (Fig. 5).

Ethics and integrity are part of moral judgments about right and wrong and an integral part of corporate responsible behaviors. From employee to top manager, all are confronted with ethical decision-making in the course of their professional lives. Recent scandals show how unethical behaviors may harm a carefully built reputation of organizations. To create a corporate ethically responsible culture, it is therefore crucial to train people on a regular basis. The "To BE or not to BE?" Game [5] is designed to train people in ethical judgment and decision-making. This game is making people aware of the importance of business ethics, being able to recognize the ethical dimensions of everyday business decisions, develop an understanding of one's own moral compass and strategies to deal with ethical dilemmas. The game is designed for professionals with a training need in business ethics and corporate responsibility. After playing the game trainees will receive objective feedback on how they tend to deal with ethical dilemmas. Do they adhere to rules and regulations? Do they take multiple perspectives into account when resolving the dilemmas? This information helps them to reflect on their leadership behavior and put them in an excellent position to improve. The business ethics game is currently implemented in a corporate that takes corporate responsibility very seriously. Each employee has to take the mandatory training, including several e-learning modules. The Business ethics game is now running for a year in an additional blended learning training program and well received by the trainees. Trainees' feedback centers around (1) level of engagement: fun and motivating to do and better fitted to the needs of Millennials, (2) its effectiveness: higher retention rate compared to classic classroom training and e-learning, (3) its efficiency: short (hours instead of days) and to-the-point scenarios, (4) its level of realism: high quality learning content reflecting the leadership challenges of the organization, (5) the fact that it is objective: standardized and automatically generated feedback on ethical judgment and decision-making skills. Again, subject matter experts used the authoring tool to define the scenarios.

Fig. 5. Business Ethics Game.

4   Game Analytics

Game analytics [9–11] is about maximizing the value of player data. It should deliver actionable insight to improve skills and human performance. The Conceptual Assessment Framework CAF-Model [12] provides an interesting starting point to map latent traits (student model) to the operationalization of assessment tasks (task model) (see Fig. 6). CAF is part of the Evidence-Centered Design method applied for assessment design. The mapping in CAF results in a so-called evidence model. The evidence model quantifies the leadership construct based on observable in-game player actions. The simplest deterministic evidence model is based on a single player action that linearly increases or decreases a relative score on one or more of the student model dimensions. For example, dealing with a dilemma requires a decision that the player must take; IF a yes decision is made THEN the scoring on e.g., leadership dimension A is increased by x and the scoring on dimension B is decreased by y. More complex non-linear or probabilistic evidence models can be top-down defined as well, for example, using Bayes nets that depends on various combinations of available observed variables.

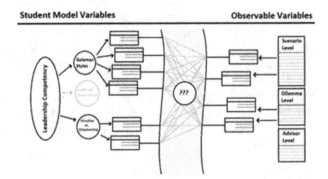

Fig. 6. Conceptual Assessment Framework CAF-Model Linking student model variables to observable variables (taken from 13).

The evidence model we implemented in the authoring tool for DILEMMA is based on yes/no decisions that the player has to take for each dilemma that needs to be solved. This evidence model is defined with a domain expert who should be able to argue what the effect of a yes/no decision is on the three leadership dimensions considered, thereby ensuring a certain level of fidelity, accuracy and validity of the evidence model.

From the exportable game logs simple descriptive statistics can be generated such as the time needed to answer dilemmas, the number of dilemmas answered, the number of times advices of various team members were indicated as important, et cetera. This is done on an individual level and provided as feedback to the player. Further statistical analysis can be done on aggregating levels based on all game logs. This makes it possible to hypothesize and examine team and group behaviors across a number of other parameters as well e.g., level of expertise, gender, country, culture, business domain, etc.

In another explorative study [13], we were interested in utilizing data mining techniques to predict the scoring on student model dimensions. The idea was that if we could predict player behavior on-the-fly, then we are able to recommend didactical suggestions to the player maximizing the learning experience. It is beyond the scope of this paper to discuss data mining and predictive analytics extensively [see 14]. The key in predictive analytics is to create a model from the available set of observed variables to predict an outcome (class). For illustration purposes suppose the class is Figurehead (one of the student model dimensions in the Mayor game) that we discretized in two categories; low and high scoring on this dimension. The question than is can we find a model based on in-game observed variables that predicts a scoring on the Figurehead dimension for each of the two categories low and high? Such a model is derived via a learning algorithm. There is a variety of learning algorithms based on different methods, for example, decision tree-based methods, rule based methods, neural networks, Bayesian belief networks, et cetera. Depending on the method used the algorithm will learn specific patterns that map the observable variables ('predictors') to the class for which you know the class answer. The model captures these relationships for a 'training data set' and can then be used to get predictions on a 'new data set' for which you do not know the class answer. The Waikato Environment for Knowledge Analysis (WEKA) [15] toolbox and several embedded machine learning techniques are helpful to predict the classes. Although our results were promising, we learned that a variety of experts are needed to examine this type of data analysis. The learning architect and trainer need to indicate what they want to know. The game developer needs to implement the data models to capture the set of meaningful predictors, the data analytics experts needs to understand the learning goals and game play constructs and available observable variables to come up with meaningful classifiers.

5 Conclusion

We presented the DILIMMA game engine and authoring tool. The standardized decision room game format provides non-game experts the tools to develop their own scenarios. The games are accessible via a Learning Management System for training or research purposes. One can elucidate various pre-defined student models or can use predictive analytics techniques to predict certain behaviors. Our future work is focusing on the unobtrusive measurement of competency where we explore the combination of several top-down (e.g. Bayesian networks) and bottom-up data mining techniques to analyze and predict human behaviors. Further, we will focus on player profiles in terms of actions, tactics, and strategies.

References

1. Kahneman, D.: Thinking Fast and Slow. Macmillan (2011). ISBN 978-1-4299-6935-2
2. Gonzalez, C., Lerch, J.F., Lebiere, C.: Instance-based learning in dynamic decision-making. Cognit. Sci. **27**, 591–635 (2003)
3. Schell, J.: The Art of Game Design: A Book of Lenses, 2 Edn, by A K Peters/CRC Press (2008)
4. van der Spek, E.D., Sidorenkova, T., Porskamp, P., Rauterberg, G.W.M.: The effect of familiar and fantasy aesthetics on learning and experience of serious games. In: Marsh, T., Sgouros, N., Pisan, Y., (Eds.), Entertainment Computing – ICEC 2014: 13th International Conference, ICEC 2014, Sydney, Australia, October 1–3, 2014: Proceedings, pp. 133–138. Lecture Notes in Computer Science, No. 8770. Berlin: Springer (2014)
5. www.txchange.nl
6. de Heer, J.: How do architects think? A game based microworld for elucidating dynamic decision-making. . In: Auvray, G., et al. (eds.), Complex Systems Design & Management, pp. 133–142. © Springer International Publishing Switzerland 2016 133 (2015). https://doi.org/10.1007/978-3-319-26109-6_10
7. de Heer, J., Porskamp, P.: Human behavioral models from microworlds. In: Nicholson, D. (Ed.), Advances in Human Factors in Cybersecurity, Advances in Intelligent Systems and Computing, vol. 593, pp. 173–184. © Springer International Publishing AG 2018 (2017). https://doi.org/10.1007/978-3-319-60585-2_17
8. Théron, P.: Informing business strategists about the cyber threat: why not play serious games? In: Hills, M. (Ed.), Why Cyber Security is a Socio-Technical Challenge: New Concepts and Practical Measures to Enhance Detection, pp. 129–158. Northampton Business School, University of Northampton, UK (2016). ISBN 978-1-53610-090-7
9. Shute, V., Ventura, M.: Stealth Assessment: Measuring and Supporting Learning in Video Games. MacArthur Foundation, The MIT Press, Cambridge, Massachusetts. ISBN 978-0-262-51881-9 (2013)
10. Loh, C.S., Sheng, Y., Ifenthaler, D.: Serious Games Analytics: Methodologies for Performance Measurement, Assessment and Improvement. Springer International Publishing Switzerland. ISBN 978-3-319-05833-7 (2015)
11. El-Nasr, M.S., Drachen, A., Canossa, A.: Game Analytics: Maximizing the Value of Player Data. Springer, London (2013). ISBN 979-1-4471-4768-8
12. Almond, R., Mislevy, R., Steinberg, L., Yan, D., Williamson, D.: The conceptual assessment framework. In: Bayesian Networks in Educational Assessment. Statistics for Social and Behavioral Sciences. Springer, New York, NY (2015)
13. de Heer, J., Porskamp, P.: Predictive analytics for leadership assessment. In: The 9th International Conference on Applied Human Factors and Ergonomics – 3rd International Conference on Human Factors in Management and Leadership. 22–26 July, Orlando, Florida, USA. © Springer International Publishing AG, part of Springer Nature 2019 Kantola, J.I., et al. (Eds.), AHFE 2018, AISC 783, pp. 516–523 (2018). https://doi.org/10.1007/978-3-319-94709-9_51
14. Daniel, T., Larose, C.D.: Data Mining and Predictive Analytics, 2nd Edn (2015). ISBN: 978-1-118-11619-7
15. Waikato Environment for Knowledge Analysis (WEKA). https://www.cs.waikato.ac.nz/ml/weka/

A Gamified Approach Towards Identifying Key Opportunities and Potential Sponsors for the Future of F1 Racing in a Declining Car Ownership Environment

Evangelos Markopoulos[1(✉)], Panagiotis Markopoulos[2],
Mika Liumila[3], Younus Almufti[1], Chiara Romano[1],
and Paulina Vanessa Benitez[1]

[1] HULT International Business School, 35 Commercial Road, Whitechapel,
London E1 1LD, UK
evangelos.markopoulos@faculty.hult.edu,
{yalmufti2015, cromano2015,
paubenitez2015}@student.hult.edu
[2] University of the Arts London, Elephant and Castle, London SE1 6SB, UK
p.markopoulosl@arts.ac.uk
[3] Turku University of Applied Science, Joukahaisenkatu 3,
20520 Turku, Finland
mika.luimula@turkuamk.fi

Abstract. This research work aims to propose new approaches towards identifying key opportunities and potential sponsors for the future of F1 racing in an environment with declining car ownership, without resorting to endless licensing agreements. The paper presents a gamification approach on which an innovative and disruptive operations framework can be developed to help, without operational complexity and commitment, F1 teams gain new customers (fans) and recapture essential markets and targets groups. The paper also contributes on establishing a base for effective strategy development based on the user's/player's engagement and behavior. Furthermore, this work extends towards the analysis of the game's operations and the marketing initiatives needed to succeed. The proposed approach varies from OHH (out of home advertising), interactive marketing, celebrities, F1 drivers' endorsements, and other related supportive initiatives such as search engine optimization on online research platforms and other promotion and marketing dissemination initiatives.

Keywords: Gaming · Gamification · Serious games · Esports ·
Entertainment games · Monetization · Sponsorship · F1 · Formula 1 ·
Blue ocean strategy

1 Introduction

F1 (Formula One) can be considered as one of the most attractive, challenging but also most expensive sports in the world. The F1 industry can be seen as a multidisciplinary area which integrates finance, business management, marketing, technology,

© Springer Nature Switzerland AG 2020
T. Ahram (Ed.): AHFE 2019, AISC 973, pp. 179–191, 2020.
https://doi.org/10.1007/978-3-030-20476-1_19

innovation, manufacturing, material engineering, and many more. The overall cost and complexity of operations of the glamorous and amazing world of F1 constantly increases while the fans and followers decrease. This situation results to the decrease of interest and increase of the risk on the viability of the sport which extends much beyond racing.

This research work aims to propose new approaches towards identifying key opportunities and potential sponsors for the future of F1 racing in an environment with declining car ownership without resorting to endless licensing agreements. The paper presents a gamification approach on which an innovative and disruptive operations framework can be developed that can contribute to F1 teams on gaining new customers (fans), recapture essential markets and targets groups, without operational complexity and commitment. The paper can also contribute on establishing a base for effective strategy development based on the user's/player's engagement and behaviour.

The proposed approach utilizes the F1 history though its drivers, manufacturers and time periods, to bring up the wealth of memories, emotions, but also challenges, needed towards reinventing a sponsorship model for all type of organizations involved in the operations of F1. To illustrate this approach a game called 'F1 Legends' has been designed and supported with the operations activities needed for successful strategy execution on sponsors engagement. The proposed model can be adopted by any F1 organization seeking such benefits. The game can also be customized as a different game for different organizations such as the 'Ferrari Legends', the 'McLaren Legends' the 'Williams Legends' etc. The same concept can also be adapted by the F1 sponsors who support the sport at different times. In such a case the game can be called the 'Shell Legends', the 'Marlboro Legends', etc., where the F1 car becomes the legend of the sponsor.

The overall framework of the gamification approach, which can be execute via an on-line platform, can increase brand awareness and interest of younger audiences, and provide access to emerging and gaming friendly markets where F1 records low popularity. The game emphasizes on providing a highly innovative online platform using competitive racing among players, tournaments with high rewards and ladder boards, in an attempt to operate as an eSport open to all kind of players and not necessarily to the professional games. The paper extends towards the analysis of the game's operations and the marketing initiatives needed to succeed. The proposed approach varies from OHH (out of home advertising), interactive marketing, celebrities, F1 drivers' endorsements, and other related supportive initiatives such as search engine optimization on online research platforms and other promotion and marketing dissemination initiatives.

2   The e-Sports Gaming Phenomenon

The e-sport industry had been invented since the arcade era, however esports did not become mainstream until very recently. The worldwide phenomenon of esports had a quite humble start with simple tournaments hosted around arcade machines by the amateur players of that period. The evolution of the esports over the time was achieved with a predictable slow pace related to the development of the gaming industry, until the introduction of the multiplayer games with Quake to be considered as the first competitive game with multiplayers. The very first unofficial esports tournament took place on 1972 with the Space Invaders game, and the first official esports tournament was on 1997 with the Red Annihilation Quake game.

The advancement from the first competition to what esports are considered today is tremendous. Esports will soon be a billion-dollar business sector and a global audience of over 300 million fans [1]. In 2018 the revenue from esports only in North America reached the $345 million (Fig. 1).

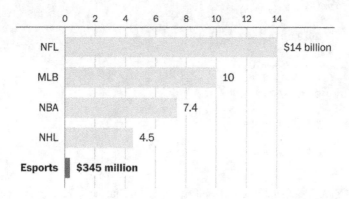

Fig. 1. Annual revenue for esports (North America only) and US major league sports

Tremendous growth is also indicated at the esport prize money. Today esport athletes surpass what established sports provide to their athletes. The champions of the DOTA-2 International received in 2018 the amount of 25.5 million$ (Fig. 2) [2]. The total prize pools for selected esports and traditional sports tournaments indicate a significant drive in the growth of the esports.

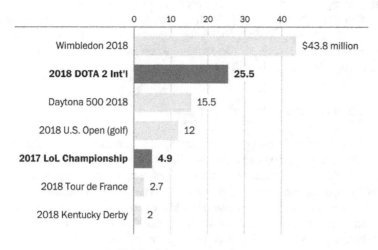

Fig. 2. Prize money at esports.

The growth of esports extend also to the followers they have. Over the time the amount of people interested in competitive games grew significantly, and in turn the esport industry grew as well. The grabbing eyeballs on esports has gone beyond expectations reaching 106.2 million only on the 2017 championship (Fig. 3).

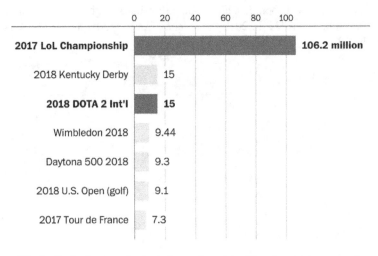

Fig. 3. Peak viewers of selected esports and traditional sport tournaments.

Esports radically redefined the terms 'athlete' and 'sport'. The idea of a virtual athlete has become a reality with the U.S. government giving P1 Visa to esports professionals and recognizing them as professional athletes. South Korea is also a leading county in professional video game competitions, calling them esports, by creating organized leagues, training well-financed professional teams and filling giant stadiums with frenzied fans to cheer on their favorite players [3].

With gaming becoming a trend in recent years more people express interest in esports as they prefer to see professional players play their favourite games. The ranking system in most games also helps the popularity of esports since professionals and casual players can play on the same ranking ladder with the only difference between them to be purely their skills and strategy.

3   esports as a Blue Ocean Strategy

The term blue ocean and blue ocean strategy has been invented by W. Chan Kim and Renée Mauborgne as a marketing theory which presents strategic moves that can create a leap in the value of an organization, its buyers, employees, while generating new demands by making the competition irrelevant. The term practically reflects to the name itself as Blue Ocean are new markets while red oceans are saturated markets [4]. The key characteristics of the blue ocean is summarized in Table 1.

The threats and opportunities of going blue are quite similar in any industry even and they could not be different for the gaming industry and the esports as well.

Table 1. Blue and Red Ocean strategy characteristics.

Blue ocean strategy	Red ocean strategy
Compete in existing market space	Create uncontested market space
Beat the competition	Make the competition irrelevant
Exploit existing demand	Create and capture new demand
Make the value-cost trade-off	Brake the value-cost trade-off
Align the whole system of a firm's activities with its strategic choice of differentiation or low cost	Align the whole system of a firm's activities in pursuit of differentiation and low cost

However there have been games that evolved into esports with significant success, audiences and impact to the economy band the society.

'League of Legends' (LoL) is one example which created a gaming blue ocean and turned out to be one of the most popular games in the world of esports. LoL is a MOBA (Multiplayer Online Battle Arena) type of game, where MOBA is an extremely popular game mode. LoL started simply as a community made mod known as 'Defense of the Ancients' (DotA) for a strategy game called 'Warcraft 3'. However, 'League of Legends' was the first true MOBA game which created a blue ocean to the gaming industry via MOBAs. LoL rose to popularity amongst gamers very fast as it was free, it was a computer game able to run on most systems and because it has a very competitive nature. That same competitive nature turned it in to an esport with the most followers in the world (Fig. 4). The game increased its player base from 12 million in 2011 to 119 million in 2017 [5, 6].

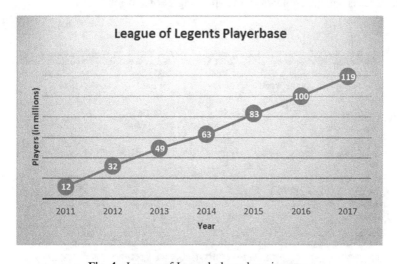

Fig. 4. League of Legend player base increase.

eSports can certainly lead to blue oceans but there are pre and post conditions that need to be satisfied towards achieving and sustaining this goal.

One of the preconditions is the popularity of the game before entering the esport challenge. In the League of Legends example there were many factors that helped but its popularity was a significant one. Even upon release the game LoL had already a decent player base mostly from the DotA fans, ready to download and play it. While the tournaments started the small but loyal fanbase kept promoting the game. The game developers also encouraged the momentum of the fans by adapting a referral system that was rewarding players with in-game awards. Most successful esports derived from companies with very loyal followers, such as Blizzard, otherwise the effort and time needed to reach the top of the list is massive that can turn the overall attempt and investor very risky.

4   The F1 Economics

Formula One (F1) began in the early 1920 as the European Grand Prix till the late 1930's. The concept evolved to Formula One with the Fédération Internationale de l'Automobile's (FIA) which standardized the competition and racing rules. A few years later, in 1950, the competition was named World Championship of Drivers and the first race took place at Silverstone. The term Formula One is composed with 'Formal' meaning the 'set of rules and regulations drivers have to deal with', and the 'One' meaning the top category of racing.

Since then, Formula One has evolved into a very impressive bur also expensive sport. The cost of keeping Formula One's wheels turning has been revealed in 2018 from the financial statements of 10 teams, indicating the total annual spend of $2.6 billion. [7]. However, profitability is marginal for only a few teams while most of them recorded loses. Figure 5 indicates the spending per team for 2016.

F1 TEAM FINANCES

	Revenue	Costs	Net profit/loss
Ferrari*	£382.0m	-£464.0m	-£82.0m
Force India Formula One Team	£81.4m	-£109.1m	-£11.6m
Haas Formula	£100.0m	-£94.9m	£4.1m
Haas Formula Italia	£6.9m	-£6.2m	£0.4m
McLaren Racing	£190.3m	-£193.6m	-£3.2m
Mercedes-Benz Grand Prix	£289.4m	-£275.1m	-£3.8m
Mercedes AMG High Performance Powertrains	£140.4m	-£126.9m	£1.5m
Red Bull Technology	£247.8m	-£238.1m	£8.4m
Renault Sport Racing (UK)	£119.1m	-£121.0m	-£3.3m
Renault Sport Racing (France)	£191.1m	-£197.5m	-£8.8m
Sauber Motorsport**	n/a	n/a	n/a
Scuderia Toro Rosso	£127.3m	-£124.8m	£1.5m
Williams Grand Prix Engineering	£177.7m	-£170.8m	£5.9m
TOTAL	**£2,053.4m**	**-£2,122.0m**	**-£90.9m**

* All data is derived directly from 2016 financial statements, except in the case of Ferrari which is only partially derived from financial statements.
** Sauber is based in Switzerland and therefore does not file publicly available accounts.

Fig. 5. Formula One team finances.

The F1 economics in terms of costs can be categorized on the component costs of formula one cars, the sponsorship cost per area on a car, and the team budget operations.

Figure 6 indicates the 2015 cost of a formula one car, to be £6 million on an average, with the gearbox cost at £750.000, the exhaust £172.000, the frond wing £150.00, the brakes £149.000, and even the steering wheel at £50.000, among the rest.

The main revenue stream of the F1 comes from the sponsorships. The total revenue that can be generated from sponsorships on a car can be £66.7 million on an average (shown in Fig. 7), with the sponsorships on the airbox, sidepod and rear wind to be at £17million each [8]. The revenues from the sponsors are 11 times more than the cost of a car, but it is not enough to cover the overall operations of an F1 team.

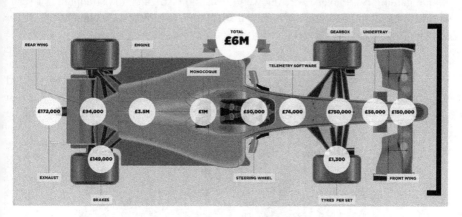

Fig. 6. Component cost and total cost of a Formula One car

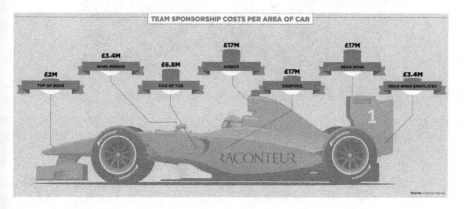

Fig. 7. Sponsorship costs per area on a Formula One car

The cost for the total operations of an F1 team can reach the £158 million. From this, the research and development costs are nearly £41million, the salaries £42 million, the production £39 million, and the operations 36£ million.

5   The F1 Esports Blue Ocean

With marginal profits for a few F1 teams, but mostly loses for the rest, it becomes difficult to foresee the future of F1. New business strategies and ecosystems need to be developed to link the traditional sport activity with the eSport concept, expanding the follower and sponsors base, after all both types of the sport can learn from each other. Biathlon is an example of such an approach which combined the actual biathlon with mobile gaming [9]. Players were allowed in Biathlon X5 game to participate in a real-world competition hosted by International Biathlon Union World Championships 2015 and compete with other players in the same time (Fig. 8).

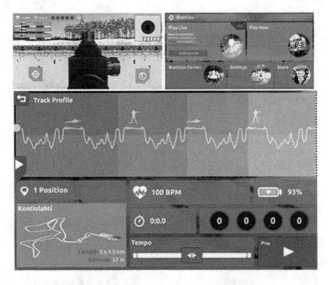

Fig. 8. Screenshots from the Biathlon X5 game.

This application consisted of two modes namely "Play Live" and "Play Now". The first one could be played only when the real-world competition was in progress.

In this project, we learned that this type of combination of real-life sport and gaming can be seen as an ecosystem consisting of various screens and applications to be used in the same time while watching or actually participating in the competition. In the F1 case spectators have already various screens which can be chosen based on everyone's preferences. Therefore, it is possible that experiences collected in IBU World Championships can be utilized in the F1 gaming sponsorship engagement as well, but there are challenges on that. In esports the requirements to create a blue ocean are not only related with the innovation of the new product or service and the market it will serve but also on the followers that can attract which will form the critical mass for effective operations and return on the investment.

Strategy games, First-person shooters, Fighting, and MOBAs are the common game genres featured in esports. However, a game genre that is very popular but was completely absent from the world of esports is racing games. Up until recently, racing game competitions had limited followers without attracting the attention of the average viewer, but lately this is starting to change. Project CARS managed to become the first proper racing game esport. The game attracted the attention of many prestigious car manufacturers such as McLaren, Renault, Ginetta and others, and hosts collaborative esport tournaments with them [10]. Furthermore, the McLaren Shadow esport competition provides a leading and inspirational structure as incentive for the players and the followers [11]. Though the McLaren Shadow Project the company searches for the brightest and the best racing talent in esports, for the next phase in McLaren's esports strategy [12].

It appears that the era for racing sports has moved into a different dimension. Methods and practices have been successfully applied, popularity is growing, and the leading car manufactures are seriously engage in it. Racing game fans who are very often fans of real cars and real racing events have the opportunity to enjoy more from the connection of the esport to the real sport.

6 A Multiplayer esports Gamification Approach for Achieving F1 Sponsor's Engagement

Racing esports can increase the followers of the car manufacturers and the overall followers of the competitions they are take part of. Esports, and specifically in the F1 can contribute in sponsor engagement which can be translated into economic relief for the F1 teams. Such a sponsor engagement-oriented approach can be achieved with the development of an online multiplayer Formula One game. This proposed solution is targeted not only to achieve new sponsorships for F1 but also to expose the sports to new and younger age groups. Creating a popular game is an efficient way to reach millennials and generate a strong brand connection with them. The proposed game enables F1 teams to create a large portfolio of virtual sponsorships and apply direct and indirect marketing practices.

The architecture of such a game has been developed and various allocated areas for sponsorships have been included that can increase the overall revenue streams. The game has been designed to have two modes. The first one is a multiplier mode which consists competitive online races and a levelling up system that unlocks stages and elements after the user completes part of the marketing campaign. The second one is the campaign mode and has been designed to assure brand awareness for the F1 team as it revolves around the history of the team.

The multiplayer mode provides a strong sponsorship model that supports the exposure for the sponsors in a consistent manner the same way it they get in the actual sport. Furthermore, the player gains benefits from tournament participation while playing a competitive experience that is monitored by the F1team itself.

The game can support F1 teams to achieve more sponsors with the method of unlockable items and screens loading. Unlockable are the items a player can use to personalize his/her car and driver (only for aesthetic purposes) and to differentiate themselves from other drivers. The player obtains these through playing the campaign, leveling up the driver, winning a tournament or by purchasing various game elements or abilities with actual currencies or game currency. All unlockable elements can be sponsored and contain different rarities according to the cost of the sponsorship. On the other hand, loading screens are used on idle time for the player. The game has four different loading screens displayed when the player race is loading. The loading screens can promote and utilize best the legend of the car manufacturer along with a sponsor.

In order to increase awareness and engage more sponsorships, the game is designed to be played through frequent tournaments where sponsors can change on each. Different unlockables are available per tournament and per sponsor. The scenery of the race includes potential sponsors as well. Along the racetrack the game provides various billboards and signs available for sponsors. Cities and countries can also sponsor F1 to have the races simulated in their location. Campaign and cutscenes have been designed for more sponsorship opportunities. When a player meets a campaign milestone a cutscene appears with a paid sponsor. The races offered in this game can also be sponsored along with a variety of other game before, during and after the race. The sponsorship process can be customized based on the items, tournaments and frequency the sponsor desires to appear in various elements or races.

7   Sponsorship Benefits Through Gaming Platforms

Gaming platforms can provide the base for multidimensional sponsorship practices in product awareness and branding which can be repeated over time intervals with different players and on different target groups. Some of the benefits that can be achieve with such an approach are briefly listed:

Increase consumer loyalty: Customer, loyalty and trust can be generated due to the exposure a brand has with such sponsorship strategy.

Generates awareness: Promotion in alternative and interactive ways, different from traditional marketing campaigns, helps the public to learn in detail the brand, its values, identity and objectives, and develop activities that makes the brand more identifiable.

Reach new target groups: Games can reach larger and specific target globally with less effort needed for the conventional marketing campaigns.

Improved brand positioning: sponsorship is a good way to improve branding and thereby improve positioning among the same target, as well as among its competitors.

Spread the publicity "by word of mouth": Innovatively interactive and immersive games create fans only with a 'thumbs up'.

Promotion of sales: Direct and indirect sales or brand promotions during the various stages of the game.

Motivate B2B marketing: New business relationships or strategic alliances with sponsors are formulated.

Encourages engagement: Consumers can feel part of the brand. The sponsor generates engagement with the client and within the company and therefore the sense of belonging is transmitted to the target.

8   Risks and Considerations. SWOT and BCG Matrix Analysis

In essence, a SWOT examination is intended for consistency of clear and shrouded factors that may impact a proposed strategy. By analyzing the proposed approach through SWOT, potential challenges will be evaluated in order to be resolved.

According to Grand View Research, Inc., the video game industry will grow by 171.96 billion USD by 2025 [13]; therefore, by taking advantage of the statistics, the proposed online game will be able to provide F1 a new strategy to gain sponsors.

Regarding the strengths, F1 has an excellent worldwide audience of more than 550 million; hence, it already has a strong brand identity which can help the game be recognized faster than other less known competitors. Furthermore, the game is proposed to be launched in the market with a freemium pricing strategy, which consists in providing the service without any charge; however, fees can still be charged for extra highlights, features, or virtual products. The game also provides weekly tournaments which can have a different sponsor every week on all elements. These strengths generate the opportunities for potentially new and more sponsor engagement. On the other hand, weaknesses can be considered the high cost to develop the game and the users' expectations that might be higher from what can be launched on version 1.0. This could lead to possible threats that the game might face, such as not having enough users to attract the sponsors.

Placing the characteristics presented in the SWOT analysis into a Boston Consulting Group growth-share matrix, the proposed approach would be considered as a Question mark. With a high market growth in the fast-growing esports sector, but low market share in the interest of F1 in the esports, the proposed approach is capable for the best which can be the fast adaptation, and the worse which is low adaptation. However, the market does exist, and the opportunities are open.

9   Areas of Further Research

The proposed approach to address a key issue of the F1 market is a simplified way to resolve sponsorship-oriented challenges based on awareness and dissemination of the manufacture's brand at new target groups. The way gamification is introduced in this case aims to demonstrate its application in an integration of serious and entertainment scope of games. The research will be extended with the development of a full-scale game hopefully with the participation of an industry manufacturer in order to adjust it into to actual needs, targets and goals. This will provide reliable metrics that can validate the effectiveness of the proposed approach in the real market and under real conditions with a real partner. The outcome of such progress will generate results that can be published and impact the way new sponsorship models can be achieved for gamified products and strategies in very niche markets.

Applied gamification for economics and sponsorship engagement are research areas that can grow together. Such an integration can provide financial, development and operations models that can contribute to the adaptation of gamification as an effective strategy and approach in F1 and not only.

10   Conclusions

As firms extend their operations into global markets, it is becoming increasing apparent that they have to create high levels of brand awareness which can contribute towards higher revenues and profit margins. Additionally, with increasing competition in most industries, firms are compelled to adopt relative strategies aimed to differentiate themselves from competitors and to better connect and form relationship bonds with their target audiences.

Serious games and corporate gamification can be considered more of a strategy than a technology. The adaptation of such strategies can be applied even in the most niche markets. F1 is a very conservative industry to anything that is not directly related with the car, the engineering performance it represents and the driver of course. However the sport if facing a clear decline on its followers while the operation costs increase. The need for alternative sponsorship models is very important for the viability of the sport and the overall economy build around it.

The introduction of gamification is an approach to resolve the challenge of achieving new sponsor funding by involving a new customer base. The young generation and the tech generation compose a significant part of the worlds population which needs their own communication channels to engaged not only in F1 but in anything that needs their attention. The proposed approach creates a new customer base for the F1 sponsors without the direct involvement of the F1 in it. It utilizes the history, the records and the brand F1 still has, in order to flip the fan group, revitalize its existence, and be applied in other similar industries such as the automotive sector, aerospace, defense etc. Gamification has no limits on the industry as it exists to create the pleasure of exploration while unconsciously contributes towards achieving corporate and organizational strategic goals.

References

1. World Economic Forum. The explosive growth of eSports. https://www.weforum.org/age nda/2018/07/the-explosive-growth-of-esports/
2. The Washington Post. The massive popularity of esports, in charts. https://www.washingto npost.com/business/2018/08/27/massive-popularity-esports-charts/?noredirect=on&utm_ter m=.2757af149d10
3. The New York Times. For South Korea, E-Sports Is National Pastime. https://www.nytimes. com/2014/10/20/technology/league-of-legends-south-korea-epicenter-esports.html

4. Kim, W.C., Mauborgne, R.: Blue Ocean Strategy: How to Create Uncontested Market Space and Make the Competition Irrelevant. Harvard Business School Press, Boston (2004)
5. Is League of Legends Finally Dying In 2019. https://www.lol-smurfs.com/blog/is-league-of-legends-dying/
6. Forbes. Riot Games Reveals 'League of Legends' Has 100 Million Monthly Players. https://www.forbes.com/sites/insertcoin/2016/09/13/riot-games-reveals-league-of-legends-has-100-million-monthly-players/#5a9a5a45aa8b
7. The $2.6 Billion Budget That Fuels F1's 10 Teams. https://www.forbes.com/sites/csylt/2018/04/08/revealed-the-2-6-billion-budget-that-fuels-f1s-ten-teams/#240e71cc6595
8. Raconteur. The Economics of Formula 1. https://www.raconteur.net/infographics/the-economics-of-formula-1
9. Besz, A., Gornicki, M., Heinonen, T., Kiikeri, T., Ratamo, I., Luimula, M., Suominen, T., Koponen, A., Saarni, J., Suovuo, T., Smed, J.: Three apps for shooting sports: the design, development, and deployment. In: Chorianopoulos, K., Divitini, M., Baalsrud, H., Jaccheri, L., Malaka, R., (Eds.), Entertainment Computing – ICEC 2015, Lecture Notes in Computer Science 9353, pp. 329–342. Springer (2015)
10. Project CARS X Ginetta. https://www.projectcarssports.com/ginetta.html
11. PC2 Project CARS X McLaren. https://www.mclaren.com/formula1/2018/mclaren-shadow-project/
12. McLaren Shadow ProJet: Explained. https://www.mclaren.com/formula1/2018/mclaren-shadow-project/mclaren-shadow-project-explained/
13. Grand View Research, Inc., Gaming Market Size Worth $171.96 Billion by 2025 | CAGR: 6.5%: Grand View Research, Inc. PR Newswire: News Distribution, Targeting and Monitoring. www.prnewswire.com/news-releases/gaming-market-size-worth-17196-billion-by-2025–cagr-65-grand-view-research-inc-671617663.html

Cybersickness Among Gamers:
An Online Survey

Stanislava Rangelova[1,2(✉)], Daniel Motus[1], and Elisabeth André[2]

[1] BMW Group, Knorrstrasse 147, 80788 Munich, Germany
{Stanislava.Rangelova,Daniel.Motus}@bmw.de
[2] Human Centered Multimedia, Augsburg University, Universitaetsstrasse 6a,
86159 Augsburg, Germany
andre@informatik.uni-augsburg.de

Abstract. In recent years a trend of head-mounted displays (HMDs) appears on the consumer market and it expands the entertainment dimension by adding a new segment to it called virtual reality (VR) gaming. However, VR games not only bring enjoyment to the players, but they also bring discomfort called cybersickness. In this study, an online survey among gamers was used to identify whether and which symptoms of cybersickness occur while playing VR games. The results showed that the most experienced symptoms during or after VR exposure regarding gaming are nausea, fatigue, and general discomfort. Additionally, the paper aims to give insights about what gamers do against cybersickness associated with VR games. The findings showed that the side effects of VR games are widespread among gamers and an appropriate solution has not been found yet. This paper is a starting point for more extended research on cybersickness induced by VR games.

Keywords: Cybersickness · Virtual reality · Video games · Simulation sickness · Human factors

1   Introduction

Virtual reality (VR) is a computer-simulated environment where one interacts within the environment similarly as he or she would in the "real world." The trend of VR came back a few years ago after more than twenty years of disappearance on the market. Video gamers, as well as researchers, have had the opportunity to enter a virtual world through a head-mounted display (HMD) at rather low cost. The HMDs offer a stereoscopic 3-dimensional (3D) environment with a wide field of view (FOV). However, VR games not only bring enjoyment to the players, they also bring discomfort called cybersickness.

Cybersickness is a condition that can occur during or after exposure to a virtual environment and it can include symptoms such as *headache*, *eyestrain*, or *disorientation* [1]. Despite the similarities with motion sickness and simulation sickness, cybersickness seems to differ in incidence, severity, and symptoms [2]. The cybersickness outbreak is unregulated; it can appear suddenly or later after the VR simulation is over. The cause of cybersickness is not soundly accepted. Three theories

© Springer Nature Switzerland AG 2020
T. Ahram (Ed.): AHFE 2019, AISC 973, pp. 192–201, 2020.
https://doi.org/10.1007/978-3-030-20476-1_20

hypothesize the different origin of the sickness: cue conflict theory, postural instability theory, and rest frame theory. According to cue conflict theory, a so-called "mismatch" could arise between visual, vestibular, and muscular proprioceptive systems, all of which can incite a perceived increase of discomfort as a response from the body [3]. The postural instability theory states that persons get sick from a prolonged postural instability induced by the outside environment [4]. The rest frame theory states that if a person perceives a direction as upward and this direction is different from the actual upward due to gravity, then cybersickness could occur [5]. Symptoms such as *general discomfort, eyestrain,* and *difficulty concentrating* are more likely to be experienced while a user is using a fully immersive virtual environment such as HMDs [6]. For example, *eyestrain* could be experienced due to the close distance between the eyes and the HMD's screen, which could also induce headache. These symptoms, also known as oculomotor symptoms, are the primary difference between cybersickness and motion sickness, where the nausea discomfort prevails [7].

The fast-growing adoption of VR technology expands to other sectors outside of the laboratory research. With an estimation of 171 million active VR users by 2018, the VR market is set to increase with a remarkable rate in the coming years [8]. However, cybersickness may significantly interfere with the acceptance of the VR technology [9]. Despite the growing popularity of VR, there are just a few studies on the commercial VR games and how they affect the users [10–13].

This paper aims to investigate the occurrence of cybersickness among video gamers who use VR in their leisure time. Additionally, it is investigated which mitigation technique they use. Furthermore, relationships between motion sickness history, vision correction, and physical activity and cybersickness are addressed. The respondents as gamers gave a different perspective on cybersickness induced by HMDs. In the literature and the press, there are vast spread ideas of remedies which supposed to have worked against cybersickness. However, it is interesting if these remedies were already tested by the gamers and if they had had an effect.

The paper is structured as follows. Section 2 provides related work on the research topic. Section 3 describes the structure, the data collection, and the analysis of the survey as well as the respondents. Section 4 reports the results. The last section concludes the paper with a discussion of the results and shows a way toward future research.

2   Related Work

A trend of using an HMD as a medium to play video games has an uprising and VR games have become more and more popular. With fascinating graphics and high immersion, the VR game has the advantage to become a new gaming environment. However, the cybersickness related to video games could be induced from VR games as well. Therefore, there is still strong ongoing research on methods to reduce cybersickness outbreak. Some of the techniques which are already suggested are: dynamically reducing the FOV [14], changing the viewpoint [15], adding of a virtual nose [16], virtual hand-eye coordination task [17], and reducing the time spent in the VR [18]. For example, in a study evaluating cybersickness while playing commercial console games with an HMD, Merhi et al. [19] reported that cybersickness is induced more often in the standing playing position. The study compares sitting vs. standing playing position and sitting position with two console games. Moreover, the author stresses that the postural instability during the standing position precedes the cyber-sickness onset as it is shown in previous research [20].

In the literature, 50 factors which are related to cybersickness were reported [21]. However, not all of them are well researched. For example, video games experience, physical activity, and vision correction are some of them. Video games experience showed a trend to be one of the predictive factors of cybersickness outbreak [22]. It is assumed that users who play video games (e.g., console or PC games) daily are more resilient to the malaise due to the accommodation to virtual stimuli. Motion sickness history is another factor which could use as a predictive element of cybersickness [23]. Users with higher motion sickness susceptibility are more prone to experience dis-comfort in VR [24, 25].

Additionally, a positive correlation between vision correction and cybersickness was reported [22] as well as a positive correlation between physical activity and user performance in VR [26]. It might be assumed that there is a correlation between physical activity and cybersickness. This assumption is based on the reported rela-tionship between body-mass index (BMI) and cybersickness [27]. It is a requirement to be in good health to participate in cybersickness studies, but it was not tested whether the level of physical activity could affect cybersickness.

3   Methodology

This study aims to examine the occurrence and severity of cybersickness among video gamers. The correlation between motion sickness history, vision correction, and physical activity are investigated. Age and gender are not presented as a co-variable due to unbalance distribution between the respondents. Furthermore, this paper tries to understand how gamers manage with the induced cybersickness and their perspective of what possible could help reduce the discomfort. Such a perspective on cybersickness and VR games is investigated by analyzing the responses of an online survey. The survey included qualitative and quantitative questions. This section gives an overview of the survey structure, followed by data collection and analysis methods. The section ends with a description of the respondents.

The survey included six sections:

- Introduction section including the purpose of the study
- Demographic questions
- Motion sickness susceptibility questions
- Physical activity questions
- VR experience and cybersickness induced by VR questions
- Suggestions for reducing cybersickness

Due to space constraints, only the items relevant to this paper are described in the results section.

3.1   Data Collection and Analysis

The survey link, with short information about the survey, was spread through a few web forums (e.g., Oculus Forums, Reddit Oculus Rift, and Virtual Reality). Due to the specific target group, the forums were chosen based on their popularity among video gamers and in particular, among VR gamers. The survey was written in standard English. The respondents were informed how long it would take the survey, as well as, each section and that all answers would be analyzed anonymously. The participation was entirely voluntary and no rewards were given.

An adapted version of Virtual Reality Symptom Questionnaire (VRSQ) [28] was used. The questionnaire consists of 13 items located in two groups, general body symptoms and eye-related symptoms (Table 1). The online survey used a scale ranging from 0 to 3 (e.g., none, slight, moderate, and severe).

The short version of the Motion Sickness Susceptibility Questionnaire-short (MSSQ-short) [29] was used. The questionnaire contains 18 items, representing different types of transportation, separate into two groups (childhood and adulthood). For each item, the respondents had five choices to answer: "Not Applicable – Never Traveled," "Never Felt Sick," "Rarely Felt Sick," "Sometimes Felt Sick," and "Frequently Felt Sick." Each choice, except "Not Applicable – Never Traveled," responds to a value from 0 to 3 ("Never Felt Sick" – "Frequently Felt Sick").

Since the physical activity can contribute to better body balance, it was added to the online survey. The International Physical Activity Questionnaire (IPAQ) [30] was used to obtain data on the health-related physical activity of the respondents. The questionnaire contained seven questions on which respondents had to answer how much time they spent on a particular activity in minutes or hours per day. The total score is presented in minutes per week.

3.2   Respondents

The survey collected 73 responses, from which four were female and 69 males. The mean age was 34 years. Forty-one of the respondents had completed some bachelor's degree; three of the respondents had completed a doctoral degree, and ten a master degree. Overall, the respondents were mainly young educated males.

The respondents were asked how often they wear glasses or contact lenses. Thirty-three of the respondents reported wearing glasses or contact lenses often, more than five times per week; however, 25 reported wearing no glasses or contact lenses. The respondents were asked as well how often they play video games (Fig. 1.) Therefore, the respondents were considered video gamers based on the frequency of playing video games.

4   Results

4.1   Cybersickness Correlations

According to the literature on cybersickness, the susceptibility to cybersickness can be predicted based on the motion sickness history and other factors such as gender and

Table 1. The list of items from the VRSQ divided into two main groups, general and eye-related symptoms.

General symptoms	Eye-related symptoms
General discomfort	Tired eyes
Fatigue	Sore/aching eyes
Boredom	Eyestrain
Drowsiness	Blurred vision
Headache	Difficulty focusing
Dizziness	
Difficulty concentrating	
Nausea	

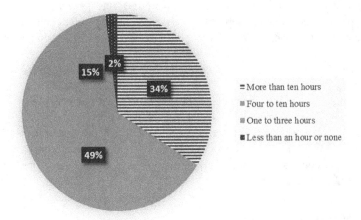

Fig. 1. Graphical representation of the respondents separated by video game play measured in hours per week.

previous experience with VR environments. Due to the unbalanced sample size regarding gender, this variable was excluded from the analysis. Statistical analysis was conducted to investigate whether there is a correlation between motion sickness history, physical activity, vision correction, and cybersickness.

A Spearman's rank-order correlation was run to determinate the relationship between motion sickness susceptibility during the childhood and cybersickness. There was a very weak correlation between the two variables, which was not statistically significant (r (73) = 0.19, p = 0.118.) Increases in motion sickness susceptibility score from the childhood were not correlated with increases in cybersickness onset. The same correlation coefficient was computed for motion sickness susceptibility during the adulthood. There was no correlation between these variables (r (73)= 0.16, p = 0.184) The increased score from the MSSQ-short calculated from the section related to the adulthood was not correlated with the increased total score from the VRSQ regarding VR experience. Overall, the analysis indicated that increases in motion sickness susceptibility were not correlated with increases in cybersickness (r (73) = 0.18, p = 0.128.)

A Pearson chi-square test was performed to examine the relationship between vision correction and cybersickness. The correlation between these variables was not significant, $X^2(1)$ = 0.001, p = 0.972. No association between wearing regularly eyeglasses and cybersickness was observed. Furthermore, a Pearson product-moment correlation was run to determinate the relationship between the physical activity of the respondents and cybersickness onset. There was a positive correlation between physical activity and cybersickness, which was statistically significant (r = 0.26, n = 73, p = 0.029). Higher physical activity was correlated with increases in VRSQ total score.

4.2 Cybersickness Associated with VR Games

Seventy-one of the respondents had previous experience with HMDs. The most frequently used brand was Oculus Rift CV1 (90%), followed by HTC Vive (9%) and Sony PlayStation VR (9%). Fifteen respondents took some of the following substances before starting the VR simulation: medicine, alcohol, or drugs. Only two of them reported that it helped them reduce cybersickness. Most often VR games were used as VR applications, and 71 of the respondents already have played VR games. The most played VR game was Robo Recall[1], followed by The Elder Scrolls V: Skyrim VR[2], and Echo VR[3].

From all respondents, 79% experienced to some degree at least once cybersickness induced by VR simulation. The most experienced symptoms from the general symptoms are *nausea*, *general discomfort*, and *fatigue* and from eye-related symptoms are *tired eyes* and *eyestrain* (see Fig. 2.)

Thirty-one respondents stopped the VR application immediately after they felt sick. Cybersickness disturbed the overall VR experienced for the majority (55%) of the respondents. However, that did not prevent them from using VR again (89%). 33 respondents answered affirmative to the question, *"Did you take any actions during the game to reduce cybersickness (e.g., reduce head movements, and close your eyes)?"* A follow-up question revealed that the taken actions helped 27 respondents to reduce the discomfort. Additionally, only 19% of the respondents tried ginger or products containing ginger as a remedy against cybersickness.

4.3 Suggestions for Reducing Cybersickness

The most used remedies are to stop the simulation and take a break as well as adaptation. Some of the gamers did not do anything against cybersickness and pushed through the moments of discomfort. The suggestions by the respondents are organized into three groups which are associated with the user, the software, and the hardware, accordingly.

---

[1] https://www.epicgames.com/roborecall/en-US/home.

[2] https://store.playstation.com/de-de/product.

[3] https://www.oculus.com/experiences/rift/1369078409873402.

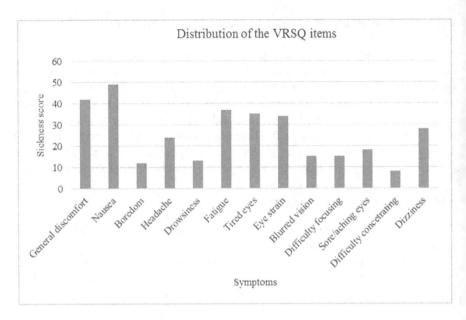

Fig. 2. Summation of VRSQ scores in over the thirteen symptoms.

The respondents were asked to share their experience with tested methods of reducing cybersickness. Forty-one out of all respondents answered. Five of them did not experience cybersickness and their responses were removed from the analysis. The most used method (n = 11) was to stop or to take a break and to continue later with the VR simulation. Five respondents adapted to the VR with time, three took anti-motion sickness medicaments, and two drank water. The other methods were diverse and were given by only one respondent. Some of the methods were to close the eyes, to take a mint bonbon, to adjust the inter-pupal distance, to take ginger, and not to play games with artificial motion. Nevertheless, six respondents did not do anything to reduce the cybersickness onset and continue to play.

Additionally, the respondents were asked to give their own opinions on what can be changed to feel less sick while using HMDs. That gave respondents the chance to make suggestions using their own words. Forty-three out of all respondents answered. The suggestions are edited and separated into three categories: user related, software related, and hardware related (Table 2). One of the respondents suggested following the developer guidelines. In particular, to add a cockpit with solid geometry (if that applies); to add a tunnel effect during fast head movements, also to make these options available to toggle on and off; to add free and to snap rotation options. Another respondent made a point regarding the stance reducing cybersickness during the VR experience, as written: "*Stance. Bending my knees send to help my body think it is on a platform when my avatar moves.*" Other respondent answered the same question that the players should gradually adapt to the VR games movements. For example, he wrote the following order: "*Touch Demo > Robo Recall > Elite Dangerous > Project Cars > Sprint Vector > Skyrim.*"

Table 2. Suggestions on how to reduce cybersickness by the online survey respondents split into three categories: user, software, and hardware related.

User-related	Software related	Hardware related
Closing eyes	High framerate	Comfortable HMD
Adaptation	Reduce vection	Motion cues feedback
Mint tablets	Roll cages	Higher resolution
Stop	Teleport movements	Wider FOV
Stance	Simple graphics	
Take a break	Locked framerate	
Short sessions	Frames of references	

5 Discussion and Conclusion

This paper reported the results of an online survey conducted through an online forum regarding the occurrence and severity of cybersickness among gamers. The results showed a positive correlation between cybersickness and physical activity and no correlation between cybersickness and motion sickness history. Furthermore, the correlation between the vision correction and cybersickness did not show statistical significance. This do not support the previous research where the correlation was significant [22]. An explanation for that might come from the different distribution between the users with corrected vision and those with normal vision in the two studies. Surprisingly, there was a statistically significant positive correlation between physical activity and cybersickness. It was assumed that gamers who might have a less active lifestyle are less susceptible to cybersickness. The results supported that assumption. Further research is needed to investigate this relationship. Opposite to previous research [24], there was no correlation between motion sickness susceptibility and cybersickness. This could be due to the lower score on the MSSQ-short, which represent a lower susceptibility level, and few respondents never experienced motion sickness.

The results showed that 79% of the video gamers who play VR games experienced cybersickness to some extent. Some of the most experienced symptoms are *general discomfort*, *fatigue*, and *eyestrain* aligned with previous research [6] on cybersickness as most experienced symptoms. Regarding the mitigation techniques suggested by the respondents, some of them are described in the literature such as adaptation [31], independent visual background [32], point-and-teleport locomotion [33], and adding motion cues [34].

This work clearly has some limitations. First, the sampling was limited to one online platform which constrained the variety of video game users. Second, the absence of an interviewer, who can inquire and clarify the questions in person, can lead to less valid data. Nevertheless, the paper showed that cybersickness is widespread among gamers and an appropriate solution has not been found yet. The results from this study can be used as valuable insights to create a more accessible, user-friendly and enjoyable VR environment for everyone. This paper could be a starting point for more extended research which could investigate the effect of different mitigation techniques such as frames of reference or ergonomics of the HMDs on cybersickness. Furthermore, physical activity and vision correction might be further investigated for their relationship and possible causality to cybersickness.

References

1. Joseph, J., LaViola, J.: A discussion of cybersickness in virtual environments. SIGCHI Bull. **32**, 47–56 (2000)
2. Kennedy, R.S., Drexler, J.M., Compton, D.E., Stanney, K.M., Lanham, D.S., Harm, D.L.: Configural scoring of simulator sickness, cybersickness and space adaptation syndrome: similarities and differences. Virtual Adapt. Environ. Appl. Implic. Hum. Perform. **247** (2003)
3. Reason, Brand, J.J.: Motion Sickness. Academic Press Inc. (1975)
4. Riccio, G.E., Stoffregen, T.A.: An ecological theory of motion sickness and postural instability. Ecol. Psychol. **3**, 195–240 (1991)
5. Prothero, J.D.: The role of rest Frames in Vection, Presence and Motion Sickness, pp. 169. University of Washington (1998)
6. Jinjakam, C., Hamamoto, K.: Simulator sickness in immersive virtual environment. Biomed. Eng. Int. Conf. (BMEiCON) **2012**, 1–4 (2012)
7. Cobb, S.V.G., Nichols, S., Ramsey, A., Wilson, J.R.: Virtual reality-induced symptoms and effects (VRISE). Presence **8**, 169–186 (1999)
8. https://www.statista.com/statistics/426469/active-virtual-reality-users-worldwide/
9. Biocca, F.: Will simulation sickness slow down the diffusion of virtual environment technology? Presence: Teleoper. Virt. Environ. **1**, 334–343 (1992)
10. Davis, S., Nesbitt, K., Nalivaiko, E.: Comparing the onset of cybersickness using the Oculus Rift and two virtual roller coasters. In: 11th Australasian Conference on Interactive Entertainment (IE 2015), pp. 3–14. ACS (2015)
11. Shafer, D.M., Carbonara, C.P., Korpi, M.F.: Factors affecting enjoyment of virtual reality games: a comparison involving consumer-grade virtual reality technology. Games Health J. **8**(1), 15–23 (2018)
12. Pallavicini, F., Ferrari, A., Pepe, A., Garcea, G., Zanacchi, A., Mantovani, F.: Effectiveness of virtual reality survival horror games for the emotional elicitation: preliminary insights using resident evil 7: biohazard. In: International Conference on Universal Access in Human-Computer Interaction, pp. 87–101. Springer (2018)
13. Walch, M., Frommel, J., Rogers, K., Schuessel, F., Hock, P., Dobbelstein, D., Weber, M.: Evaluating VR driving simulation from a player experience perspective. In: Proceedings of the 2017 CHI Conference Extended Abstracts on Human Factors in Computing Systems, pp. 2982–2989. ACM, Denver, Colorado, USA (2017)
14. Fernandes, A.S., Feiner, S.K.: Combating VR sickness through subtle dynamic field-of-view modification. In: 2016 IEEE Symposium on 3D User Interfaces (3DUI), pp. 201–210. (2016)
15. Farmani, Y., Teather, R.J.: Viewpoint snapping to reduce cybersickness in virtual reality (2018)
16. Whittinghill, D.M., Ziegler, B., Case, T., Moore, B.: Nasum virtualis: a simple technique for reducing simulator sickness. In: Games Developers Conference (GDC) (2015)
17. Curtis, M.K., Dawson, K., Jackson, K., Litwin, L., Meusel, C., Dorneich, M.C., Gilbert, S. B., Kelly, J., Stone, R., Winer, E.: Mitigating visually induced motion sickness. Proc. Hum. Factors Ergon. Soc. Ann. Meet. **59**, 1839–1843 (2015)
18. Curry, R., Artz, B., Cathey, L., Grant, P., Greenberg, J.: Kennedy SSQ results: fixed vs. motion-base ford simulators (2002)

19. Merhi, O., Faugloire, E., Flanagan, M., Stoffregen, T.A.: Motion sickness, console video games, and head-mounted displays. Hum. Factors **49**, 920–934 (2007)
20. Stoffregen, T.A., Smart Jr., L.J.: Postural instability precedes motion sickness. Brain Res. Bull. **47**, 437–448 (1998)
21. Rebenitsch, L.R.: Cybersickness Prioritization and Modeling. Michigan State University (2015)
22. Rebenitsch, L., Owen, C.: Individual variation in susceptibility to cybersickness. In: Proceedings of the 27th Annual ACM Symposium on User Interface Software and Technology, pp. 309–317. ACM, Honolulu, Hawaii, USA (2014)
23. Stanney, K.M., Hale, K.S., Nahmens, I., Kennedy, R.S.: What to expect from immersive virtual environment exposure: influences of gender, body mass index, and past experience. Hum. Factors: J. Hum. Factors Ergon. Soc. **45**, 504–520 (2003)
24. Wright, R.H.: Helicopter Simulator Sickness: A State-of-the-Art Review of Its Incidence, Causes, and Treatment (1995)
25. Matas, N.A., Nettelbeck, T., Burns, N.R.: Dropout during a driving simulator study: a survival analysis. J. Saf. Res. **55**, 159–169 (2015)
26. Basu, A., Ball, C., Manning, B., Johnsen, K.: Effects of user physical fitness on performance in virtual reality. In: 2016 IEEE Symposium on 3D User Interfaces (3DUI), pp. 233–234 (2016)
27. Taha, Z.H., Jen, Y.H., Gadzila, R.A.R., Chai, A.P.T.: The effect of body weight and height on incidence of cyber sickness among immersive environment malaysian users. In: Proceedings of 17th World Congress on Ergonomics (2009)
28. Ames, S.L., Wolffsohn, J.S., Mcbrien, N.A.: The development of a symptom questionnaire for assessing virtual reality viewing using a head-mounted display. Optom. Vis. Sci. **82**, 168–176 (2005)
29. Golding, J.F.: Motion sickness susceptibility. Auton. Neurosci. Basic Clin. **129**, 67–76 (2006)
30. Booth, M.: Assessment of physical activity: an international perspective. Res. Q. Exerc. Sport **71**, 114–120 (2000)
31. Regan, C.: An investigation into nausea and other side-effects of head-coupled immersive virtual reality. Virt. Real. **1**, 17–31 (1995)
32. Duh, H.B.-L., Parker, D.E., Furness, T.A.: An independent visual background reduced simulator sickness in a driving simulator. Presence: Teleoper. Virt. Environ. 13, 578–588 (2004)
33. Frommel, J., Sonntag, S., Weber, M.: Effects of controller-based locomotion on player experience in a virtual reality exploration game. In: Proceedings of the 12th International Conference on the Foundations of Digital Games, p. 30. ACM (2017)
34. Yoon, H., Lee, S., Park, J., Choi, Y., Cho, S.: Development of racing game using motion seat. In: 2017 International Symposium on Ubiquitous Virtual Reality (ISUVR), pp. 4–7. IEEE (2017)

Gamifying the Rice Industry: The 'Riceville' Paradigm

Evangelos Markopoulos[1]([⊠]), Ho Fung Kisum Chan[2],
and Lincoln Lee Ming[2]

[1] HULT International Business School, 35 Commercial Road Whitechapel,
London E1 1LD, UK
evangelos.markopoulos@faculty.hult.edu
[2] University College London, 66-72 Gower Street Bloomsbury,
London WC1E 6BT, UK
kisumc@gmail.com, mingt.lee@gmail.com

Abstract. The global agriculture industry remains underdeveloped due to a lack of education among farmers about best practices and technologies. Similarly, corporate gamification and the rise of serious games have demonstrated their effectiveness in education within our era. The implementation of gamification principles through accessible technological platforms has massive potential in building best practices in the agriculture industry. Rice agriculture has a resistance to the adoption of new methodologies and technologies. This paper introduces 'Riceville', a game used to simulate a virtual farming environment promoting the use of effective best-practices introduced by leading rice organisations like the International Rice Research Institute (IRRI). By gamifying best practices, the social norms of low adoption and perception of risk can be altered to boost the reception of novel practices within the industry. This will not only serve to modernize the rice industry but also improve yields for farmers while helping agriculture companies increase exposure.

Keywords: Gaming · Gamification · Rice · Agriculture · Technology · Management · Production · Management · Serious games · Agriculture

1 Introduction

Many farmers around the world are still using outdated traditional agricultural practices. Despite, a large amount of resources from global charities, governmental and commercial agricultural organisations allocated to educating the farmers, the resistance occurs when trying to convince them of the effectiveness of novel technologies. The scarcity mindset within farmers in one of the driving factors contributing to low rates of adoptions, ultimately stunting the growth of industries, especially ones typically characterised with low levels of education. Mainstream platform technologies such as smartphones are now able to successfully penetrate rural farming communities such as the case of high smartphone adoption of farmers in Myanmar.

This platform presents an exciting opportunity to catalyse the education of farmers. One such means is through gamification that rewards best farming practices, most

© Springer Nature Switzerland AG 2020
T. Ahram (Ed.): AHFE 2019, AISC 973, pp. 202–214, 2020.
https://doi.org/10.1007/978-3-030-20476-1_21

effective technologies and even proven agronomy. This paper suggests a farming game that targets rural rice farmers, enabling them to simulate farming and get rewarded for using the most route to grow their farm. In addition, to raise competition and stickiness, the incorporation of a ranking system can be a further form to gamify the platform. It may accelerate the adoption of best practices and technologies whilst providing a great platform to change the entire mindset of whole rural village communities that are predominantly farming focused.

2 The Gamification Era

Gamification as an academic field and also as management and operations practice is still in its infancy, treated as an original idea. A definition that is frequently cited in relative works presents gamification as the incorporation of game elements into non-game contexts [1]. The word gamification could refer to games created with the purpose of turning a tiresome and hard task into an engaging activity, while the incorporation of educational features is desirable. Furthermore, gamification may refer to the evolvement of an existing structure, like a website, an enterprise application or an online community, to an educational tool by applying some of the techniques and ideas that make games appealing. In other words, gamification is the strategy which uses game mechanics and techniques in order to drive user behaviour by increasing self-contribution.

Gamification is a popular topic among business professionals besides the academia and is exercised in sectors such as engineering, medicine and military. It is described as serious games, pointification, behavioural games and games with a purpose, with the aforementioned terms being similar, yet different. The work of Seaborn and Fels [2] is proposed where several definitions of gamification and the related concepts are categorized and elucidated. Gamification is considered by industries as a tool for supplementing branding initiatives or a business strategy tool [3, 4].

3 Corporate Gamification and Serious Games

The term "serious game" has been used since the decade of 1960, long before the introduction of computer and electronic devices into entertainment [5]. It was used to define gamified processes without the use of technology as a scenarios-based model operating metaphorically as a game of strategy with probabilities, possibilities and skills on handling information, conditions, decisions and results.

Many references define serious games or applied games as games designed for a primary purpose other than pure entertainment. The "serious" adjective is generally prepended to refer to products used by industries like defence, education, scientific exploration, health care, emergency management, city planning, engineering, and politics [6]. This, kind of biased, characterizations can be unfair to the entertainment games that do have serious scenarios, technology, graphics, sound, animation, effects and other elements that can turn entertainment games into unique experiences. Serious games are successful only if they are designed based on entertainment games design principles. A good serious game must be entertaining or at least so immersive that players are addicted to play. What is serious game, and what is not serious cannot and shall not be determined by the type of its user's target group, functionality or operations, but solely on its quality, effectiveness and benefits to those using it for a specific purpose, any purpose.

4　Examples of Industry Serious Games

Gamification is often correlated to digital game-based learning (DGBL), which is defined as the use of "game-based mechanics, aesthetics and game thinking to engage people, motivate action, promote learning and solve problems" [7]. In more specialized topics in engineering, gamification includes game systems in CAD-type environments. Broughet al. [8] developed Virtual Training Studio where users train to perform assembly tasks. Li, Grossman and Fitzmaurice [9] presented GamiCAD, a gamified tutorial for AutoCAD, based on missions, scores and rewards. Additionally, a gamification approach was designed by the RWTH Aachen University and tested in cooperation with a German car manufacturer, in order to enhance training strategy for workers in low volume assembly systems and increase ramp-up performance, with promising results [10]. Furthermore, Hauge and Riedel [11] tested two serious games, namely COSIGA and Beware, in order to evaluate gamification for teaching engineering and manufacturing.

Applying game technologies in engineering is a quite unique business area, and even quite unique research area). An efficient use of existing digital content such as 3D technical CAD drawings together with game technologies is one of the key elements of gamified industrial applications. IndustrySim (Fig. 1) is an example with CAD drawings of coal fired power plant. IndustrySim has characteristics which can be found from IS research framework presented by Hevner et al. [12].

One of the most successful games in production management is Plantville. The online game was release on 2011 by Siemens and revolutionized production gamification. Taking part of its name after the popular online game Farmville, Plantville enables players to immerse themselves in an production environment, make decisions to change and improve it, and interact with other colleagues.

The game simulates a plant manager's experience and challenges players to increase productivity and sustainability of an industrial plant. During the game, players are faced with the challenge of maintaining the operation of their virtual plant while trying to improve key areas of manufacturing (Fig. 2). The players are measured against on several Key Performance Indicators (KPIs), that go beyond the production management process as they approach the plant's operations in a wholistic way covering safety, on time delivery, quality, energy management and employee satisfaction targets [13].

Plantville players have the option to select one of the three different plants the game provides, in three different sectors (vitamin plant, bottle production plant and train building plant). Plant engineering experts from Siemens worked closely with gaming experts from Pipeworks Software Inc., developer of software and technology for Xbox 360, Nintendo Wii, Playstation 3 and other gaming consoles. This collaboration delivered a visually rich environment which made the game a big success.

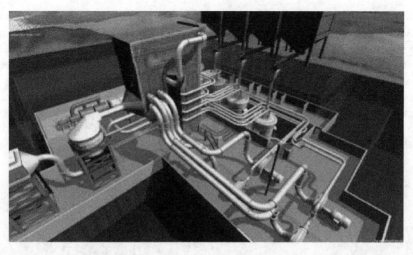

Fig. 1. Screenshot (with the GUI disabled) of an example coal-fired power plant built inside the IndustrySim prototype.

Fig. 2. Screenshot from the Plantville game [14].

A similar example in popularity and success of an industry serious games is Quest for Oil, a real time strategy resource management game in the area of oil exploration has been developed by MAERSK in 2013. The game is composed from three mini games related to production management, drilling management and seismic manage- ment (Fig. 3). The players can experience off-shore drilling being on an oil drilling rig in the pursuit of oil in the North Sea and off Qatar. The players go on a subsurface journey, exploring the geographical underground of the oil industry at both locations. The goal of the games is to test the players analytic skills on searching for oil based on a seismic map while they face challenges surrounding the oil exploration [15].

Fig. 3. Screenshot from the Quest of Oil game [16].

IndustrySim, Plantville and Quest for Oil are indicative examples of serious games that made an impact in the adaptation of games in production and operations man- agement, challenging more industries to follow this path.

5 Serious Games in Agriculture Management

Agriculture management in our modern day requires much planning, allocating proper resources, dealing with unorthodox weather patterns and financial constraints. It also requires managing a good deal of risk as profits are more often than not tied to a good crop. Furthermore, globally agriculture is being disrupted by new smart farming machinery, precision agriculture and digitization. Managing an already complex pro- cess while integrating new potential challenges can make agriculture management difficult especially for small holder farmers who still produce majority of the world's food supply.

Serious games offer a way to make the managing this complex process easier. Balancing resources, integration of new events and planning environments are a common part of many famous game plays. For example, sim city which teaches basic urban planning would be a concept easily transplantable onto a farm environment. It would offer a way to quickly learn and pick up basic farm agronomics. Furthermore, it offers a quick way to learn and pick up the dynamic nature of the changes affecting agriculture that perhaps older farmers had never seen before due to climate change and urbanisation. Agricultural games designed with education in mind have long been available. An example of this is AgVenture, a computer game designed to educate farmers and students on basic farming strategy [17].

In recent years, there have also been agricultural streams that have a greater reach inciting interest from a more mainstream player base. The farming simulator series of games is a prime example of this. Released in 2012, the video game has sold over four million copies combined just two years after it's release. The players in this multiplayer game are involved with planting crops, mowing wheat, hiring manual labour, and other activities [18] (Fig. 4).

Fig. 4. Screenshot from the Farming Simulator 14 multiplayer game.

There is a clear demand for such a game from younger generations evident from the main player demographic for this series being young children and farmers. Furthermore, the agricultural niche is becoming ever more mainstream demonstrated through electronic sport (e-sport) competitions that are starting to appear including the Farming Simulator Championship with a total of 250.000€ in prizes (a 150.000€ increase since the competition's first year, one year ago) [19]. This opportunity has not gone unnoticed as several research institutions have already begun to recognise its potential and begun developing serious games targeted at the agriculture sector [20, 21]. In fact, in a study done on serious games as a tool for sustainable development, agriculture was recognised as a sector for focus [22].

6 Challenges in the Rice Industry

Rice is one of the 3 main crops in the world but the most important one for majority of middle- and low-income countries. Over 500 million tons of rice are produced and consumed each year. Its importance cannot be understated not just because it's a staple of over half the world but because any disruption to the industry would cause massive repercussions for food security and global income. For example, in 2008 when the price of rice tripled, the world bank estimates 100 million people went below the poverty line as a result [23].

Over 70% of the world's rice is produced by smallholder farmers living in low to middle income countries. Their farm sizes are typically no larger than 2 acres. These farmers face a variety of issues due a lack of education, climate change and rapid urbanisation. They normally use traditional and improper agricultural practices that result in low yields and low income as a result. Penetration of new technologies including machinery and best practices such as proper usage of pesticides are slow because they are rarely understood by the farmers [24]. However, when a new practice or technology is adopted, it has the potential to spread quickly but farmers typically implement it without proper understanding which is risky as it might result in ruined crops. With low incomes, a lack of understanding of best practices, this puts the world's food security of rice at risk.

However, penetration of smart devices is surprisingly high in most smallholder populations. Despite gaps in their understanding of these devices, items like smartphones have become a daily tool that farmers use for a variety of tasks. Hence this offers a touchpoint for serious games to be able to deliver educational content to rice farmers in particular. A serious game has the potential to provide a method to teach farmers best practices in the industry as well as allow them to practice managing scenarios that they otherwise would not risk in real life. This could potentially allow them to make better decisions and implement best practices for their actual crops. Finally, such a game would also serve as a platform for agricultural companies who have historically found it hard to market to rural smallholder farmers to engage with them.

7   Potential of Rice Industry in Asia

The rice industry in Asia, in particular the rice farmers of Asia which number a population of more than 200 million [25], would be particularly suited for a serious game. Most of the worlds rice production takes place in the Asia Pacific region (Fig. 5) [26] which is also the region with the most games as more than 900 million were recorder in 2016 (Fig. 6) [27].

Furthermore, serious games set in the agriculture content have been shown to attract farmers not just regular players. Farming simulator, a serious game for agriculture which has over a million users that is targeted to western countries like the United States and the United Kingdom, reported that over a quarter of their users have a background in farming with up to 10% full time professional farmers. Interviews with these farmers demonstrate that professional farmers use these serious games to learn more about machinery that they don't have access to in real life. To test out various scenarios on their crops and simulate their real-life processes or even just for the satisfaction of being able to try something new without real life risk [28]. It demonstrates that professional farmers are already utilizing serious games to increase their own sense of knowledge and fulfill creative desires that would be too risky to implement in real life.

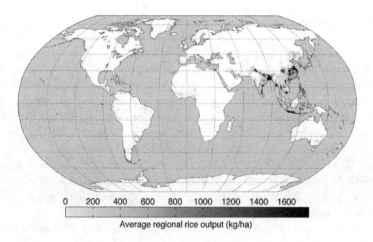

0   200   400   600   800   1000   1200   1400   1600

Average regional rice output (kg/ha)

Fig. 5. Screenshot from the Farming Simulator 14 multiplayer game.

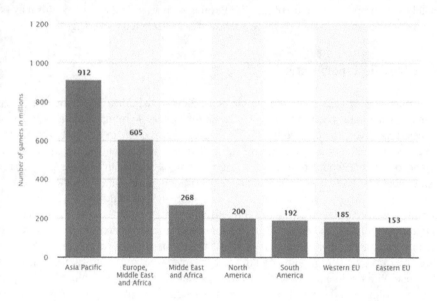

Fig. 6. Number of video gamers worldwide in 2016, by region (in millions)

This provides an exciting opportunity for serious games in the rice industry because majority of rice farmers are located in the Asian Pacific Region, while majority of gamers also reside in the same location demonstrating a large overlap of potential demand. Furthermore, a serious game in agriculture from another region has shown to be able to attract significant interest from the farming population and been successfully leveraged for its benefits. The same could happen in Asia as well for the small holder rice farmers.

8 The Riceville Case

A key issue mentioned above regarding the rice industry is the need to accelerate adoption of new agriculture technologies and practices in the farmer demographics. Additionally, there is an exceptional rate of smartphone adoption within the same populations of low economically developed countries including Myanmar. This presents an exciting opportunity to leverage new technology adoptions to simulate a variety of basic agricultural farming tasks to promote farmer education. This can be done by gamifying and rewarding behaviours which reflect real life best farming practices. Examples of which best farming practices can be modelled after could include IRRI's Best Management Practices for differing land and weather conditions. In addition to Best Management Practices, the game could also be a platform to simulate and educate basic agronomics as well [29].

It would be a basic simulation game that simulates relevant factors that are involved in farming conditions. Factors including cost of farming inputs, growth rates, weather conditions, soil conditions, fertilizer effectiveness and etc. can all be simulated. Ultimately with proper simulation of real-life farming scenarios for the farmers, this may be a form of education, through gamification, can accelerate their learning.

9 Riceville Operations

The foundations of the game are simple and follow a traditional city-builder/resource management strategy game such as Cities Skyline or the Farming Simulator series of games respectively to begin with [30]. The player takes the role of a farmer who has the objective of making more profit for his farm. A variety of mechanisms are available for this to occur that would reflect real life opportunities so that the farmer playing the game would be able to learn and implement learnings to real life.

The game begins with access to a basic plot of land. Additionally, with some starting resources, that are initially limited, which may include: money, experience, influence, reputation and other. They can use these starting resources to purchase seeds, fertilizer amongst other farming inputs. All these can be used to progress through a three-part cycle: Planting and sowing, crop maintenance and crop selling. Each part designed with the intention to allude to and promote the use of agricultural techniques and technologies reflective of real-life best practices. The planting and sowing cycle would include processes such as choosing the best types of seeds for the best soil types, choosing which day to plant the seeds to avoid bad weather conditions or which machines to use to decrease the amount of time required for planting. Crop maintenance would involve selecting the best pest control substances and techniques to use within the game. Finally, the crop selling stage would involve the process of selling different varieties of crops during the best time to maximize return on investment of initial resources.

During the different stages, different resources could be rewarded for various actions done within game. An obvious example would be a reward of money during the crop selling stage. Another example is experience that can be gained in the planting and crop maintenance stage. This will be a pivotal part in the progression of the game for the player through the upgrade mechanism for different assets in the farm. The resources gained can be used to purchase/add value to existing farming inputs or enhance existing processes/assets used by the player. Enhancements and the usage of different assets/technologies within the game will all have an influence on the final crop that is to be sold and increase efficiencies of other processes. This creates a platform in which farmers have a safe environment to test out/simulate new technologies on a virtual farm. By seeing the positive effects, it may also encourage them to upgrade their real-life existing processes/technologies.

To facilitate the engagement, the gamification element of a ranking system is incorporated to boost competitiveness between players. A ranking system based on score given to combined achievements or experiences the player has. This also creates a platform to highlight model farmers/farms. An interface has been designed where players can visit other players' farms to see ideal farm practices that will be positively reinforced. Furthermore, the implementation of daily rewards from consecutive playing of the app would also be appropriate to increase the stickiness of the game.

To once again boost farmer engagement, the ranking system selects players who could be awarded rewards in real life for example via partnerships with companies/governments/NGOs to further emphasise that despite the simulation being a game, there are real life benefits through both implementing the structure and also possible materialistic rewards. An example is a partnership with an agricultural company where top ranking farmers can get subsidized rates for that company's fertilizer. Non-agriculture related organisations may also be appropriate such as a partnership with telecom companies for telecommunication benefits.

One of the most important partnerships would be with a knowledge partner such as a rice research organisation which constantly updates and refines a 'best practices' set of guidelines to instruct farmers. Obtaining this input can turn the game into an engaging, yet highly educational resource for farmers to have fun whilst they learn the best agricultural practices in the world. Even with the possibility of gaining real life benefits.

10   Riceville Contribution to the Rice Industry

Benefits to the farmer are obvious, education on modern agriculture practices will greatly increase the productivity of the farmers. This will also raise the overall productivity of the rice industry in tandem since the farmers are at the beginning of the supply chain, directly increasing the volume of production. This also provides an exciting opportunity to influence the decision of which rice variety types a farmer should plant that is desired by a nation's rice industry.

Disaster scenarios such as a drought is predicted can be prepped against by setting those weather conditions for the players and farmers to play through. Successful players who heed in game warnings will be rewarded by having more produce to sell at the end of a game cycle. On an individual farmer level, this educates them on what to do to prepare for certain conditions whereas on an industry level, overall rice output could be maximized given whatever weather conditions take place.

Finally, Riceville could act as a bridge between existing rice institutions/businesses and farmers. It provides a win-win solution for many stakeholders in relation to farmers. For the policy makers, they can simulate new policies in game to gain data on how new policies could affect the farmers whilst farmers can get up to date information and prepare how to address new policies appropriately to maximize their gains. Even for corporations, that provide products/services for farmers, Riceville is another route to engage with the farmers directly whether through paid sponsorship or to increase familiarity of their business concepts.

11   Areas of Further Research

Further research on the gamification of the articular sector will be extended to the design of Riceville under Virtual and Augmented Reality (VAR) technology in order to achieve more immersion and provide more effective results and benefits to the players who could be the actual or the potential rice farmers. VAR can enable virtual manipulation and provide exposure to new technologies - by being able to visualise and handle virtually, it can boost learning of how new machinery can work. VAR can also help train farmers to recognise negative signs regarding their crops: e.g. to spot certain plant diseases. A VAR version of Riceville could offer exciting opportunities for farmers such as enabling the virtual manipulation to tools or technology that they haven't bought yet or it not available in their region. Alternatively, it could also provide them training for new machinery that they do not fully know how to use without the risk of damaging their crops.

Other areas to move into would be using serious games to not just teach farmers best practices of rice but for other crops as well. They would be able to experiment with different harvest seasons and simulate in advance what would be the best type of agriculture expansion for investment. For example, a rice farmer would be able to decide if he should invest in purchasing a cow or perhaps try planting maize as well as rice for the next harvest. It doesn't have to just stop at farming either, educational content of relevance to small holder farmers could be taught as well for example, home economics or simulations of other business industries.

12   Conclusions

The adaptation of serious games is a growing challenge and opportunity for all sectors to primarily achieve employee engagement and operations efficiently. However, there are sectors like the agriculture that present a decline of the people involved in it, and especially the farmers, require any possible help from technology in order to restructure its workforce in size, skills and commitment. Riceville is not only a serious game attempted to gamify the agricultural industry but a shared value social innovation technology aiming to improve the lives of hundreds of thousands rice farmers and revitalize international the rice farming community. The proposed technology emphasizes more on the social contribution to the rice farmers and the rice industry, than the operations efficiency and profitably by the optimization of the farming process.

Riceville is a game that can prove that gamification can be aligned with the UN sustainable goals, improve the lives of the people, reduce hunger and poverty, and assure decent work and economic growth.

References

1. Deterding, S., Khaled, R., Nacke, L.E., Dixon, D.: Gamification: toward a definition. In: CHI 2011 Gamification Workshop Proceedings, Vancouver, BC, Canada (2011)
2. Seaborn, K., Fels, D.: Gamification in theory and action: a survey. Int. J. Hum Comput Stud. **74**, 14–31 (2015)
3. Zichermann, G., Linder, J.: Game-based Marketing: Inspire Customer Loyalty through Rewards, Challenges, and Contests. Wiley, Hoboken, NJ (2010)
4. Werbach, K., Hunter, D.: For the Win: How Game Thinking Can Revolutionize Your Business. Wharton Digital Press, Philadelphia, PA (2012)
5. Clark C. Abt.: Serious Games, Viking Press. (1970)
6. Damien Djaouti, D., Alvarez, J., Jessel, J.-P.: Classifying Serious Games: the G/P/S model IRIT. http://www.ludoscience.com/files/ressources/classifying_serious_games.pdf
7. Radoff, J.: Game On: Energize your Business with Social Games. Wiley (2011)
8. Brough, J.E., Schwartz, M., Gupta, S.K., Anand, D.K., Kavetsky, R., Pettersen, R.: Towards the development of a virtual environment-based training system for mechanical assembly operations. Virt. Real. **11**(4), 189–206 (2007)
9. ESA-Entertainment Software Association. 'Essential Facts About the Computer and Video Game Industry'. http://www.theesa.com/wp-content/uploads/2014/10/ESA_EF_2014.pdf
10. Kampker, A., Deutskens, C., Deutschmann, K., Maue, A., Haunreiter, A.: Increasing ramp-up performance by implementing the gamification approach. Procedia CIRP **20**, 74–80 (2014)
11. Hauge, J.B., Riedel, J.: Evaluation of simulation games for teaching engineering and manufacturing. Procedia Comput. Sci. **15**, 210–220 (2012)
12. Hevner, A.R., March, S.T., Park, J., Ram, S: Design science in information systems research. MIS Q. **28**(1), 75–105 (2004)
13. Siemens Canada Internet. https://www.siemens.com/content/dam/webassetpool/mam/tag-siemens-com/smdb/regions/canada/press-archive/2011/english-2011/05-07-2011-siemens-launches-plantville.pdf
14. Plantville Siemens. http://www.computer-blog.net/tag/plantville-siemens/
15. Ecogamer. Quest for Oil. https://www.ecogamer.org/energy-consumption/quest-for-oil/
16. Serious games market. Serious Games for Oil Exploration Hands-On Experience https://www.seriousgamemarket.com/2013/06/serious-games-for-oil-exploration-hands.html
17. Cross, T.: AgVenture. A Farming Strategy Computer Gam. J. Nat. Resour. Life Sci. Educ. **22**(2), 19, 103–107 (1993)
18. Eurogamer. Farming Simulator 14 plows onto iOS and Android. https://www.eurogamer.net/articles/2013-11-18-farming-simulator-14-plows-onto-ios-and-android
19. Farming Simulator. Introducing the Farming Simulator League. https://www.farming-simulator.com/newsArticle.php?lang=en&country=us&news_id=104
20. SEGAE: A serious game project for agroecology learning. In13th European IFSA Symposium 2018 Jul 1 (2018)
21. Smart-AKIS. Serious Games for Smart Farming? - Smart-AKIS. https://www.smart-akis.com/index.php/2017/01/20/gates-serious-games-for-smart-farming/

22. Liarakou, G., Sakka, E., Gavrilakis, C., Tsolakidis, C.: Evaluation of serious games, as a tool for education for sustainable development. Eur. J. Open Distance E-Learn. **15**(2) (2012)
23. Ricepedia. Challenges. http://ricepedia.org/challenges
24. Schreinemachers, P., Afari-Sefa, V., Heng, C.H., Dung, P.T., Praneetvatakul, S., Srinivasan, R.: Safe and sustainable crop protection in Southeast Asia: status, challenges and policy options. Environ. Sci. Policy **54**, 357–366 (2015)
25. Ricepedia. Who grows rice. http://ricepedia.org/rice-as-a-crop/who-grows-rice
26. Kuenzer, C., Knauer, K.: Remote sensing of rice crop areas – a review. Int. J. Remote Sens. **34**(6), 2101–2139 (2013)
27. Statista. Number of gamers worldwide by region 2016. https://www.statista.com/statistics/293304/number-video-gamers/
28. The Guardian, Lane, R. Meet the real-life farmers who play Farming Simulator. https://www.theguardian.com/games/2018/jul/24/meet-the-real-life-farmers-who-play-farming-simulator
29. IRRI. Best Management Practices for rainfed lowland rice in Cambodia - IRRI Rice Knowledge Bank. http://www.knowledgebank.irri.org/agronomy-guides/item/best-management-practices-for-rainfed-lowland-rice-in-cambodia
30. Rock Paper Shotgun. Why road-building in Cities: Skylines is a pleasure. https://www.rockpapershotgun.com/2017/02/10/cities-skylines-roads/

Cultural Playing: A Videogame Case Study of Never Alone and the Teaching of Culture

Hélio Parente de Vasconcelos Neto[✉]
and Maria Aurileide Ferreira Alves

University Centre UniFanor Wyden, 60191-195 Fortaleza, Ceará, Brazil
hpn.helio@gmail.com, mafa@campus.ul.pt

Abstract. The present article proposes to study the game design as an effective instrument of teaching and propagation of culture. It has also been taken in consideration the aspects that are established as necessary for a game to be an effective educational game and how they are present in our object of study. For such study, we analysed the video game "Never Alone", produced by the North American company Upper Games and how it relates to and transmits the folkloric and cultural elements of the Iñupiaq people through its own unique game design, taking in consideration its content, playability and visual, or textual information. In addition, we have also discussed the importance of the propagation of unknown culture in new popular media while being respectful and accurate to its source and people. All data was collected via playing the game whilst observing its contents under the scope of the analysis.

Keywords: Culture · Never alone · Game design · Education · Iñupiaq

1 Introduction

It is no news that, in our contemporary society, videogames acquired their place in the daily lives and leisure hours of the younger generation, after all, its capacity of maintaining the player's concentration in one single task for hours is one of its main characteristics. Its due to this specific aspect that the pedagogical area of knowledge has taken interest in studying the possibility of inserting this new media in the process of teaching.

Another important aspect to be taken in consideration is the growing power of this sector in the media and entertainment industry. In Brazil, it is notable the strength of this industry, as we can verify in the following extract of Sav's article:

"Approximately 23% of Brazilian population are assiduous or casual players, which correspond to around 45 million players" [13]. With such reach and concentration power, it is natural that an interest surges in applying these characteristics in the teaching process. "Therefore, there today areas dedicated to the application of games in the learning process called Digital Game-Based Learning – DGBL." [16] apud [13, 21]. (Author's free translation).

© Springer Nature Switzerland AG 2020
T. Ahram (Ed.): AHFE 2019, AISC 973, pp. 215–223, 2020.
https://doi.org/10.1007/978-3-030-20476-1_22

By the means of a fast analysis of data collected during the using of the videogame, the present article seeks to expose a few possibilities of its utilization in the teaching area.

2 Methodology

To write this essay it was elected to use an experimental and empiric methodology via case study of the mentioned videogame, which was played by the author so that the same could comprehend its functioning, meet its characters and plot. For 30 days, during the month of May of 2018, the videogame was played, and observations made e registered regarding the functioning of the game, as well as the conceptual and stylistic elements and its playability so that it was possible do assess the way culture is being transmitted to the player or user. Bibliographic fonts and the informative material supplied by the developers of the game, via internet, was also used in the making of this essay.

3 Results and Discussion

3.1 Game Design

Before speaking about the use of videogame in education, it is convenient to enlighten the importance of game designing in the process of creation of games, and of which stages this area acts on and what are its main characteristics.

Game design is easily mistaken with other stages of game production. In an emergent market like digital game production, it is hard to conceptualize in a clear and objective way in what does a game designer act upon professionally. According to Tavares [17]:

The majority of parts of the areas involved, specially music or visual arts, although do have some of their specific knowledges, most commonly the technical ones, redirected to the acting of a game like these, such knowledge is not native to the area. The exception is the game designer, who must master specific concepts, although it must also drink from other areas, direct or indirectly related to game design [17]. (Author's free translation).

This means that the process of game designing is an interdisciplinary process, that appropriates not only the technical information present in the area of design but as well as every other knowledge of any area that is present during the production, be it indirectly of directly. Game design, therefore, does not only works with the visual aspect of the game, but with the work as a whole, including the psychologic aspects as well as the visual, artistic, aesthetic and functional ones. Teaching area.

3.2 Game Design and Education

Today, the introduction of game design in education and studies that seek to better understand the relation between both areas are, more and more, present the academic world. Many universities and schools promote the use of digital educational games in the classroom as an alternative method of teaching. We can confirm this assumption if we that the following excerpt from Savi into consideration, [16]:

Now, in institutions, instead of closing their doors to games there is an increasing interest between researchers and professors in discovering in which ways digital games can be used as a resource to support the learning process and which are its benefits. [16] pud [9].

If there is indeed a knowledge and research production in this area, what are the results of such studies? Which are the characteristics of a videogame that makes it desirable by the pedagogical area? Which characteristics must a game possess to be considered an educational game?

Digital games provide "attractive and inovative educational practices, in which the student has the opportunity to learn in a most active, dynamic and motivational way" [16]. For the game to be effectively applied in the pedagogical area it is necessary for the game designer to have in mind the following five principles formulated by [12]:

Learnability: the game must be of easy learning. Icons and symbols must be of great clarity and easy understanding. Efficiency: the game must be efficient, in such way that the user, after having learned to play it, may obtain a high level of productivity. Memorability: the game must be easy to remember, in a way that the casual user can return to the game, after some time not using it, without having to learn everything once more. Errors: the must have a low error rate, so that the user, when committing errors can easily solve them [12].

Once aware of Nielsen's principles, another challenge emerges: how to develop educative games that do not lose its dynamic modern character in favor of its didactic. According to Savi, [16]:

This is the current problem of companies that develop such games: because, many times, when giving emphasis in the didactic of the game we end up losing the leisure provided by the game, a fundamental characteristic of this educational tool. "Digital game must attend to pedagogical requirements, but it is also necessary to be careful not to make the game into a didactic product, making it loose is pleasurable and spontaneous character [16] apud [5]".

Continuing on this theme, Savi concludes, [16]:

Therefore, we conclude that having leisure components inserted in the study process is important because "when the student is relaxed, usually there is a grander reception and disposition for the learning process". [16] apud [7, 15].

3.3 Educative Game Design and Culture

Once the characteristics of game design in education are presented, it becomes necessary to identify the relation between propagation of culture and the game, as well as the relation between culture and teaching. Culture is an essential part of individual knowledge, through it the student characterizes the knowledge acquired and interacts with the world around it. Through the contact with different cultures an amplification of the individual's perspective about the world occurs, making the student capable of producing more based and elaborate knowledge. Aside from this, culture, in the educative process is foreseen in Brazilian legislation. The article 26 of Law number 12796/2013, altering the Directive's Law and the Basis of National Education (LDB, 9394/96), foresees culture as necessary and obligatory in the educative process.

In the article "Culture in The Process Of Teaching And Learning In Children's Education", Geraldo (2010) [6], brings us the following conclusion:

Analysing what the many authors have to say we can perceive how fundamental the cultural elements to children's education are. We can even say that they are the center of their education. These elements make possible the acceptance of the most diverse ways of thinking, of questioning and of acting in the modern world, where knowledge is increasingly superfluous, the consequence of this is a provisory society. [6].

Knowing that culture is an essential part of the process of teaching, in which way are videogames a part of it? By being a type of media that, usually, is linked to a form of culture, be it by the teachings present in it, or, simply by being an artistic production of members of a specific society. Digital games are cultural tools of a people:

According to Pierre Levy, an intellectual technology, in most cases, exteriorizes, objectivises and virtualizes a cognitive function, a mental activity. In this line of thought, we understand digital games as a cultural tool, comprehending it as an exteriorization, objectivation and virtualization of a cultural expression [2].

Continuing on this theme:

The relation between digital games and cultural artifacts has been studied for some years. De Paula points that if we do not treat this cultural artifact (digital games) as an unique language and give emphasis to its possibilities, we shall continue to see them only as transmissions of the northern American way of thinking [2].

Therefore, we can deduct that the pedagogical area of knowledge utilizes digital games as an alternative way of teaching and that the aforementioned area produces and consumes studies aimed at the perfecting of this teaching method. The media called videogame is a cultural representation of a people. The present case study now has as an object of analysis the real application of the theories before mentioned, as well as the characteristics of educative games present in the game Never Alone, to this we elected to use the empiric methodology.

By the means of a fast analysis of data collected during the using of the videogame, the present article seeks to expose a few possibilities of its utilization in the teaching area.

By the means of a fast analysis of data collected during the using of the videogame, the present article seeks to expose a few possibilities of its utilization in the teaching area.

4   The Game

The idea of the game was to produce a platform that promotes both leisure and education, as well as propagation of the Native American culture of the people of Alaska, known as the Iñupiat. Developed in 2015, by Upper One Games, the game

Never Alone, or Kisima Innitchuna, was produced to many diverse platforms such as the following: Steam, App Store, Mac App Store, GOG and Google Play. Upper Games is the first North American company to belong to Alaskan native people.

According to its original idea, the developers, when elaborating the art and the game's story, interviewed forty elders of the Iñupiat community, as well as artists, story-tellers and other members of the local group so that the same could contribute to the making and elaboration of the concepts. The result was a unique game, with a faithful representation of that people's culture, both because of the art used and because of its adapted story.

4.1   The Plot

In Never Alone, we play as a young girl named Numa and an arctic fox while we search for the cause of an endless storm that envelopes the girl's village and its adjacent lands. During the game, the user is present to the famous folkloric characters and to an inspiring ambience both cruel and beautiful, which is faithful to the typical Alaskan weather, where temperatures reach -20 °C.

Play as young Iñupiat girl and an arctic fox as they set out to find the source of the eternal blizzard, which threatens the survival of everything they have ever known. Guide both characters in single player or play cooperatively with a friend or family member as you trek though frozen tundra, leap across treacherous ice floes, swim through underwater ice caverns, and face numerous enemies both strange and familiar in the journey to save the girl's village [18].

For the game story-telling a group of local story-tellers, from the Inupiat people, were hired, and it is necessary to be said that a large part of their culture is passed through in an oral way, making, therefore, the figure of the story-teller an essential part of their culture. The game is available in 16 languages, but the plot telling made during the gameplay is made only in their native language, a group of dialects called Inñupiaq. According to the Alaskn Native Language Center:

Iñupiaq is spoken by the Iñupiat on the Seward Peninsula, the Northwest Arctic and the North Slope of Alaska. It is closely related to other Inuit languages across the Arctic in Canada and Greenland. It is divided into two main dialects: Seward Peninsula and Northern Alaskan. About 2,144 speak the language, most of them elderly [10].

Inñupiaq is a language that faces extinction, with around 2,144 speakers, of which the majority are elders, media production that utilize this language are imperative so that we can conserve it an propagate it. This language is also present during the game in the section called "Ancient Wisdoms".

4.2   The Art

As said before, native artist was hired to make the graphic production of the game, besides that the developers were very attentive to have, as an referential material, that people's traditional artistic production, examples being paintings, ivory carved objects, clothing, drawings and sculptures (Fig. 1).

Fig. 1. Example of traditional art

This reference reflects itself not only in the character's design, but as well as in the ambience design of the game, be it through the clothing used by the main characters, be it in the way that the mythological creatures present in the game were drawn or even in symbols present (Fig. 2).

Fig. 2. Example of traditional clothing

During the cut scenes its notable, in a clearer way, the reference of the *scrimshaw* art, defined by Silveira (2016) [2] as the word used to designate the art of the carving of ivory and sperm whale jaw bones and it an intrinsic part of the Iñupiat culture, making this one of the main forms of folkloric story-telling (Figs. 3 and 4).

Fig. 3. Example of traditional scrimshaw

Fig. 4. Example of the game's characters.

4.3   The Gameplay

Never Alone is not a game in which we have a life point, item-acquiring or monetary system. There is, however, the possibility of Game Over.

The game is categorized as an atmospheric Puzzle-Game, it possesses few commands, being them simple and easy to be memorized, which constitutes one of the basic principles of educative game design stablished by Nielsen (1994) [12]. The playability is simple and intuitive, which helps the gamer to absorb intensively the atmosphere and ambience of the game and promotes social interaction, as the game can be single-played or played in pairs.

Progressing with the plot, the gamer unlocks a tool in the menu called "Ancient Wisdoms". Working as a kind of production backstage, the "Wisdoms" shows the player the development of the game, with stretches about the artistic and sociocultural aspects of the videogame, for example: the animation process, the character's design and soundtrack making (artistic aspects) and interviews in the native tongue of the Inñupiaq, where the player is exposed to the folklore, the classic tale used as a basis to the game's plot, the traditional art, the language, the community, the way of life of that people and much more (sociocultural aspects). There are more than thirty minutes of video, in 28 videos that share the knowledge and culture, all of which was passed through generation to generation in the community, orally or via scrimshaw (Fig. 5).

Fig. 5. Image of one of the Ancient Wisdoms

The videogame is very well made both in the plot, in the gameplay and also in the graphic part, making it not only an educative game, but interactive and dynamic, which is necessary in the area of educational games.

5   Conclusion

The area of educative game design is a growing presence in the academic world; its main challenge is to develop good quality games that promotes good leisure time as well as education. In what concerns the teaching through culture and its relation to digital games, Never Alone is a modern example of an educative game that, when seen as a design case study, serves to the purpose of showing what are basic qualities that a game must have so that it its considered educational and what are the characteristics of a functional digital game, that propagates culture of a remote community in a respectful, educative and dynamic way.

Therefore, we believe this game to be one good example as to how this media can be explored and used to educative means and to the transmission of information so subtle and so important to the preservation of the memory of different people, such as culture. In this sense, after the perspective here exposed, we believe that game design has a lot to contribute to a positive insertion of videogames in the academic and educational area.

References

1. Filho, A.C.M., Oliveira, M.P.: Folclore E Videogame: Jogando, Aprendendo E Valorizando A Cultura Popular. Curitiba: Intercom - Sociedade Brasileira De Estudos Interdisciplinares Da Comunicação 40° Congresso Brasileiro De Ciências Da Comunicação (2017)
2. Silveira, G.C.: Jogos Digitais Como Ferramenta Cultural: Uma Proposta Interdisciplinar. In: SBC - Proceedings Of Sbgames, São Paulo (2016)
3. Conditt, J.: Never Alone Review: Into The Storm (2014). https://www.engadget.com/2014/12/10/Never-Alone-Review-Into-The-Storm/
4. Dalla, M.I.H.Z.: Planejamento: Análises Menos Convencionais. Porto Alegre (2000)

5. Fortuna, T.R.: Sala De Aula É Lugar De Brincar? Porto Alegre. Cadernos De Educação Básica **6**, 147–164 (2000)
6. Geraldo, A.F., Carneiro, N.P.: A Cultura No Processo Do Ensino E Aprendizagem Da Educação Infantil. (2000)
7. Hsiao, H.C.: A brief review of digital games and learning. In: The First IEEE International Workshop On Digital Game And Intelligent Toy Enhanced Learning, Los Alamitos, CA, USA, DIGITEL 2007, pp 124—129. IEEE Computer Society (2007) http://doi. Ieeecomputersociety.Org/10.1109/DIGITEL.2007.3
8. Dias, T.J., et al.: Estudo Sobre Os Diversos Gêneros De Jogos E Sua Aplicabilidade No Ensino. Vila Jacuí (2014). http://gestaouniversitaria.com.br/artigos/estudo-sobre-os-diversos-generos-de-jogos-e-sua-aplicabilidade-no-ensino
9. Kirriemuir, J., Mcfarlane, A.: Literature Review In Games And Learning. Futurelab, Bristol (2004). http://www.Futurelab.Org.Uk/Resources/Publications_Reports_Articles/Literature_Reviews/Literature_Review378
10. Krauss, M: Alaskan Native Languages (2011). https://www.alaskanativelanguages.org/
11. Fardo, M.L.: A Gamificação Aplicada Em Ambientes De Aprendizagem. CINTED-UFRGS, Santa Catarina (2008)
12. Nielsen, J., Loranger, H.: Usabilidade Na Web. Elsevier Brasil (2007)
13. Orrico, A.: Mercado Brasileiro De Games Já O Quarto Maior Do Mundo E Deve Continuar A Crescer. O Estado De São Paulo, São Paulo (2012). http://www1.Folha.Uol. Com.Br/Tec/1165034-Mercado-Brasileiro-De-Games-Ja-E-O-Quarto-Maior-Do-Mundo-E-Deve-Continuar-A-Crescer.Shtml
14. Oliveira, M.P.; Filho, A.C.M.: Folk-Game: Como A Cultura Popular Pode Ser Adaptada Para Os Jogos Eletrônicos De Computador - O Caso De Never Alone. Ponta Grossa: RIF Artigos/Ensaios, pp. 87–103 (2017)
15. Prensky, M.: Digital Game-Based Learning. Mcgraw-Hill, New York (2001)
16. Savi, R.: Jogos Digitais Educacionais: Benefícios E Desafios. CINTED-UFRGS, Santa Catarina (2008)
17. Tavares, R.: Fundamentos De Game Design Para Educadores. Salvador, Bahia (2005)
18. Upper One Games, LLC.: A Great Story Has The Ability To Move The Audicience In Many Ways (2014). http://www.Upperonegames.Citci.Org/
19. Upper One Games, Never Alone. Inspirational Art (2016). http://Neveralonegame.Com/Inspirational-Artwork/
20. Upper One Games, Never Alone. Never Alone (2016). http://Neveralonegame.Com/Game/
21. ECK, R.V.: Digital game based learning: it's not just the digital native who are restless. Educause Rev., pp. 16–30 (2006). http://Net.Educause.Edu/Ir/Library/Pdf/ERM0620.Pdf

The Design of a WebGL-Based 3D Virtual Roaming System for the "Batu Hitam" Shipwreck

Gantian Bian and Baosheng Wang[⊠]

School of Design, Hunan University, Changsha, Hunan, China
gantianbian@foxmail.com, walterwang840217@gmail.com

Abstract. This paper details a practical project which we developed to explore the approaches to designing for the storytelling of the virtual reality systems related to cultural heritage. We reconstructed the "Batu Hitam" shipwreck and the cultural items it carried. Then we designed a WebGL-based 3D virtual roaming system applied to PCs. Users can interact with the ship (zoom in/out, rotate, and explode) in this system or board the ship and walk around. For the fluency of storytelling, we designed and rationally arranged nine virtual scenarios composed of animated virtual characters and cultural items, providing an ideal path for visitors. We also use multimedia (text, image, audio, and video) to provide immersive experiences. The system shows the history and structure of the ship as well as life on board. This work suggests that interactivity, fluency, and immersion contribute to the storytelling of the virtual reality systems related to cultural heritage.

Keywords: Virtual reality · Cultural heritage · WebGL · Storytelling · Interactivity · Fluency · Immersion

1 Introduction

Cultural heritage is the masterpieces of human beings with outstanding value. However, due to natural degradation and man-made destruction, most cultural heritage is isolated from the general public in terms of time, space, and cultural sensitivity [1]. Thus, we need to provide accessible cultural heritages and introduce them to laymen in an easy-to-use way [2].

Fortunately, the development of modern photogrammetric modeling techniques has opened up many possibilities for the preservation, research, and reconstruction of these damaged or even lost cultural relics without causing further damage to them [3]. Virtual reality technique enables people to move and navigate in a virtual environment related to cultural heritage, creating a platform for people to learn and develop their understandings about the heritage [4]. The development of WebGL provides easy access to cultural heritage from anywhere in the world without the need for plug-ins [5].

Recently, the digitization of cultural heritage has mainly focused on modeling or scanning reconstruction. Many historical items and sites have been given new digital existence. The utilization of virtual reality technique can also make it possible to

© Springer Nature Switzerland AG 2020
T. Ahram (Ed.): AHFE 2019, AISC 973, pp. 224–232, 2020.
https://doi.org/10.1007/978-3-030-20476-1_23

examine in depth, perceive details, and even interact with these reconstructed historical items and sites [6, 7]. Although some projects have already started to integrate storytelling into the digitization of cultural heritage [8], there are still no systematic approaches to designing for the storytelling of cultural heritage digitization. In addition, 3D reconstructions of ships for historical purposes are quite rare, with most being static, such as the reconstruction of Pepper Wreck [9].

The vacancy of similar works motivated us to initiate this project. We reconstructed the "Batu Hitam" shipwreck, including the people and items it carried. Then, we designed a WebGL-based 3D virtual roaming system focusing on the interactivity, fluency, and immersion aspects of storytelling. Our purpose is to convey the historical information about the "Batu Hitam" shipwreck to the users so as to provide them with an understanding of the life on board. The benefits of this system lie in its accessibility to large audiences and the appealing quality of the diffusion of cultural heritage.

2   Historical Context

The "Batu Hitam" is a Tang Dynasty Shipwreck found in the waters off Belitung Island, Indonesia in 1998. Based on the timber species, construction methods, and the hull form of the excavated wreckage, experts speculated that the ship was an Arab Dhow.

The shipwreck is approximately 21 feet (6.4 m) wide and 58 feet (18 m) long. It carried a commercial quantity of Tang dynasty ceramics as well as rare items of imperial quality. They are more than 67,000 pieces of ceramics from different kilns, including the Changsha kiln, Yue kiln, Xing kiln, and Gongxian kiln. There are also some gold and silver wares and bronze mirrors [10]. These cargoes are large in number and well preserved and are of great research value. From the Changsha kiln ceramics with Arabian-inspired patterns and the rare items of imperial quality, it can be inferred that the original mission of the ship was to transport Chinese goods from Southeast Asia to West Asia and North Africa for trade or diplomacy; unfortunately, it struck against the rocks and sank.

The discovery of the "Batu Hitam" shipwreck is an empirical study of direct maritime commerce between China in the Tang Dynasty and West Asia (covering Turkey, Arabia, and other Gulf regions). The reconstruction of the "Batu Hitam" and its cultural items will enable both professionals and non-professionals to gain a deeper understanding of shipbuilding and handcraft techniques from the Tang dynasty, as well as the cultural exchange between the Western and Eastern world through the Maritime Silk Road.

3   3D Modeling

3.1   Survey and Modelling

According to the interpretative and reconstructive hypotheses supported by critical study of the shipwreck site, bibliographic and iconography references, and typological comparisons with similar Arab dhows of the same period [11], we used 3D Studio Max to reconstruct the hull and the internal structure of the shipwreck directly at 1:1 scale. The ship space was laid out reasonably, which can help realize the needs for boarding and roaming.

To show more details, some representative life appliances and precious cultural items have also been modeled, including typical Changsha kiln ceramics, precious gold and silverware, and bronze mirrors. Considering that we need more accurate and realistic models, these cultural items were modeled by photogrammetry. The original cultural items were supplied by a local collector.

All of the selected cultural items (more than 40) were captured with a Nikon D90 camera. Because of the short camera-object distances, we used small apertures (f20) to maintain a proper focal length. To ensure that the adjacent photographs have an overlapping area of 70%, the modelling of each cultural item required over 60 photographs taken at four angles (15°, 45°, 75°, 90°) from eight cardinal directions. In case of insufficient light, we used three professional illuminators to maintain a homogeneous light and texture of the object, avoiding shadows and reflections (Fig. 1). The photographs of the selected cultural items were processed by a community called Altizure for realistic 3D modeling.

Fig. 1. On-site photogrammetry

3.2 Simplification

Because memory is a constraint on the complexity of the content that a system can run the simplification of these models requires a trade-off between resolution and rendering speed. Considering the number of vertices that is acceptable to ensure a good rendering speed, we decided to use a multiresolution scheme for different presentation needs of the models to allocate memory reasonably. The reconstructed "Batu Hitam" wreck, for example, has a complex internal structure consisting of 1,260,000 vertices, the invisible part of which can be simplified to the greatest extent. Therefore, the final shipwreck model that was imported into the virtual roaming system has only 180,000 vertices. However, since cultural items require a detailed presentation they are still maintained at a high resolution. All of the high-resolution models obtained through manual and photogrammetric modeling were simplified by Blender, an open source software.

4   System Design

4.1   Introduction

The primary task of this project is to provide easy access to the "Batu Hitam" ship-wreck from anywhere in the world. Therefore, we designed a virtual roaming system based on WebGL, which makes it possible to run the system on the localhost or on web browsers (Mozilla Firefox 4+, Google Chrome 9+, Safari 5.1+, Opera Next and Internet Explorer9+) without needing plug-ins. The virtual roaming system was developed by Unity 3D, a multi-platform integrated game development tool. Its WebGL build option allows us to publish Unity content on a web browser using the WebGL rendering API.

In the virtual roaming system, the ship is floating on the ocean and has a certain depth of draught. Ambient lighting comes from the natural light of the sun and the sky is clear. The "Batu Hitam" ship has two floors, the deck and the cabin. The hatch is located in the middle of the deck. It provides convenient access to cabin areas by stairs. The cabin is used to hold cargos and provide living space. The definition of space and all its components is determined by their functionality.

Fig. 2. Four interfaces that appear after clicking each button

There are four buttons in the upper-left corner of the interface: Help, Panorama Mode, Roaming Mode, and Exploded View. In the panorama mode, users can view the model of the "Batu Hitam" in the simulated ocean environment. In the roaming mode, they can board the ship and walk around from a first-person view. The two mode buttons provide free switching between the panoramic mode and roaming mode; the

Help button provides the system operation instructions that can be viewed at any time through a 2D pop-up panel; the Exploded View button can clearly show the structure of the ship (Fig. 2).

4.2   Storytelling

Stories provide a framework for understanding events and conveying cultural values [12]. Information conveyed by storytelling can be more easily absorbed by users [13]. The "Batu Hitam" shipwreck contains a wealth of information. To spread this information in a more attractive and innovative way and to bridge cultural, time and space divides, our design focused on the interactivity, fluency, and immersion aspects of storytelling.

Interactions. Interactions run through the whole storytelling process. In panorama mode, users can learn the details of the "Batu Hitam" ship through the mouse. Specifically, they can use the scroll wheel to zoom in and out or drag the mouse to visualize the 3D model from a different point of view. They can also click the Exploded View button to separate the interior from its hull and observe the structure of the ship. Through these interactions, users can gain a whole image of the ship. Considering the complexity of the ship and its surroundings, we chose to fix all of the objects in the center of the view. Then, we set up a free look camera that can move freely and controlled the rotation angle of the camera to achieve the free rotation of the objects. Zooming in/out of the objects can also be achieved in this way.

In the roaming mode, visitors can board the ship and walk around. This helps them establish a sense of space. The roaming mode is achieved by a first-person controller, which allows users to control the view and navigation. The interactions are completed through the arrow keys. Visitors can use the up and down arrow keys to move forward or backward, or the left and right arrow keys to turn left or right.

Fluency. Any story is inseparable from the characters and a coherent plot. Because the space in the "Batu Hitam" ship is very complex and articulated, to convey the historical information completely, we designed nine virtual scenarios of life on the ship, such as making food, drinking tea, the sailors playing dice for entertainment, and so on. Virtual characters are core elements of every scenario, including Arab sailors, merchants, and a Chinese man who might be a merchant or diplomat. They wear costumes of that period. With clues of people's life on board, all of the reconstructed cultural items were reasonably distributed in the nine scenarios. Through these scenarios, we initially established the character image and the plot.

The purpose of the storytelling is to make space lively, as the virtual reconstruction that is rendered by static objects can only provide a limited user experience [14]. To solve this problem, we introduced animation into this system to provide simulations of the daily activities of virtual characters. We created an invisible, bone-like structure (skeleton) inside the characters and linked it to the external mesh (skin) using the open-source software Blender. The animation of the skeleton can thus induce the corresponding deformation of the skin. This allows the virtual characters to not only present a realistic appearance but also have a vivid figure. Their body language can convey more information to users and facilitate the development of the plot. These actions include giving directions, moving, and using tools, and so on.

For example, one of the Arab sailors is rolling a roller on a grindstone to make food on the deck. A variety of other stone utensils are scattered on the table, including a greater in the form of a fish, a mortar, and a pestle. There is also a wooden rolling pin. In the cabin, two sailors are checking the cargos (Fig. 3). Users who experience this method of storytelling in different scenarios can better understand the shapes and usage of various items.

Fig. 3. Making food by special utensils and checking cargos

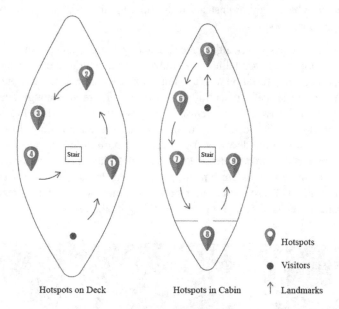

Fig. 4. The hotspot distribution map

In the specific context of the virtual roaming system, the mobility of the users challenges the fluency of storytelling. As we have mentioned, the space in the ship is very complex and articulated. More importantly, the nine virtual scenarios are laid out across the whole space of the ship. To ensure the integrity of the plot, we decided to guide the visitors along an ideal path. In this way, the plot is synchronized with the visitor's path, which will form a consistent overall storytelling. We placed an eye-catching red hotspot in each scenario and numbered them. Combined with the arrow-shaped landmarks attached to the ground, visitors can easily conduct a complete visit (Fig. 4). There is also an Arab sailor standing beside the hatch to show users the direction of the stairs, which provide access to the cabin areas. It should be noted that we needed to set the appropriate motion speed to ensure a smooth experience and the coherence of the plot.

Another factor that limits the fluency of storytelling is the performance of the system. To solve this problem, we need to optimize the memory and performance in terms of design and technology. From the perspective of design, on the basis of ensuring the integrity of the plot, reasonable selection and allocation of characters and items in each scenario as well as the reduction of unnecessary items can save memory to a large extent. From the perspective of technology, in addition to model optimization, another effective way to reduce the memory footprint is texture compression and optimization. We can use 16-bit compressed textures instead of 32-bit textures to release most of the memory bandwidth. Generating mipmaps for textures used in a virtual environment can also be useful. It enables the smaller triangles to get a lower resolution texture. This step is fundamental to speed up loading time, reduce memory footprint, and significantly improve rendering performance. As for lighting performance, we chose bake lighting instead of dynamic lighting, which is a simple trick to reduce the memory footprint.

Immersion. On the basis of boarding the ship and walking around, in order to enhance the immersion aspect of storytelling, we utilized multimedia to create experience.

An important manifestation is the simulation of the marine environment. The infinite ocean plane is realized with a projected grid. There are enough details in the mesh for small scale waves. Then, we used the open source ocean shader to achieve the animation effect of the waves. We also provided a seamless transition from above to below the water with an underwater fog effect. The purpose of this work is to ensure that the ocean surface looks as realistic as possible. There are sounds of waves and seagulls. All of these sounds are binaural sounds and the time interval between each sound file is random. These sounds are naturally incorporated into the animation of the waves, drastically enhancing the immersive feeling of the users.

Fig. 5. The triggered images when approaching a hotspot

Another important aspect of the immersive design is embodied in each scenario. When a visitor gets close to the hotspot of a scenario, the image and audio with descriptive information of the scenario will instantly be triggered (Fig. 5). We also deployed an animated video in one of these scenarios to show the sinking of the "Batu Hitam". All of the triggering of the multimedia information are achieved by collision detection. The multimedia components are hidden until the visitor enters the trigger area. This design brings a multi-sensorial and immersive experience to the user.

5   Conclusions

In this paper, we have presented the approaches to designing for the storytelling of a WebGL-based 3D virtual roaming system, highlighting the three aspects of story-telling: interactivity, fluency, and immersion. Interactivity runs through the process of storytelling to establish a meaningful relationship between users and the system. It is manifested in many aspects, such as the interactive model display (rotation, zooming, and exploded view), the virtual roaming from a first-person view, and the exploded view. The fluency of storytelling depends on design and technology. Depending on the historical information to be conveyed, different design choices are to be made at the fluency stage of storytelling. In a virtual reality system for free exploration, the design and reasonable layout of scenarios contribute to initially establishing the characters and the plot. But the mobility of users challenges the fluency of storytelling. The best solution is to synchronize the storytelling with the visitor's path. As for the design of scenarios, appropriate animation elements are conducive to creating vivid characters and facilitating the development of the plot. For example, we can convey more information to users by simulating characters' daily activities. From the view of technology, we have underlined the most significant challenge to the fluency of sto-rytelling—memory and performance optimization. Here are some useful suggestions based on the work we have done. Firstly, a multiresolution scheme can be used to simplify complex 3D meshes or point clouds according to different display require-ments of different models. Secondly, texture compression and optimization is an effective way to reduce memory footprint. 16-bit textures over 32-bit textures can be used, as well as always enabling mipmaps for textures used in a virtual environment. Finally, we can use bake lighting instead of dynamic lighting to reduce load time. The immersion of storytelling depends on the combination of text, image, audio, video, and other media. This multimedia information can bring a multi-sensorial and immersive experience. For example, the animation effect of the waves combined with the sound of waves and seagulls can make the marine environment more real and give users a feeling of immersion. All in all, this work has provided a new paradigm for applying open source technologies to the diffusion of cultural heritage on the web.

There are still some deficiencies in our work, such as that the existing interactive mode can only allow users to observe but not participate in the activities. In the future, our work will focus on more flexible ways of interaction, such as steering the ship, pouring water, carrying goods, and so on.

References

1. Faye, L.: VR Cultural Heritage Sites: A Case Study on Creating Immersive Visual & Auditory Experience. Research Gate (2016)
2. Gonizzi Barsanti, S., Malatesta, S.G., Lella, F., Fanini, B., Sala, F.: The Winckelmann300 project: dissemination of culture with virtual reality at the Capitoline Museum in Rome. Arch. Photogramm. Remote Sens. Spatial Inf. Sci. **XLII-2**, 371–378 (2018)
3. Scianna, A., La Guardia, M., Dodero, E., Petacco, L.: 3D virtual CH interactive information systems for a smart web browsing experience for desktop PCs and mobile devices. Arch. Photogramm. Remote Sens. Spatial Inf. Sci. **XLII-2**, 1053–1059 (2018)
4. Kersten, T.P., Tschirschwitz, F., Deggim, S.: Development of a virtual museum including a 4D presentation of building history in virtual reality. Int. Arch. Photogramm. Remote Sens. Spatial Inf. Sci. **XLII-2**, 361–367 (2017)
5. Parthenios, P., Yiannoudes, S., Oikonomou, A., Mallouchou, F., Ragina, L., Christaki, A.: Using WebGL to design an interactive 3D platform for the main monuments of crete. In: First CAA GR Conference, pp. 206–212. Ubiquity Press, London (2014)
6. Carlo, I., Alfonso, I.: Conservation, restoration, and analysis of architectural and archaeological heritage. IGI Global, Pennsylvania (2018)
7. Carrozzino, M., Bruno, N., Bergamasco, M.: Designing interaction metaphors for Web3D cultural dissemination. J. Cult. Herit. **14**(2), 146–155 (2013)
8. Pietroni, E., Forlani, M., Rufa, C.: Livia's villa reloaded: an example of re-use and update of a pre-existing virtual museum, following a novel approach in storytelling inside virtual reality environments. In: 2015 Digital Heritage, pp. 1104–1138. IEEE Press, New York (2015)
9. Wells, A.E.: Virtual Reconstruction of a Seventeenth Century Portuguese Nau, Master's Thesis, Texas A&M University, College Station (2008)
10. Krahl, R., Guy, J., Raby, J., Wilson, K.: Shipwrecked: Tang Treasures and Monsoon Winds Smithsonian Institution, Washington (2010)
11. Jeremy, G.: The archaeological contribute to the knowledge of the extra-european shipbuilding at the time of the medieval and modern Iberian-Atlantic tradition. In Proceedings, International Symposium on Archaeology of Medieval and Modern Ships of Iberian-Atlantic Tradition, pp. 69–76 (2001)
12. Bruner, J.: The narrative construction of reality. Crit. Inq. **18**(1), 121 (1991)
13. Gershon, N., Page, W.: What storytelling can do for information visualization. Commun ACM **44**(8), 31–37 (2001)
14. Barreau, J.B., Nouviale, F., Gaugn, R., Bernard, Y., Llinares, S., Gouranton, V.: An immersive virtual sailing on the 18th-century ship Le Boullongne. Presence Teleoperator Virtual Environ. **24**(3), 201–219 (2015)

Designing Experiences: A Virtual Reality Video Game to Enhance Immersion

Ricardo Navarro$^{(\boxtimes)}$, Vanessa Vega, Sergio Martinez,
María José Espinosa, Daniel Hidalgo, and Brian Benavente

Grupo Avatar PUCP, Pontificia Universidad Católica del Perú, Lima, Peru
{ricardo.navarro,vanessa.vega,smartinezp,
mjespinosa}@pucp.pe, dhidalgoc@uni.pe,
brianbenaventet@gmail.com

Abstract. The objective of the present study is to identify the influence of a virtual reality video game on the experience of immersion. To this end, a virtual reality video game was designed taken into account all the elements needed to the design of an immersive virtual reality video game, based on psychological evidence and immersion studies. The participants of the study were ten university students between 18 and 25 years' old who had never played a virtual reality video game. Semi-structured interviews were conducted, delving into the sensation of immersion experienced when playing the video game. The results are analyzed using the evidence collected on the concepts of immersion, motivation, and psychology. Recommendations for the design of virtual reality video games are proposed.

Keywords: Human factors · Video games · Immersion · Virtual reality · Motivation

1 Introduction

The use of technological tools (e.g., ICT or e-learning) has become a relatively common and important initiative in recent years [1, 2]. This interest in the use of new technologies originates from the benefits that its use causes, and that can influence the development of social and cognitive skills.

In line with the above, one of the technological developments that have attracted the most attention in recent years is video games. These have become a common product for people and society, which has led to video game companies looking for developers who can create better video games and thus be able to increase the audience of players for their latest releases [2].

The development of cognitive and social skills are benefits of video games that have caught the attention of researchers, which are key concepts that help explain the popularity of video games [3, 4]. Also, some studies report that an important element of a successful video game is the ability to involve the player with the elements and contents of the video game [5]. When that happens, video game players report to experience a sense of involvement or an experience of "being in the game", where they do not realize the things around them, such as the amount of time that has passed or if

© Springer Nature Switzerland AG 2020
T. Ahram (Ed.): AHFE 2019, AISC 973, pp. 233–242, 2020.
https://doi.org/10.1007/978-3-030-20476-1_24

someone is talking to them [2, 5, 6]. This experience of complete concentration in the game environment and a distraction-free self-absorption is known as immersion [5–7].

Therefore, immersion is an important aspect of the gaming experience [2], and it is necessary to understand its characteristics if you want to obtain insights on how to increase this experience in desirable contexts, such as in educational games, or how to inhibit it in dangerous contexts, as in cases of video game addiction [6].

Contrary to what would be desired, however, immersion is a complex term whose definition can vary depending on the author, which implies that various characteristics can be taken into account or not. However, the definition used in this study understand immersion as a state of cognitive, emotional and motivational involvement of the player with the elements of the video game [6, 9]. In this line, since the interaction with the video game generates an immersive experience, this can influence the cognitive processes of the player, which can affect the attention and perception of the player's reality [6, 9].

As for the possible degrees of immersion, a study by Brown and Cairns [5] indicates that there are different levels. The first is the Engagement, where there is a minimum immersion state, enough for the player to devote his time to playing a video game, although he does not necessarily enjoy it. The next level of immersion is the Engrossment, where there is an interest of the player to continue playing, but there is an emotional connection with the game and its elements (characters, narrative, history). The maximum level is Total Immersion, where there is a strong connection with the game, and the player "feels in the game." At this point, the attentional and perceptual processes are strongly altered and guided towards the elements of the video game.

Jennet et al. [6] identify five areas that would make up the immersion experience. The areas are the following: Cognitive Involvement, Emotional Involvement, Real World Dissociation, Control, and Challenge.

The first area, Cognitive Involvement, refers to the attention to the stimuli that are presented in the video game. Second, Emotional Involvement, is understood as the player's interest in continuing to play, as well as its relationship to the events that occur within the game. Real world Dissociation is understood as the alteration of the perception of time and space, a phenomenon that occurs when the player thinks he has played a short time but, in fact, it has happened much more time than it calculated. Fourth, Control is understood as the facility perceived in controls; that is if the controls are simple to use. Finally, Challenge refers to the perception of the difficulty of the video game; if the game is quite easy or difficult, it is likely that a fully immersive experience will not be given.

This immersion structure, proposed by Jennet et al. [6], allows to understand the state of immersion and allows to identify characteristics that can be applied to game design since it identifies essential aspects of immersion that can be materialized in the design of video games. For example, the feeling of Challenge that the player experiences must have a balance. If the player feels that the game is not challenging or too difficult, it may not generate a state of immersion. Likewise, if the reward obtained after passing a situation in the game does not go according to the difficulty, then the feeling of immersion can also be affected. It is here that the link between immersion theory and game design becomes a field of study.

There are several approaches to the game design that provide a procedural framework to develop entertaining and gameful experiences [10, 11]. These approaches present work methodologies aimed at enjoyment, immersion and the satisfaction of particular needs of the player (e.g., competence, autonomy, gregariousness) [3], which is essential in the development of video games.

In that line, it is important to delve into the characteristics of the elements of a video game that influence the player's immersion. While the classification varies depending on the author consulted, it can be found four essential characteristics that should be taken into account when designing a video game [10, 11]: narrative, game mechanics, art and rules.

The narrative is understood as the story in which the video game will be immersed [10]. The narrative should not be understood as solely textual; on the contrary, the narrative can be developed from the art of the video game, the mechanics and the sections in which a story can be explicitly presented in a visual, oral or both.

On the other hand, the mechanics refer to the actions that the player can perform within the game [10]. That is if the character can jump, swim, or fly, if he has powers or uses weapons, what kind of weapons he uses. Likewise, the mechanics refer to all possible actions that the player can perform within the game. This element is one of the most important aspects of Game Design in general, so its role in the development of an educational video game is crucial. The mechanics of the video game, like all the mentioned aspects, must contribute to the learning objective [11], although not necessarily directly. For example, Navarro et al. [11] used the classic mechanics of a platform video game; however, they adapted them to promote specific objectives for their research (oriented to learning). Thus, a classic mechanics such as picking up an object on the road became significant for the design of the video game, giving it a different meaning but one that responded to the needs of the studio. It is important to highlight that the mechanics must be designed to achieve the game's objectives (what I want the player to be able to do within the game).

Art is a visual component that provides visual information. These must disclose essential aspects for the player, such as the theme of the game, the relevance of some aspects within the game to the player, or "feel" that they want to generate in the player.

The rules make up the universe of the video game and delimit the actions of the player. Some authors refer that the rules delimit the mechanics that a player can perform within the game, and in the way that the game has been programmed. For example, the player can jump within the game but cannot fly and can only jump in a certain way and a certain height. Thus, rules can be understood as the structure of the video game.

Currently, game developers have studies focused on understanding the characteristics of mechanics [12], narrative [13], art [14] and the structures and rules of games [10]. These studies work as an important conceptual framework that allows improving playful experiences (such as immersion). However, we still need to delve into the relationship that specific designs have in the player's experience (particularly in immersion).

Since immersion is an experience that involves both cognitive, emotional and motivational activation with the content of the video game [6, 11], there are factors associated with autonomy and competence that generate a level of greater involvement

[3, 15]. In that sense, it must be understood how such involvement can be generated from the elements of the video game and how it is based on the degree of autonomy and competence experienced by the player. Although there are studies that approach this type of phenomena with video games [13, 16], there are still no studies focused on virtual reality video games. This is extremely important because it is an emerging technology that is becoming increasingly common. Its long-term sustenance will require the creation of the most optimal experiences possible, with mechanics, controls, narrative and clear objectives that can be understood by the player and, in turn, generate immersion.

Therefore, the objective of this study is to identify the sensation of immersion experienced when using a virtual reality video game. To do this, a video game was designed in virtual reality, taking into account that the movements made by the player are as intuitive as possible.

2   Method

2.1   Participants

The participants were four university students whose ages ranged between 20 and 23 years. Of them, two were men and two women.

None of the participants had used a virtual reality video game nor was familiar with the hardware used in the study. This is important since the perception of this type of situations responds to a phenomenological methodology of the experience. Also, the participants had little experience with video games.

2.2   Materials

Video game "The Submarine": A video game was designed and developed in virtual reality to identify the immersion in the participants. Within the development before the creation of the project, a study of similar experiences within the field was carried out: experiences such as "Job Simulator," "Expect you to Die," among others, were explored. Interviews were conducted with individuals who tried such video games in situ to gather their reactions and opinions. Based on this information, the elements of interest for the user were defined. Due to the limitations of the resources, a basic experience was proposed that responds to the objectives of the study. In the initial interviews, the participants indicated that they were interested in exploring unknown places. This is how the idea of exploring with a submarine originates (the theme of the submarine was also identified in the initial interviews). Thanks to the definition of the main idea, an initial physical prototype was developed, whose use was focused on testing the proposed mechanics. In this way, the prototype was validated through continuous iteration and meticulous analysis of the feedback obtained. The video game takes place in the cockpit of a small submarine, where the player receives relevant information about the game's objectives through the voice of an NPC. The player must search and photograph fish from the bottom of the sea, and his partner (the voice that gives the information) is talking about personal aspects of the player and their objectives. There are two basic mechanics: move the submarine and take pictures. The first mechanics required that the player moved a lever that was in front of him. The second mechanism was performed by pressing a button on a camera inside the cab. The experience lasts 5 min and ends when a giant fish devours the player.

- Semi-structured interview. An interview was designed that allowed to deepen on the participant's perception of the video game. Because we wanted to delve into the effectiveness of some aspects of it, the interview had the following areas:
- Immersion: the questions asked addressed the participant's perception of immersion. In that sense, the questions revolved around the impression of the time the participant had played (which differed over the actual playing time), the feeling of being in the game, and the identification of elements (striking and non-striking) within the game.
- Rules: for this area questions were asked about what the player believed were the basic rules of the game and particularly if he had understood the objective of the game.
- Mechanics: this area referred to the actions that the participant could perform within the game. Thus, he delved into the movements that the player could make and the elements within the game with which he could interact.
- Narrative: Questions concerning the narrative of the video game were asked to the participants. The purpose of these questions was to identify if the video game allowed the player to understand the history of the game, as well as the characters who appeared in it.
- Art: Questions concerning the art of the video game were asked, and if it supported the objective and mechanics of the game; also, if it was articulated with the other elements of the video game.

2.3   Procedure

We proceeded to separate a suitable environment to install a computer with the video game in virtual reality, as well as the HTC Vive components. Thus, a space of 3×5 m was used, where a chair was placed at a prudent distance so that the sensors could identify the controls of the HTC Vive and the movements of the player. The position where the chair was placed was marked so that participants are in the same space in all applications.

An open call was made to university students, indicating the place and time where they should attend to participate in the study. At the beginning of the intervention, the participant was helped to place the HTC Vive viewer. They were given necessary information on how to use the controls and where to position themselves. Finally, they were seated in a chair and told to follow the goal that the game would point to at the beginning. The application lasted 5–7 min. Afterward, a semi-structured interview was conducted about what happened in the video game.

In the following section, the results obtained are presented and discussed.

3   Results and Discussion

The results of the interviews with the participants are presented below. The first analysis that is carried out in this section is the one concerning the participants' sensation of immersion. Thus, first we will delve into the details perceived by the participants about the game and how they relate to immersion; then, it will delve into the game elements that could generate immersion in the participants. Finally, the limitations of the game that could have an adverse effect on the feeling of Total Immersion are mentioned.

3.1   Immersion

The participants were able to identify the details of the game (music, art, narration) from the beginning. Likewise, signs of the first and second level of immersion were identified in the responses of the participants. This implies that specific characteristics of the game allowed to submerge the participant in the world that was presented to him; however, there was no evidence of a deep level of immersion, or Total Immersion, although it was reported that the game generated some similarity with the real world (which could be a determining factor of immersion). This can be seen in the following quotes from the interviews:

> "I felt that I was turning to the left and there was sea, to the right and there was sea, up and there was the submarine's hull, there were also sounds, those little bubbles that make you feel (...) The narrator's voice was very clear, maybe with more noise in between." (participant 1, female, 22 years)
> "The environment, the depth, the sea, the bubbles, the seabed, inside the submarine was the radar, almost everything conditioned as if it were like that. (...) The sound also (...) the atmosphere, the music, it felt real." (participant 2, male, 21 years)
> "The sound, the realistic images helped you feel under the sea (...) Elements of the submarine not particularly (...) When you are under the water it feels like an echo, and it looked very well reproduced by the game." (participant 3, male, 23 years)

This information accounts for a considerable degree of immersion, as is engrossment [6]. Thus, there is a relationship between the player and the video game, generating interest in continuing to play and, at the same time, a certain degree of enjoyment.

While there is no evidence of a higher degree of immersion, which could generate Total Immersion, evidence was found that the game can generate strong emotions, such as fear. This may give indications that the player could be able to experience Total Immersion:

> "I got scared with the big fish at the end, it was so sudden, they didn't give you a warning that it was close, I tried to move, but then there was an impact." (participant 1, female, 22 years)

Thus, the unexpected appearance of a threatening stimulus within the game, and the player's inability to do anything generated a particular emotional response, in this case fear. This could be because levels of emotional and cognitive involvement were high, which generated a stronger sense of immersion and, consequently, the generation of strong emotional responses [2, 6].

Based on the previous evidence, it is essential to analyze these results from the perspective of game design: how mechanics and narrative could have influenced the generation of immersion, and what characteristics of these game elements were most beneficial to achieve this.

3.2 Game Elements

Although the game was new and was not related to any previous experience that the participants had had, the rules were understood.

> *"The objective was to find a new species of fish and take a picture."* (participant 3, male, 23 years)
> *"The objective was to find a specific fish and take a picture of it. (...) But I did not know how to locate that fish, how to look for it."* (participant 4, female, 20 years)

Even the rules were understood easily: although one of the participants did not pay attention to the narration - where the objective of the game was indicated - he was able to understand what he should do.

> *"I did not pay attention to what I said, but to drive, pay attention not to crash and keep the ship (...) I felt comfortable, I expected some things to appear that would help the environment more or be a guide."* (participant 2, male, 21 years)

This information could indicate that the game was intuitive enough for players to understand the objectives and, if necessary, deduct them at the time of play, which could be due to factors related to art and narrative.

> *"The game started with the submarine descending and I had to take a picture, then they told me there was a massive object, and that it was coming to me."* (participant 1, female, 22 years)
> *"Two friends were investigating, one was out of the water, they were talking about their life and that (...) It was not a big story, but it was interesting."* (participant 3, male, 23 years)
> *"The little information I had was that I was in a submarine, with a person who was in a plant, I might be a biologist or something, looking for a species."* (participant 4, female, 20 years)

Visual details (buttons, stickers, photography) were received positively, as they supported the submarine's environment. That is why all the participants managed to recognize the space where they were, which could help them to perform actions within the game in a more fluid way.

Likewise, it was observed that although some participants had a limited understanding of the game narrative, this seems to have not affected the interest that the game generated in them. However, this could be due to the interest people had in experiencing virtual reality.

However, there were negative aspects that may have affected the level of immersion of the participants. Although, the design of the game took into consideration to facilitate the movements and actions of the players, there were drawbacks with some game mechanics.

> *"I felt that control of the lever was sometimes odd, in the same game I had more control, but I felt that the lever did not respond to my actions, to where I wanted to go."* (participant 1, female, 22 years)

"The controls were very slow (...) I felt that I was going very slowly, maybe a little faster, to improve the experience." (participant 2, male, 21 years)
"It was difficult to get used to controlling the submarine, but once you had it you could play." (participant 3, male, 23 years)
"There was little difficult to handle the lever to move. Despite moving the lever forward and wanting to go forward, I could not." (participant 4, female, 20 years)

The participants report negative emotions and ignorance in two aspects: the first was not being able to handle the main controls adequately, that is, the lever and the second was not to identify the other controls because they did not see them.

In the case of not being able to control the ship, the participants report that it is because they are not used to the controls of the VR game and the environment. For example, the players had to look around to see other screens and not just to the front, or to be able to direct the main lever with the VR controls. Due to the lack of time for exploration, or the absence of clear indications about game handling of the controls and the mechanics, participants often considered some controls to be part of the environment and not something they could use.

3.3 Immersion in VR Games and a Framework to Explain Enjoyment

One aspect to consider in this study is the importance of intrinsic motivation in the player, since it can generate a positive gaming experience and enjoyment [17]. The game elements can immediately give feedback to the player, either visually, through the narrative or even mechanically [10, 18]. This feedback allows to motivate certain behaviors of the player, which may be expected within the game. In order to generate the motivation of the player for certain behaviors, it is important to understand the specific basic needs of the human being, which must be satisfied. In studies with video games, Przybylski et al. [3] and Ryan et al. [15] point out the autonomy of the player as an essential aspect of intrinsic motivation. This assertion is based on the studies of Ryan and Deci [19], where it is mentioned that the more autonomy exists in behavior, the more intrinsic the motivation will be.

Likewise, it is also mentioned that the generation of intrinsic motivation in behavior leads to positive emotions in the person [20], which may be associated with immersion and enjoyment. That is to say that the more autonomous the behavior and sense of immersion in the player, the more positive emotions it can generate and enjoyment [3, 15, 17]. This evidence can be approached from the game elements mentioned before. If the game design takes into account the objectives of the game, the emotions, and behaviors that it wants to generate in the player, it must present them in such a way that the player feels autonomy when playing. Also, game designers must take into account the feedback given to the player through the game elements (mechanics, rules, art, and narrative), which allow the gaming experience to flow.

Based on the evidence collected, and the analysis performed, a framework of action can be established that describes the process by which players' immersion occurs (Fig. 1).

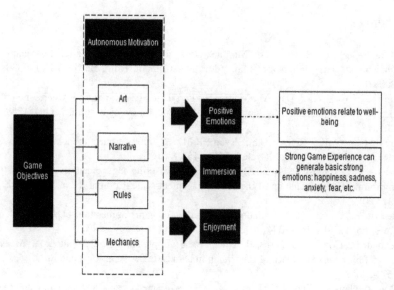

Fig. 1. Framework of game design for immersion, positive emotions and enjoyment

This framework would allow to understand the elements of the game that can influence the player's immersion in a virtual reality game. Further studies should be made to prove the validity of this framework and its relationship with enjoyment.

Also, this framework takes into account the importance of motivation in the experience of immersion and the game elements.

4 Conclusions

From the analysis, it is possible to point out that the video game has allowed participants to feel "immersed" in the game experience. Also, this sensation of immersion had led to experience strong emotions, such as fear. The participants reported having a slight degree of fear before the abrupt end of the video game. This can be explained by the existence of various graphics and audio features that favored the feeling of immersion mentioned by players, such as bubbles, screens, sound effects, among others.

The main drawback in the gaming experience and, therefore, in the immersion experience were related to the mechanics. The mechanics were well defined; however, the actions within the game presented some flaws that made handle the submarine a difficult task and, consequently, affected the participants' gaming experience.

In summary, the video game seems to positively influence the immersion of the players; however, the difficulties of handling some mechanics is a factor that reduces immersion. These characteristics should be considered when developing future VR games, because if it is taking into account in the development, it is possible to achieve higher levels of immersion. Finally, it is pertinent to include a tutorial or detailed guide that allow players to fully exploit the potential of this type of video game, especially if the gameplay time is limited.

References

1. Moreno-Ger, P., Burgos, D., Martínez-Ortiz, I., Sierra, J.L., Fernández-Manjón, B.: Educational game design for online education. Comput. Hum. Behav. **24**(6), 2530–2540 (2008)
2. Cairns, P., Cox, A.L., Day, M., Martin, H., Perryman, T.: Who but not where: the effect of social play on immersion in digital games. Int. J. Hum.-Comput. Stud. **71**(11), 1069–1077 (2013)
3. Przybylski, A.K., Rigby, C.S., Ryan, R.M.: A motivational model of video game engagement. Rev. General Psychol. **14**(2), 154 (2010)
4. Wiebe, E.N., Lamb, A., Hardy, M., Sharek, D.: Measuring engagement in video game-based environments: investigation of the user engagement scale. Comput. Hum. Behav. **32**, 123–132 (2014)
5. Brown, E., Cairns, P.: A grounded investigation of game immersion. In: CHI 2004. ACM Press, pp. 1279–1300 (2004)
6. Jennett, C., Cox, A.L., Cairns, P., Dhoparee, S., Epps, A., Tijs, T., Walton, A.: Measuring and defining the experience of immersion in games. Int. J. Hum.-Comput. Stud. **66**(9), 641–661 (2008)
7. Hou, J., Nam, Y., Peng, W., Lee, K.M.: Effects of screen size, viewing angle, and players' immersion tendencies on game experience. Comput. Hum. Behav. **28**(2), 617–623 (2012)
8. Shernoff, D.J.: Optimal Learning Environments to Promote Student Engagement. Springer, New York (2013)
9. Freina, L., Ott, M.: A literature review on immersive virtual reality in education: state of the art and perspectives. eLearn. Softw. Educ. (1) (2015)
10. Adams, E.: Fundamentals of Game Design. Pearson Education (2014)
11. Navarro, R., Zapata, C., Vega, V., Chiroque, E.: Video games, motivation and history of Peru: designing an educational game about Mariano Melgar. In: International Conference on Applied Human Factors and Ergonomics, pp. 283–293. Springer, Cham (2018)
12. Sicart, M.: Defining game mechanics. Game Stud. **8**(2) (2008)
13. Falcón Quintana, D.A.: La relación entre la inmersión en la narrativa de los Massively Multiplayer Online Role-Playing Games (MMORPGS) y las conductas pro-sociales dentro del videojuego (2018)
14. Castaño Díaz, C.M., Tungtjitcharoen, W.: Art video games: Ritual communication of feelings in the digital era. Games Cult. **10**(1), 3–34 (2015)
15. Ryan, R.M., Rigby, C.S., Przybylski, A.: The motivational pull of video games: a self-determination theory approach. Motiv. Emot. **30**(4), 344–360 (2006)
16. Cox, A.L., Cairns, P., Berthouze, N., Jennett, C.: The use of eyetracking for measuring immersion. In: CogSci 2006 Workshop: What Have Eye Movements Told us so Far, and What is Next (2006)
17. Birk, M.V., Atkins, C., Bowey, J.T., Mandryk, R.L.: Fostering intrinsic motivation through avatar identification in digital games. In: Proceedings of the 2016 CHI Conference on Human Factors in Computing Systems, pp. 2982–2995. ACM, May 2016
18. Mekler, E.D., Brühlmann, F., Tuch, A.N., Opwis, K.: Towards understanding the effects of individual gamification elements on intrinsic motivation and performance. Comput. Hum. Behav. **71**, 525–534 (2017)
19. Deci, E.L., Ryan, R.M.: The "what" and "why" of goal pursuits: human needs and the self-determination of behavior. Psychol. Inquiry **11**(4), 227–268 (2000)
20. Smit, B.W., Ryan, L.C., Nelson, C.A.: Does Autonomy Increase or Decrease Negative Emotional Displays From Service Workers? J. Pers. Psychol. (2016)

Logic Evaluation Through Game-Based Assessment

Carlos Arce-Lopera[⊠] and Alan Perea

Engineering Faculty, Universidad Icesi, Cl. 18 #122-135, 760031 Cali, Colombia
{caarce,alanperea}@icesi.edu.co

Abstract. Digital game–based evaluations may be useful to solve several problems of traditional paper-based assessments, such as students' test anxiety and evaluation rigidity. On the other hand, video games allow to record inter-action data of the thinking process that can be used later in formative assessments. A game application was developed as a tool for evaluating the logic abilities of first year university students. The game was designed as a puzzle with different difficulty levels. Experimental results showed that the game scores were not significantly different from the grades obtained with traditional paper-based evaluations. However, for most students, the game-based interaction was significantly different by lowering perceived frustration and increasing user engagement. The use of gamification on student assessment can lower test anxiety and reveal useful insights on student thinking processes. Moreover, automatic and real time feedback could drastically improve learning and guide students to understand complex scenarios.

Keywords: Human factors · Gamification · Assessment · Logic

1 Introduction

Game–based learning is linked to increase cognitive abilities and positive motivational impact in students [1]. In recent years, gamification as a tool for student evaluation have gained supporters as their use attempts to solve several problems of traditional paper-based assessments [2, 3]. One of the main problems that students face when evaluated is test anxiety. Indeed, anxiety during evaluation activities influence nega-ively student achievement and motivation [2, 4]. Moreover, the rigidity of standardized evaluation assumes that all students learn and behave uniformly. However, this assumption does not hold in most educational environments. Students learn and solve problems in different ways and at a different pace. If an evaluation is rigid, then it may hinder the motivation and the knowledge acquisition.

However, computer-based evaluations may also increase anxiety when not care-fully designed [5]. Therefore, careful design has to be taken into account when developing computer-based evaluation environments that foster motivation and student engagement [6].

One of the main advantages of using digital video games as an assessment tool is that they allow to record the interaction data. This data can be related to the thinking process of the student when confronted to each problem on the evaluation. Then, this

© Springer Nature Switzerland AG 2020
T. Ahram (Ed.): AHFE 2019, AISC 973, pp. 243–250, 2020.
https://doi.org/10.1007/978-3-030-20476-1_25

wealth of information can be used to design real-time feedback that could change the gameplay. The idea is to use each individual interaction data to maximize learning in a much more personalized and flexible way. Therefore, this approach helps to collect data as input to formative assessments and gain understanding and insight on whether students are learning. More importantly, teachers can develop personalized plans aiming to develop students′ skills. Finally, these processes can help students take control and become self-regulated learners [7] rather than having a reactive role regarding evaluations and learning.

2   Methods

2.1   Game Design

A game application was developed as a tool for evaluating the logic abilities of first year university students. The game was designed as a puzzle game with 3D graphics and seven different difficulty levels. The game was based on the push box principle, similar to the single-player combinatorial puzzle game Sokoban. The main objective was to arrange some objects into designated places with a set number of movements. This gameplay was selected as the rules were simple and visually intuitive; hence the game was very easy to understand. Moreover, the game was considered interesting and challenging which made it prone to increase the willingness to be played voluntarily.

Each level of difficulty provided different sets of problems increasing in complexity. Figure 1 shows a screenshot of one of the levels of the game. The game character was designed using the European mole, a small subterranean mammal, as inspiration. Therefore, the game story followed Norbert, the mole, in the search of the exit from a subterranean maze. The maze had seven underground layers. Each layer had a unique exit that was revealed only when some boxes are strategically placed on indicated places (see Fig. 2). Each layer had a set of barriers and passages that cannot be moved. The player moved Norbert, which can only push one box at a time, cannot pull a box, and cannot occupy the same location as a box or barrier. To guide the

Fig. 1. Screenshot of the game. The environment is dark and illuminated by torch light.

movement of the player a two-dimensional map was located at the left-bottom of the screen (see Fig. 3). This gameplay belonged to a family of motion planning problems with movable obstacles that is PSPACE-complete and it has been proved that the associated decision problem is NP-hard [8].

Fig. 2. Screenshot of the game visualizing the place where to put the objects.

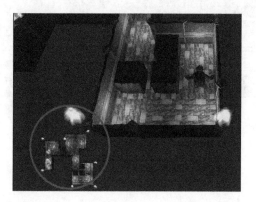

Fig. 3. Screenshot of the game showing the associated two-dimensional map that guided the player.

2.2   Subjects and Task

Volunteers participated in the evaluation of the game application. All were first year university students enrolled in the undergraduate course Formal Logic. All were naive to the purpose of the study and were instructed to play the game in a mobile phone with the game pre-installed. Participants played the game for 20 min. After the time elapsed, the game stopped. This time restriction was set to emulate the same time restrictions of short paper-based evaluations.

To be able to compare the game results with more traditional evaluations, two different paper-based tests were prepared. Both traditional tests were actual tests designed for the undergraduate course Formal Logic. One of the paper-based evaluations tested syllogism and the other abstract reasoning by association. Some examples of the paper-based evaluation are shown in Fig. 4 for syllogism and Fig. 5 for abstract reasoning.

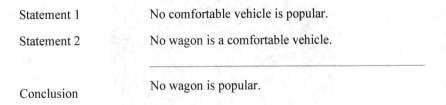

Statement 1	No comfortable vehicle is popular.
Statement 2	No wagon is a comfortable vehicle.
Conclusion	No wagon is popular.

Fig. 4. Example of syllogism. Students had to answer if the conclusion was true or false depending on the statements above.

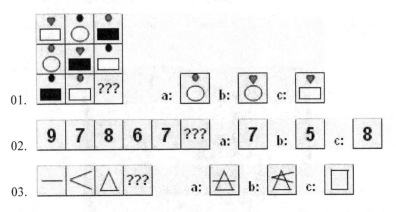

Fig. 5. Examples of the abstract reasoning questions on the paper-based evaluation. The student answered by selecting one option between a, b or c options.

2.3 Experimental Design

We designed and conducted randomized evaluations of the subject's performance for the three tasks: the game and the two paper-based tests. The NASA-TLX was used as a tool to assess the perceived workload for each test immediately afterwards. This multi-dimensional scale was selected because it is reasonably easy to use and reliably sensitive to experimentally important manipulations [9].

The NASA-TLX divide the cognitive load into six dimensions that are represented on a single page as one questionnaire. The six dimensions are: Mental demand, physical demand, temporal demand, performance, effort and frustration. All dimensions are rated using a visual scale with 100 points maximum in 5-points steps. Mental demand is the perception of the amount of mental and perceptual activity necessary to complete the task; was the task easy or demanding? Simple or complex? Physical demand is the perception of the amount of physical activity required; was the task physically demanding? Temporal demand is the perception of time during the task; was the pace slow or time pressure was felt when doing the task? Performance refers to the overall perception of success when performing the task; how satisfied were the user with their performance? Effort is described as the perception of vigorous attempt necessary to achieve their level of performance; how hard did they have to work (mentally and physically) to achieve the task? Finally, frustration refers to the perception of stress, irritation and annoy that the user felt during the task.

In addition to the NASA-TLX, for each paper-based test all answers were recorded with the associated grade. On the other hand, for the game application, all the players' movements with their associated response time were recorded for analysis. Therefore, each time a player moved the character using one of the arrows, the associated direction of movement and time was recorded.

3 Results

Experimental results showed that the puzzle game scores were not significantly different from the grades obtained with the paper-based test scores. The score of the game was set as the maximum level that the player cleared which is similar to the way the paper-based tests were scored. Moreover, Fig. 6 shows the results of the NASA-TLX for the three experimental conditions. Black bars show the results for the game-based assessment. Grey bars show the results for the abstract reasoning test and white bars for the syllogism test. Results show that on average there was not a significant difference between the three tasks in the six dimensions of the NASA-TLX. These results may indicate that the game-based interface was an adequate activity similar to the other two paper-based evaluations but without any significant advantage or disadvantage in terms of cognitive load. The only clear advantage may be that in the game-based interface the movements and response time of students are recorded. The data can be used to alert the instructor if the student need additional support or to design more specific tasks to reinforce learning or focus on specific levels.

However, after careful analysis, two different groups of students were clearly identified depending on their perceived frustration with the game interface. Figure 7 shows the results of the students that found the paper-based tests more frustrating than the game-based test. Figure 8 shows the results of the students that considered the game as more frustrating than the paper-based test. The group frustrated with the paper-based test where the largest with over 60% of the total number of students. Their frustration was particularly high for the abstract reasoning paper test (mean = 5.8, SD = 1.6). This frustration score was significantly different ($t(6) = -2.93$, $p < 0,05$) from the frustration perception of the game-based interaction(mean = 3.8, SD = 1.6). Moreover, for this group of students, all other five dimensions of the NASA-TLX and their grades for the three tests were not significantly different. This may indicate that, for this group, there was an advantage to use the game-based testing. Lower perceived frustration in this group may indicate increase user engagement which may include lower sensibility to anxiety during the interaction. Intriguingly, this group perceived the

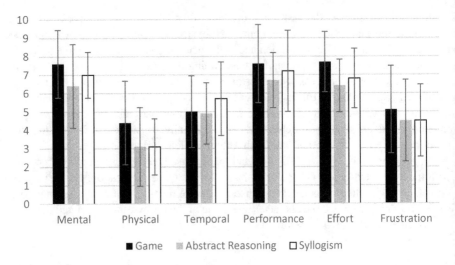

Fig. 6. Overall results for the NASA-TLX. Error bars denote the standard deviation of the results. No significant differences were found for the six dimensions.

test of abstract reasoning as highly frustrating which may indicate that they were not accustomed or not feel comfortable when doing the abstract reasoning type of evaluation. Further research is needed to understand the reason why this group reported more frustration when solving the abstract reasoning test than the syllogism test.

On the other hand, the group of students that considered that the game-based test was more frustrating (Mean = 7, SD = 2.16) rated both paper-based tests as significantly less frustrating than the game-based test ($p < 0.05$). Specifically, this group found the abstract reasoning test as the least frustrating (Mean = 1.6, SD = 1.29) with

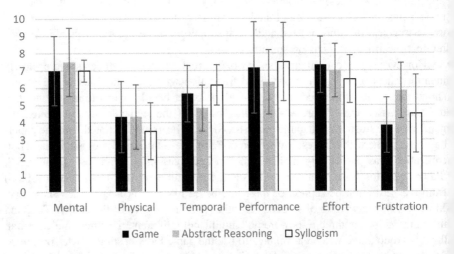

Fig. 7. Results for the NASA-TLX of students that found paper-based test frustrating. Error bars denote the standard deviation of the results.

significantly lower mental demand (t(3) = 7.83, p < 0,05) in comparison with the game interface. Also, this group rated significantly lower (t(3) = 2.48, p < 0.1) the effort necessary to complete the abstract reasoning task (Mean = 5.5, SD = 0.57) in comparison with the game-based test (Mean = 8.3, SD = 1.7). Interestingly, no significant differences were found on perceived performance or the real performance, i.e. the grades of the three types of test. These results showed that this smaller group of students may be reluctant to game-based interfaces by considering them frustrating and requiring more effort demanding more mental resources.

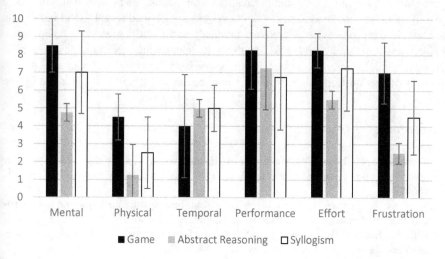

Fig. 8. Results for the NASA-TLX of students that found the game as frustrating. Error bars denote the standard deviation of the results.

To be able to further understand if we could predict the grades of the paper-based tests by using data interaction from the game interface, a linear regression was conducted using the events recorded by the system. The test on syllogism was predicted with a correlation coefficient of 0.97 by using only three variables from the game interface, namely the total score, the number of times the player pressed the push button and the time response. However, using the same technique, the abstract reasoning grade prediction showed only a correlation coefficient of 0.49. These results indicate that the interaction with the game interface was correlated with the results of the syllogism test. However, further research is needed to understand the difference between the interaction with the game and the results of the abstract reasoning.

One difference of the game interface and the paper-based tests was that players had a real-time feedback by the system, i.e. they were aware of the status of each response. This was not the case for the paper-based tests. Also, the game interface allowed multiple repetitions of the same level until it was cleared. These game characteristics allowed players to try different approaches to solve the problems. In the paper-based test only one trial was possible without real time feedback on performance. The advantage of giving real time feedback and the possibility to reset the interaction to the initial status of the game interface may provide the opportunity for students to develop different types of resolution strategies, such as try and error or even brute force. Further research is needed to clarify how this type of strategies can be related and associated to logic evaluation and learning.

4 Conclusion

We introduced a game-based evaluation for logic reasoning targeting university first year students, based on the game mechanics of the popular game Sokoban. The use of gamification on student assessment can lower test anxiety and reveal useful insights on student thinking processes. Moreover, automatic and real time feedback could drastically improve learning and guide students to understand more complex scenarios. However, careful interface design for the game-based tests must be guaranteed as to prevent frustration from students that are not accustomed to game dynamics and rules.

References

1. Connolly, T.M., Boyle, E.A., MacArthur, E., Hainey, T., Boyle, J.M.: A systematic literature review of empirical evidence on computer games and serious games. Comput. Educ. **59**, 661–686 (2012)
2. Mavridis, A., Tsiatsos, T.: Game-based assessment: investigating the impact on test anxiety and exam performance. J. Comput. Assist. Learn. **33**, 137–150 (2017)
3. Kiili, K., Ketamo, H.: Evaluating cognitive and affective outcomes of a digital game-based math test. IEEE Trans. Learn. Technol. **11**, 255–263 (2018)
4. Hancock, D.R.: Effects of test anxiety and evaluative threat on students' achievement and motivation. J. Educ. Res. **94**, 284–290 (2001)
5. Kolagari, S., Modanloo, M., Rahmati, R., Sabzi, Z., Ataee, A.J.: The effect of computer-based tests on nursing students' test anxiety: a quasi-experimental study. Acta Inform. Medica. **26**, 115–118 (2018)
6. Eseryel, D., Law, V., Ifenthaler, D., Ge1, X., Miller, R.: An investigation of the interrelationships between motivation, engagement, and complex problem solving in game based learning. J. Educ. Technol. Soc. **17**, 42–53 (2014)
7. Nicol, D.J., Macfarlane-Dick, D.: Formative assessment and self-regulated learning: a model and seven principles of good feedback practice. Stud. High. Educ. **31**, 199–218 (2006)
8. Dor, D., Zwick, U.: SOKOBAN and other motion planning problems. Comput. Geom. **13**, 215–228 (1999)
9. Hart, S.G.: Nasa-Task load index (NASA-TLX); 20 years later. In: Proceedings of the human factors and Ergonomics Society Annual Meeting, vol. 50, pp. 904–908 (2006)

Design of Virtual Public Resilience Spaces of Traditional Cultures

Jingjing He[1], Chaoxiang Yang[2(⊠)], and Michael Straeubig[3]

[1] School of Digital Art, Shanghai University, Shanghai, China
[2] East China University of Science and Technology, Shanghai, China
yangchaoxiang@qq.com
[3] University of Plymouth, Plymouth, UK

Abstract. China's contemporary society is fragile, mainly because of the lack of cohesion within the society or the low level of social resilience. This article is based on the topic of resilience (Manzini 2013), adapted to the design of virtual public space in the context of regional cultures. By analyzing the characteristics of the existing virtual works and resilience design, including the relationships between designers, users and virtual environments, we can explore the design theory of virtual public resilience space of traditional cultures and methods for cultural innovation. My research is based on concepts of local community, cultural resilience, participation in design and active design. It reorganizes and superimposes existing spaces while focusing on the analysis of activities and needs of the participants. With the design of narrative virtual environments and by imagining the possibilities of future design as theoretical and practical foundations, we can drive cultural innovation and achieve educational transformation.

Keywords: Digital traditional culture · Virtual cultural space ·
Virtual narrative environments · Resilience design · Regional culture

Since the 1990s the Chinese people's living patterns and social methods have undergone tremendous changes and people have been increasingly individualized and alienated. With the development of information network technology and the improvement of people's living standards "loose flexible and temporary social networks are gradually replacing our original communities which has led to an increasingly fragile social system. In the face of existing crises and contradictions the complexity of society is constantly increasing and contemporary society is fragile" Davies [1]. Zolli [2]. Taleb [3] "mainly because of the lack of cohesion within society or the low level of social resilience" Keck [4]. Therefore the "Cultural Resilience" (CoR) project of the University of the Arts of London (2014.01–2016.07) and the "Design for Resilience City" project (2015) of the School of Design and Creativity Tongji University Shanghai China all attempted to introduce the concept of "Resilience" into the construction of community culture. The project uses a variety of

Project Source: "Speculation and Evidence: Theory and Teaching Mode of Intelligent Design for Product Combined with Augmented Reality" (No.: C19100, Shanghai University). Education Research Project, Shanghai Municipal Education Commission, 2019.

© Springer Nature Switzerland AG 2020
T. Ahram (Ed.): AHFE 2019, AISC 973, pp. 251–260, 2020.
https://doi.org/10.1007/978-3-030-20476-1_26

methods to trigger greater participation such as community workshops student design interventions co-creation projects and public engagement activities. Mancini finally boils down to proactive design "creating meaningful forms of participation and a flexible sustainable way of living and working is the goal of community building" Ni Yuqing [5].

In August 2018 the author's project team conducted a public opinion survey and interview with residents of Jiading District Shanghai China, on the topic of "intention to participate in regional cultural tourism activities combined with AR technology" with 522 valid data. The project hopes to consider the regional cultural innovation development path combining AR technology with the concepts of local community cultural resilience participation design and active design Manzini [6]. Among them data analysis shows that 86.2% of the population expressed interest in cultural activities combined with virtual reality technology 82.1% expressed their willingness to participate in the experience and content design provide design related materials feedback etc. 49.8% believe that the content design is novel and interesting helping to understand traditional culture and stimulate the fun of learning new knowledge 52.3% believe that they can interact more with others (friends strangers relatives etc.) 46.8% think it is more interesting than the traditional display form. In addition the project team conducted a survey and measurement of the audience on the display effect of the AR work "Shan Hai Jing: Shen You" (2017.12) exhibited at the "China Creation Myth – Internet Art Exhibition" at the Liu Haisu Art Museum in Shanghai. Analysis and effective investigation of 921 people. The total number of valid investigators is 921. After two data comparisons the conclusions are basically the same.

Therefore this paper builds a topic based on the resilience virtual public space of traditional culture with the theme of regional cultural design incorporating AR technology and Anthony Dunne's speculation theory Dunne [7]. by analyzing the spatial characteristics of existing virtual cultural works including the relationship between designers (rule makers) users (individuals) space environment and design features of the space of running rules cognition publicity possibility and resilience. The experiment explores the design ideas of the resilience virtual public space of traditional culture and lays a theoretical foundation for driving cultural innovation and the transformation of cultural education and communication.

1   Imitate the Divergent Thinking of the Human Brain

Digital culture is a new art form in which people use digital technology for cultural innovation. It is an open virtual space composed of digitalized cultural information. Like a black hole, it constantly absorbs the cultural information created by history. Meanwhile, It also constitutes a new virtual world of digital art information. Its vertical is covered with historical imprint and horizontal link to modern science and technology, in the form of digital code in the virtual space. In the mobile Internet, this organic network will spread and exchange cultural information with people in the virtual space in a more advanced, richer, more convenient and more active way. People who enter the virtual space are also more likely to get closer to culture, appreciate and evaluate culture and participate in cultural creation. Based on a virtual space of a network-like digital culture that combines multiple cultural information data and multiple terminal nodes, it seems to mimic the divergent thinking of the human brain. With its rich information resources and technology, it promotes the creation of more new cultural forms. "If the net of the spider knot is flat, then the net of the virtual space is three-dimensional, diversified, and self-propagating" He [8].

2 Operating Rules of Virtual Cultural Space

After the digital culture transforms form and meaning into codes consisting of 0 and 1 in the virtual world, the form and meaning appearing in the codes are an anarchic digital stream or manipulated false imaginary? Although the designers of the program are still making money on them, the digital tools that rely on the program are designed to break away from the designer and immediately set new rules for the digital world. First of all, this new rule is not controlled by the designer or some individuals, because the program is not unique, can create a variety of rules for the digital world, and even contradictory rules, the digital world will open up isolation space for these contradictions (just a piece of code that can be run), rules are easily created or abolished, this is just a migration in the digital world. Secondly, the designer of the rules must also rely on the digital world to reflect the rules, that is to say, the designer is also the image under the rules. After the rules are designed, they have achieved a position above the digital image, but this position is not stable, may be abolished at any time. In addition, the image of the individual in the digital world depends on the rules, and the abolition of the rules means that the image is destroyed, but the elimination of one image is only a small excuse for the creation of another image. From the perspective of the entire virtual world, the image is not unique. The simulation image is inhabited under various rules.

Just as traditional culture is digitized, when a hand-painted work is created by an artist and digitized, everything will be out of control. Because the work has been mutated in the process of digital communication, and new code has been implanted, forming a new art form (new rules). For example, the PS spoof of a work, or borrow a part of the picture to form a new art work created by another designer, the same work encounters different people with different fates (new rules), so constantly Loop, from building rules, to being ruled, to abolishing rules. The original designer can't form effective control over the completed cultural works, and the work itself will mutate, just like a digital living body that is constantly reproducing. Just as Brown Shaw pointed out in the article "Literature Space", this work is separated from the author after it is completed. The meaning of work is only presented in the reading process of someone. Similarly, rules (spaces) are only experienced by the user during use. The experience after the rules are completed is completely unique and private. Even these experiences exceed the rules themselves. Users may gain new insights from rules and images.

It can clearly be seen, the rules of cultural creation in the digital world are not binding. These rules are only a relatively independent space in the digital system, in which they exercise power, but the residents they rule are completely voluntary and mobile. And they can abandon these rules at any time. In the virtual cultural space, people can freely log in to different spaces through the network, participate in different experiences, abide by their corresponding rules, or choose not to log in or leave.

3   Cognition of Virtual Cultural Space

For the digital image that enters these spaces, from the beginning of the source, it constitutes a split (subject and object). From Lacan's Mirror Theory: infants can find many images reflected from the mirror, and use a series of gestures to "experience the relationship between mirror movement and the reflected environment, and experience the relationship between actual mixture and its replicated reality" Wang [9]. In this mirror game, the baby can recognize that the copied subject that controls the game is the baby itself. This is different from the subject of self-thinking by the French philosopher Descartes. The baby confirms the self-image from the unconscious mirror game. But Descartes' abstract self-thinking is the forgetting of the mirror game. It is abstracted from the rational reality and ignores the dependence on the object, and it is an affirmation of the existence of an individual by the degree of self-recognition of the subject. But in Mirror Theory, the self can only be confirmed by the reflection (mirror) of the object.

In fact, mirror is not just a reflection of image. In the process of constant interaction between subject and object, to some extent, this image is constructing the self (the process of cognition). The replication and manufacturing capabilities of virtual reality built by the digital world are much greater than the mechanical reflection capabilities of ordinary mirror. The reflections obtained in the digital system have mutated during the original planned reflection propagation process. As the director expressed in the film "Source Code" and "Avatar", the protagonist's separated self (subject) and mirror (object) experience different lives in two worlds, and the mirror (object) in the parallel space has formed a whole new idea through a series of new world experiences, and variation has taken place. Then, in the virtual cultural space, according to Lacan's mirror theory, the subject defines the environment of self-existence through experience, and then, under the influence of the environment, gains recognition of self and culture, just like the philosophical thought of Descartes 's "Cogito ergo sum."

4   Possibility of Virtual Cultural Space

Heidegger pointed out in "Existence and Time" that the essence of human being is that "the existence of man always refers to the existence that is going to be" Zhang [10]. The meaning of the world is manifested by the existence of human beings, the existence of others, and the relationship between things. Digital technology enriches the possibilities of human existence, and forms a variety of experiences in digital systems. The realization of possibilities (virtual reality) is more and more convenient and simpler. Just like the space created by the program (rules), it is the place to realize the possibilities.

Possibility comes from speculation, and "speculation comes from imagination, that is, the ability of truly imagine more possibilities of the world and its alternative choices" Dunne [7]. "There are many kinds of imagination, including social, professional, scientific, technical, artistic, and design imagination. Fictional literature may be the most active creative laboratory in the world's structural career" Dolezel [11]. "The virtual world of literature is constructed by words and is realized by deriving the logical relationship of language to the extreme, such as science fiction" Bal [12]. Porter Abbott [13] or virtual environment narrative. But the traditional cultural virtual world, which is integrated with digital technology, is composed of codes 0 and 1. The design imagination can virtually visualize all material or non-material objects including virtual literature. This virtual world can be a fusion of imaginative design of game design, digital media, visual effects and augmented reality. It is carried out through the frequent interaction, collision and integration between real reality, imaginary reality, simulated reality, occurring reality and fictional reality. We can immerse ourselves in thinking about the relationship between the "Form" of virtual design and the real "Meaning", as well as cherishing and enjoying unrealistic needs.

5 Publicity of Virtual Cultural Space

In general, "public space is a social space that is open and accessible to people, is the expression of regional cultural diversity and the foundation of its identity, is a place where residents invest their emotions over time. It is through the creation of these places that the physical, social, environmental and economic health of urban communities can be nurtured" Charter of Public Space [14]. The virtual public space of traditional culture is a digital place based on virtual reality technology, which is different from the traditional public space. The research content is not to explore or build new spaces, but to reorganize and superimpose existing space, focusing on the activities and needs of people in space, emphasizing design and thinking while testing. Intervene in the development of regional culture with a new and more active attitude, and imagine the possibility of the future design of traditional culture through public participation and virtual narrative story design.

6 Resilience Design of Virtual Cultural Space

"Social resilience requires the presence of a local community, a group of people who interact in a physical environment. The proximity of the state and its relationship with the local community enables the local community to organize and solve problems in a crisis" Manzini [6]. Therefore, regional cultural development requires the existence of resilience cultural space of local communities. And virtual reality technology has the advantage of creating resilience cultural space. "The concept of resilience originated from ecology, which was proposed by Canadian scholar Holling" [15]. Hall [16] and then began to intervene in different disciplines, now extended to a four-dimensional (ecological, technical, social and economic) perspective. Resilience in this paper refers to the capacity of the social-ecological system of the virtual cultural space to "absorb or withstand perturbation and other stressors such that the system remains the same regime as the original one, essentially maintaining its structure and functions" Ni [5].

According to the existing literary works, historical records, videos, photos and other archives, the author attempts to introduce resilience design ideas in the regional cultural tourism development project "RuiJin Revolutionary Site Installation" (China, 2011.03), carry out the experiment of resilience design of virtual traditional culture, and at the same time, think about how to use the form of virtual design to spread and educate people about the true meaning of traditional culture. The project mainly uses

AR technology, from daily and micro perspectives, based on the consideration of the resilience design of regional tourism culture products, including individuality, sociality, participation and initiative design, expansion and continuity.

6.1 Individuality

Individuality, refers to the self-satisfaction and self-realization of people entering the virtual cultural world. According to specific scenes, the project creates virtual character content, virtual interactive experience area, reproduces 3D dynamic historical scene (Fig. 1). Also develops a mobile app (Fig. 2): The scanning function of the AR makes the scene reappear. Including the reproduction of historical figures, historical videos and live explanations. Camera function with special effects (Fig. 3): In the option interface of special effects, visitors can take photos according to their preferences. Object analysis (Fig. 4): In the interface of object analysis, the visitor can point the camera at the object that you want to know, and the information about the object will appear.

Fig. 1. Virtual character content

Fig. 2. Mobile app

Fig. 3. Camera function with special effects

Fig. 4. Object analysis

6.2   Sociality

Sociality, refers to the orientation of mainstream value of social (the expression of meaning) under the cultural environment created by the virtual world (the expression of form). A series of follow-up activities such as online feedback, uploading new materials, participating in storytelling, "micro-intervention" Manzini [6]. And active design through the operation of APP or virtual interface are the specific expressions of people's social cognition of regional culture. At the same time, the design team can easily obtain test data about people's needs through the pre-set function of data statistics of virtual system, so as to test and think while designing.

6.3   Participating

Participating in design and active design, is an effective way for people to understand regional cultural sociality, and it is also the effects of concrete manifestation of people's understanding of cultural. The micro-intervention here is a way of participating in design and active design, which constitutes a series of follow-up activities, with the essence of short-lived. It refers to endowing intangible (the cognitions of things that short-lived, movable, and tiny) things to tangible. It also points to the design possibilities of a new and open participation, and these possibilities are the supplement of the real meaning of the content, make traditional culture contemporary, realize cultural innovation and development. Anne Eggebert writes in relation to the short period in which students can participate in the life of a local community: "In the current climate of the mass movement of people, in a world city, 3–4 years might be understood as a significant period of habitation." Eggebert [17]. That means the population is in a state of flow is a true portrayal of contemporary society. So the people involved in the design can be permanent residents, students, immigrants or tourists. It can improve the openness of the regional culture, and also a model for open communication and exchange of differences, or the possibilities for making culture richer. Of course, these possibilities are not cultural alternatives of which have been built on stable, long-lasting regional societies, but different people's understanding of regional culture and the desire for modern life. It is a contemporary regional culture that meet the needs of modern people through the micro-intervention approach, and create a new cultural ideology with the characteristics of regional traditional culture and contemporary culture. Participation or active design is the way of people to achieve individual satisfaction and the behavior of social cognitive in the public space of virtual culture.

6.4   Expansion and Continuity

The resilience design of the virtual cultural space is also reflected in the expansion and continuity of the degree of people's self-organization, learning and adaptation. Gunderson [18, 19]. The concrete manifestations are as follows:

Superimpose the information with the real historical place, and coexist with other cultural communication forms and media, and combine the Internet technology to realize the sustainability of design, participation, interaction and experience.

Through micro-intervention, meet the individual needs of people, realize self-organization, learning and social cognition of culture, and provide more possibilities for new cultural forms.

According to the test data obtained by people's participation behavior, the virtual system continuously upgrades the existing space functions, enabling people to better adapt to the virtual culture world based on their knowledge of historical culture and the skills of equipment use.

The design of APP continues the spread of culture, and will not disappear with the departure of people. On the contrary, relying on the Internet, Internet of Things, and AI technology, an intelligent cultural network group has been formed, and people in the flow state can continue communicate with the virtual cultural world of each region. This is the expansion and continuation of the virtual cultural space.

7 Conclusion

"The virtual world of digital cultural space is mainly realized by the digital simulation of culture and the emerging of new forms of digital media culture, namely the completion of Baudrillard's third sequence of simulation" He [8]. It has the following characteristics: First, the network group composed of virtual world based on digital copy has the ability of divergent thinking. Second, its operating rules determine that the results presented by the virtual world are not necessarily the same as the facts; Third, the virtual space will not be destroyed by the leaving of some individuals, nor does it require any special commitment. Individual data has already been recorded; Fourth, based on the above three characteristics, all the information entering the virtual space system has the possibility of free combination, forming a rich understanding of culture and promoting the formation of new ideology of culture; Fifth, at beginning, the resilience virtual cultural space just presents to everyone a possible virtual cultural narrative space that is more suitable for modern people to accept cultural information than traditional methods. What the designer needs is not the audience's approval of the design, the audience may also be the designer. They can create more possibilities by design and imagination through the options available in the virtual space and build their own perceptions; Sixth, as a path for the protecting, spreading and developing of regional culture, virtual cultural space can easily and realistically present a variety of possible cultural spaces that people want. But the aim is not to create a cultural display space that can be regarded as long-lasting and stable. On the contrary, it is necessary to create an ecosystem that enables various social relationships to coexist. By creating opportunities to participate in design, people brainstorm and solve problems, open up new perspectives, and realize the expansion and continuation of regional cultural space.

The same model can be applied to more applications. In terms of regional cultural education topics, this can be a new curriculum of art education. But unlike traditional way, what we want to build is an intelligent teaching model that combines advanced technologies such as virtual reality, the Internet, big data, IOT, or AI, etc. The design inspiration is derived from traditional cultural materials, forms and patterns, combined with knowledge of social sciences, to obtain information about user needs, and to carry out the teaching model of design for the present world. Meanwhile, in the face of market homogenization, we need to shift the focus of design to "design for the possibilities of the future world" Dunne [7]. Of course, the object of education is not limited to students at school. According to the results of the author's recent survey report, people generally have a high degree of interest in digital display forms of traditional culture. The influence factors such as age, education, and occupational are low. But the familiarity with AR technology is not high, indicating the penetration rate of related knowledge is low. At the same time, the general public generally expressed their expectation for the virtual form of traditional culture, and it is expected to increase their social experience and cultural quality. Therefore, the creation of virtual spaces of regional cultural can also provide educational opportunities for the public.

From another perspective, the virtual spaces of regional cultural just as an innovative digital product that meets the needs of the times. The design difficulties we face are not only technical bottlenecks, but also need a benign combination of factors such as environment, product service, social, economic, culture and government, etc. "We need to cultivate on-going relationships with a number of organizations that would understand both the limitations and benefits of what we can offer" Penty [20].

In addition, different people's design and imagination of the possibility of traditional culture is a multi-angle understanding and interpretation of culture. The result may be a modern expression that conforms to the spirit of traditional culture, or misinterpretation. From the perspective of the existence and development of cultural diversity, the correctness is not so important. From the perspective of the dissemination and development of cultural connotations, it is meaningful to involve more people in the process of design. Because traditional culture comes from human civilization, the continued development under the cognition and inheritance of human beings is what we need to pay more attention to.

References

1. Davies, J.: The Birth of the Anthropocene. University of California Press, Oakland (2016)
2. Zolli, A., Healy, A.M.: Resilience: Why Things Bounce Back. Simon & Schuster, New York (2012, 2013)
3. Taleb, N.N.: Anti-fragile: How to Live in a World We Don't Understand. Allen lane, London (2012)
4. Keck, M., Sakdapolrak, P.: What is social resilience? Lesson learned and ways forward. Erdkunde **67**(1) (2013)
5. Ni, Y., Zhu, M.: Open Your Space: Design Intervention for Urban Resilience. Tongji University Press (2017)
6. Manzini, E.: Weaving People and Place: Art and Design for Resilient Communities. Open Your Space: Design Intervention for Urban Resilience. Tongji University Press (2017)
7. Dunne, A., Raby, F.: Speculative Everything: Design, Fiction, and Social Dreaming. The MIT Press, Cambridge (2013)
8. He, J.: Digital art space in the digital age. Art Des. (Theory) **286**, 86–88 (2014)
9. Wang, M., Chen, Y., Ma, H.: The Philosophical Discourse of Postmodernity, 173. Zhejiang People's Publishing House (2001)
10. Zhang, R.: The 15th Lecture of Modern Western Philosophy, 286. Peking University Press (2003)

11. Dolezel, L.: Heterocosmica: Fiction and Possible Worlds. John Hopkins University Press, Baltimore (1998). ix
12. Bal, M.: Narratology, Introduction to the Theory of Narrative. University of Toronto Press, Toronto (1997)
13. Porter Abbott, H.: The Cambridge Introduction to Narrative. Cambridge University Press, Cambridge (2008)
14. Charter of Public Space. Adopted in Rome, final session of the II Biennial of Public Space, 18 May 2013. http://www.pps.org/referce/placemaking-and-place-led-development-a-new-paradigm-for-cities-of-the-future/
15. Holling, C.S.: Resilience and stability of ecological systems. Annu. Rev. Ecology. Syst. **4**, 1–23 (1973)
16. Hall, P.A., Lamont, M. (eds.) Social resilience in the neoliberal era. Am. J. Sociol. **120**(1) (2014)
17. Eggebert, A.: Neighboring and networks. CoR website (2016)
18. Gunderson, L.H., Holling, C.S. (eds.) Panarchy: Understanding Transformations in Systems of Humans and Nature. Island Press, Washington DC (2002)
19. Scheffer, M., Carpenter, S., Foley, J.A., Folke, C., Walker, B.: Catastrophic shifts in ecosystems. Nature **413**, 591–596 (2001)
20. Penty, J.: Does Good Thinking make good doing? CoR Website (2016)

The Design of Card Game for Leukemia Children Based on Narrative Design

Long Ren, Hongzhi Pan[(⊠)], Jiali Zhang, and Chenyue Sun

Product Design, Huazhong University of Science and Technology,
Wuhan, Hubei, China
296232879@qq.com, 51406940@qq.com, 547834629@qq.com,
419180229@qq.com

Abstract. The mental health of leukemia children triggers more and more attention. It is emergency for medical staff to help leukemia children to establish a suitable perception of treatment so that children could have faith and build trust to the doctors and nurses. This study aims to help leukemia children to establish a suitable perception through a set of card game designed with narrative method. Researchers designed the game with questionnaire method and tested it by participant-observation. The result of the test showed the playability of the game and verified the possibility of communicating the medical information to the participants.

Keywords: Leukemia children · Narrative design · Treatment perception

1 Instruction

Recently, there is an increasing interest in psychological status of children with leukemia due to the increase cure rate for childhood leukemia. Shyh-Shin Chiou and Ren-Chin Jang assessed health-related quality of life (HRQL) of childhood leukemia survivors in Taiwan. The result showed that the HRQL of childhood leukemia survivors was noted to be worse than that of community children and nonadult siblings as reflected by significantly lower scores in both the physical summary and the psychosocial summary [9]. J. Vetsch and C. E. Wakefield performed a systematic review, searching published literature in Pubmed, PsycInfo, Embase, and the Cochrane database including all publications up to December 16, 2016. The result pointed out that survivor and parent socio-demographic factors and psychological factors such as resilience and depression were also associated with HRQL [10]. Children will suffer injection, blood draw, spinal tap and stimulation of drugs during the treatment. The great pain and the restrict of behavior let children develop the bad habit. And it makes children diffident, withdraw and pessimistic. Furthermore, some of them may distrust parents and medical workers, be resistant to treatment. There have been a several studies of decreasing negative feelings. However, there lacks a universal and effective method to help children with leukemia to improve the perception of medical treatment.

Based on the above-mentioned situation, The purpose of this study was to design a card game for children with leukemia based on narrative design in order to help children to learn the leukemia in a positive way, encourage them to receive treatment

© Springer Nature Switzerland AG 2020
. Ahram (Ed.): AHFE 2019, AISC 973, pp. 261–267, 2020.
https://doi.org/10.1007/978-3-030-20476-1_27

bravely and be full of hope. Ample researches have been conducted on Play Therapy, especially on the narrative methods. Franceschini (2013), Sandro discussed the benefit of action video game to dyslexic children. They found that only playing action video games improved children's reading speed, without any cost in accuracy, and more than or equal to highly demanding traditional reading treatments. The results showed that this attention improvement can directly translate into better reading abilities, providing a new, fast, fun remediation of dyslexia [8]. Tiffany Y. Tang (2017) designed an educational protocol to evaluate the two selected applications and a commercially available application, with the use of known tools and appropriate occupational therapy interventions [7]. As for the researches on the narrative design, Miao-Hsien Chuang (2015) adapted the product semantics model of Krippendorff into the cultural product narrative communication model by integrating narrative elements with form and meaning, as well as emphasizing the personal philosophies of designers and user participation [2]. Connie Golsteijn and Serena Wright (2013) presented guidelines for using narrative research and portraiture for design research, as well as discussing opportunities and strengths, and limitations and risks [3]. Clive Baldwin (2015) discussed the narrative ethics for narrative care. He argued that people should to see their actions as both narratively based and narratively contextual [6].

This paper proceeds as follows. The second Section describes how the designers designed the card game. And the Sect. 3 showed process of the test of this card game through observation method. The result of the observation will be presented in Sect. 4 Section 5 provides the conclusion.

2   The Study of Card Game Design

2.1   Narrative Design and Card Game

This study was based on narrative design which originate from narratology. The objects, goals and restriction of design are combined in narrative way, helping users to communicate with environment on information. The card game was designed with an adventure story including the knowledge of leukemia, encouraging children to accept treatment bravely and be positive about disease (Fig. 1).

2.2   Select and Code the Symbol

Leukemia children live in an environment full of medical information. The game designers should turn the medical information into the game language. This information includes the name of drug, the figure and the name of the medical equipment and methods, some heart tips about leukemia in daily life and the symptoms of leukemia

There were three categories of the symbols according to the function of the card (Fig. 1). Designers collected and screened the names of drugs, medical treatment and symptoms, generalized typical good habits which children with leukemia should pay attention on. And then, we determined the game's design with the characters of thes concepts.

The card game consists of designed by three parts.

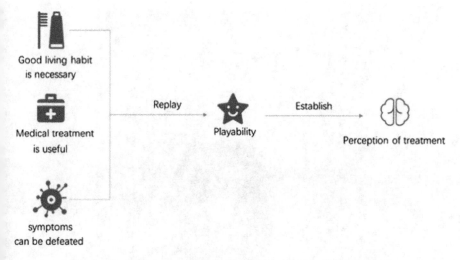

Fig. 1. How the card game help children to improve the perception

First, the most frequently occurring symptoms are designed as monster cards. These cards would stop players from advancing and winning the game.

Medical cards are designed with the concepts of medical treatment and drugs, which can help children to deal with the punishment from monster cards.

The third part is health card designed with the concept with the most frequently occurring good living habits. These cards can help players to reach the goal and win the game faster.

The main propose of this study is to explore the possibility that game can help children to follow treatment actively and learn that good living habits can help them recover from illness faster (Table 1). The cards will be designed with these symbols.

Table 1. Select and screen symbols

Category	Symbol
Heart tips	Healthy dealt, sleep on time, hand washing, oral hygiene
Medical treatment	γ-globulin, VP-16, PICC tube, glucocorticoid, paracentesis,
Symptoms	Bleed, diarrhea, fungal infection, electrolyte imbalance

In order to attract children to accept and remember the relevant medical information, the image should be anthropomorphic to improve the readability of cards. For example, the monster card BLEED was designed as a red ball with two hands handing table-knifes (Fig. 2). Designer should consider that the image must be accepted by children, which means it's better to design the image in cartoon style and make the negative characters blurred. At the bottom of the card, a line manifested the character of monster (the symptom) so that children will have a deep impression.

Fig. 2. The monster card BLEED

3 Game Test

The main method of this study is participant observation. Researchers should play with leukemia children, observing their reaction to the game to test the playability and the communication effects.

3.1 Environment

The environment was a specific word in leukemia clinic with a relatively independent large space, so that participants can play the game without disturbance and space limitation. And it ensured that the medical assistant can reach there at any time (Fig. 3).

Fig. 3. The participants and the environment of the test

3.2   Participants

There were three leukemia children randomly selected to participant this test, one was girl (C1), two were boys (C2 and C3) (Fig. 3). Before the test, doctors and nurses will check participants' (the leukemia children in the clinic) physical condition to ensure they have good enough condition in the process.

3.3   Process

There were three volunteers (designers) participant the test, two of them guided the children how to play the game, and one of them recorded the process with a camera. Designers will review the reaction of the children after the test.

Children acted enjoy in the whole process from their body language and mood. Different child acted differently in the test. C1 followed volunteers guide obediently, but showed strong desire to win at every round (Fig. 4). And C1 showed the medical equipment (PICC tube) when she drew a PICC tube card.

C2 and C3 had richer body language, they would act more obviously when they faced twists in the game. C3 danced frequently in the game and C1 would perform really excited when draw a good card. However, they had little feedback to the medical information in the cards.

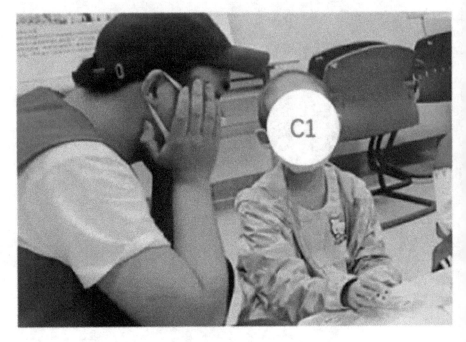

Fig. 4. C1 and the volunteer were talking about the strategy

4   Result

The process of the game test showed the practicability of this design and some deficiency of this game design.

Participants had a quick recognition when they met some cards such as PICC tube and glucocorticoid, which means this plan could trigger the association with the medical information. Secondly, participants showed great interest in the game, which verified the playability of the game.

As for the deficiency, the participants can't recognized all the medical information from the game, which means some of the symbols is too abstract to recognize and it need to be coded again. Second, designers ignored the competition of the game rules, which will decrease the playability.

5   Conclusion

The aim of this study is to help leukemia children building the perception of treatment and accepting the medical treatment with the help of card game based on narrative design.

Designers selected symbols which had high relevant of leukemia medical environment with questionnaire and coded them in narrative way. And then researchers tested the game with leukemia children in the leukemia clinic with. Through the test record, it was found that children recognized the medical information through the name of the card in most cases. And the game has playability, because participants showed great interest in the game.

Research needs to be further conducted to investigate how much the factor of volunteers influence the participants' interest. And it needs a further study to discuss how to code the symbols from name, image and description of the cards. Finally, the reproducibility should be considered in further research. More simples are needed in further test.

References

1. Dickey, M.D.: Game design narrative for learning: appropriating adventure game design narrative devices and techniques for the design of interactive learning environments. Educ. Technol. Res. Develop. **54**(3), 245–263 (2006)
2. Chuang, M.H., Ma, J.P.: Analysis of emotional design and cultural product narrative communication model. In: International Conference on Cross-cultural Design Springer International Publishing, Cham (2015)
3. Golsteijn, C., Wright, S.: Using narrative research and portraiture to inform design research. In: IFIP Conference on Human-Computer Interaction. Springer, Heidelberg (2013)
4. Koenitz, H., et al.: What is a convention in interactive narrative design? In: International Conference on Interactive Digital Storytelling (2017)
5. Marsh, T., et al.: Fun and learning: blending design and development dimensions in serious games through narrative and characters. In: Serious Games and Edutainment Applications (2011)
6. Baldwin, C.: Narrative ethics for narrative care. J. Aging Stud. **34**, 183–189 (2015)
7. Tang, T.Y., Falzarano, M., Morreale, P.A.: Assessment of the utility of gesture-based applications for the engagement of Chinese children with autism. Universal Access in the Information Society (2017)
8. Franceschini, S., et al.: Action video games make dyslexic children read better. Current Biol. **23**(6), 462–466 (2013)
9. Chiou, S.S., et al.: Health-related quality of life and cognitive outcomes among child and adolescent survivors of leukemia. Supportive Care Cancer **18**(12), 1581–1587 (2010)
0. Vetsch, J., et al.: Health-related quality of life of survivors of childhood acute lymphoblastic leukemia: a systematic review. Qual. Life Res. **27**(2), 1–13 (2018)

Human Factors in Virtual Environments

Depth Sensor Based Detection of Obstacles and Notification for Virtual Reality Systems

Peter Wozniak[1,2(⊠)], Antonio Capobianco[1], Nicolas Javahiraly[1], and Dan Curticapean[2]

[1] I-Cube - University of Strasbourg, 67081 Strasbourg, France
peter.wozniak@hs-offenburg.de,
{a.capobianco,n.javahiraly}@unistra.fr
[2] Offenburg University, 77652 Offenburg, Germany
dan.curticapean@hs-offenburg.de

Abstract. Walking interfaces offer advantages in navigation of VE systems over other types of locomotion. However, VR helmets have the disadvantage that users cannot see their immediate surroundings. Our publication describes the prototypical implementation of a virtual environment (VE) system, capable of detecting possible obstacles using an RGB-D sensor. In order to warn users of potential collisions with real objects while they are moving throughout the VE tracking area, we designed 4 different visual warning metaphors: Placeholder, Rubber Band, Color Indicator and Arrow. A small pilot study was carried out in which the participants had to solve a simple task and avoid any arbitrarily placed physical obstacles when crossing the virtual scene. Our results show that the Placeholder metaphor (in this case: trees), compared to the other variants, seems to be best suited for the correct estimation of the position of obstacles and in terms of the ability to evade them.

Keywords: Virtual Reality · Notifications · Interaction metaphor · Collision avoidance · 3D interaction · Navigation · Walking workspace · Range imaging · RGB-D

1 Introduction

After Virtual Reality (VR) had little success in the nineties, the technology is now widely available. Systems such as the HTC Vive or Oculus Rift enable room-scale VR applications with natural locomotion within virtual environments (VE). The user must be aware that objects that remain within the specified interaction area or are added later might become obstacles. While static objects (e.g. furniture) can be easily avoided, this is not so easy with objects that dynamically change their position (e.g. dogs). In normal living situations, full rooms are rather the rule than empty spaces. We present a prototype VE system for the detection of stationary obstacles by means of an RGB-D sensor and their signalling with four metaphors. In a later development stage, the detection of moving targets is also to be made possible. A preliminary study allows us to evaluate the benefits of the metaphors used and the efficiency of such an approach in terms of precision, spatial understanding and sense of presence.

T. Ahram (Ed.): AHFE 2019, AISC 973, pp. 271–282, 2020.
https://doi.org/10.1007/978-3-030-20476-1_28

2 Related Work

Research results show that walking interfaces offer advantages in navigation of VE systems over other types of locomotion [1–3]. Real walking is a natural form of locomotion, resulting in significantly higher presence than other types of locomotion [4] and reducing the probability of the occurrence of simulator sickness [5, 6] by simultaneous vestibular and proprioceptive stimulation. Real walking locomotion interfaces are based on the acquisition of the user pose and are limited to tracked space, thus raising the problem that users may encounter obstacles. Peck et al. introduced deterrents that appeared when the users were close to the edge of tracked space. They used bars as visual cues that the users were instructed to avoid [7]. The standard technique today is to display a visual cue when the user threatens to leave the previously defined empty and thus obstacle-free area [8–10]. To enable real walking in "larger-than-tracked space" VEs some interfaces use transformation techniques, like redirected walking, scaled-translational-gain, seven-league-boots, motion compression and dynamic curvature, that alter the VE or the user's motion path [7, 11]. Transformation techniques can be used to prevent collisions between users and known obstacles (e.g. walls and furniture). The principle has its limits, especially since a user's path cannot always be correctly predicted, various stop and reset techniques are necessary to signal and avoid possible collisions to users [12, 13]. Scavarelli et al. compared different collision-avoidance methods between co-located VR users but didn't use real obstacles [14]. The Magic Barrier Tape by Cirio et al. is a hybrid collision-avoidance and navigation method. The tape represents the limited walking area and can be used to shift the user position in the VE [15]. The method has been later extended to CAVE-like systems [16]. Simeone et al. builds on the idea of passive haptics [17] and explores to what extent the VE can deviate from the physical environment (PE) before the illusion breaks. A physical room and each object's bounding box were measured and modeled manually, thus possible obstacles became tangible objects within a VE [18]. In another publication Simeone outlines the visualization of non-VE users within a VE application by using a small widget, floating in the user's field of view (FOV). The pie-shaped widget represents the horizontal FOV of the Kinect and displays the positions of detected individuals in front of the sensor [19]. The Kinect space was not spatially mapped into the VE space. Similar to Nescher et al. [20], Sra et al. is using a depth camera equipped device with a SLAM based mapping and tracking, an automated 3D reconstruction of the PE plus the detection of the walkable area (WA) and also a procedural creation of the VE that uses static visual cues to prevent users leaving the WA. The WA detection is performed in advance and not at runtime. The authors also outline the need for additional warning methods to avoid collisions and the lack of detection of moving obstacles. It is also pointed out that these methods should be evaluated in terms of how much they influence the feeling of presence [21].

3   Research Approach

The possibility of locomotion in virtual reality in a natural way usually presupposes that there is sufficient space in the real environment. The current VR systems, such as the HTC Vive or Oculus Rift, can track user position and movements within a specified area but are not able to detect arbitrary obstacles. Thus, usually all potential obstacles must be removed for the duration of use to avoid collisions and stumbling. Remaining objects or objects added later represent a potential danger to the user.

It would therefore be useful if the VR system could automatically detect obstacles within the tracking area and signal them to the user if necessary. Our prototypical implementation should serve as proof of concept. For our system we use an RGB-D camera to detect the position, orientation and size of the physical obstacles on the floor and specially designed visual metaphors to allow the user to avoid the obstacles within the VE. We use an HTC Vive based VE system and a Microsoft Kinect 2 sensor. The built-in camera systems allow to capture the environment spatially. The depth information is used to calculate a bounding box per object. Each bounding box represents an obstacle in space.

Once an existing obstacle is detected in the vicinity of the user, the VE should notify him of the presence of a potential hazard. Notification can be provided in any of the following modalities: audio, video, haptic, or in any given combination of these modalities [22]. However, visual notifications seem more adapted to our purpose since they convey spatial information that can allow effective implementation of avoidance strategies [14].

In our context, a good metaphor should convey enough spatial information to allow the users to avoid potential obstacles, without being too detrimental to their subjective experience in the VE. We designed 4 visual metaphors, inspired from previous work (see Sect. 5), to compare their efficiency in the task at hand and compared them during an experimental protocol (see Sect. 6). Our objective was to assess the spatial understanding of the position of the obstacle in the environment and the impact of the chosen metaphor on presence (see Sect. 7).

For this purpose, the participants should solve a simple task within the VE and avoid randomly placed obstacles.

4   Prototypic Implementation

Our system requirements were the following:

- Interactive 3D visualization via VR headset
- VR roomscale tracking
- Unobtrusive detection of size and position of obstacles in space
- Signaling of obstacles using metaphors

The Unity 3D development environment version 2017.4.1f1 (64bit) and OpenVR API [23] was used for the implementation [24]. We choose the HTC Vive that allows room scale tracking on an area of max. 5.5 meters diagonal for a HMD, two handhold controllers and optional trackers [25].

For our preliminary study, we focused on the detection of stationary obstacles, but kept the possibility open to expand the system later on to the possibility of detecting moving obstacles. We decided on a camera-based approach using the Microsoft Kinect 2 depth-imaging sensor. This sensor can determine the distance in a range of 0.5 m and 8 m with a resolution of 512 × 424 pixels and a frame rate of 30 Hz.

The Kinect 2 sensor is based on the ToF principle. The scene in front of the camera is illuminated by modulated infrared light and the runtime of the reflected light pulses is determined. The precision of the depth data depends on a number of different factors. Besides various possible sources of error, the degree of light reflection is one of the most important factors for good measurement results. If only a small portion of the infrared light pulse is reflected back, only a small amount of light can be detected. The signal-to-noise ratio deteriorates and so does the depth measurement. The Kinect 2 sensor works reliably over a distance range of 1–4 m. Microsoft defines a distance of 1.0–2.5 m as the optimum distance. As the distance increases, the amount of light reflected decreases in the square. Usually the deviation of the depth measurement is between 1.0- and 3.5-mm. Bright surfaces provide better measurement results. For very dark surfaces the deviation is less than 1 cm. Vignetting effects of the cameras and the infrared illumination cause an additional radial drop in measurement accuracy. This results in additional deviations of ±1 mm. It is recommended to place the relevant parts of the image in the center of the imaging sensor. Furthermore, the Kinect 2 sensor requires a warm-up phase of approx. 60 min until the measurement deviations remain constant on average [26, 27]. The depth sensor's field of view (FOV) is 70.6° horizontally and 60° vertically. The depth image represents a distance measurement per pixel and can be represented as a three-dimensional point cloud. Using the Point Cloud Library (PCL) [28] we have applied the following processing scheme to the Kinect depth sensor readings:

1. Capture Kinect depth image information and generate the point cloud
2. Reduce data volume. Filter the data by min. and max. distance and down sampling (PCL VoxelGrid filter) [29]
3. Detection of floor by using RANSAC algorithm [30] to recognize the largest planar area within the point cloud
4. Objects standing on the ground are separated by segmentation of related point clusters located above the plane
5. Calculation of a bounding box per detected point cluster
6. Pass the bounding box information to Unity 3D

The position, size and alignment of each bounding box (BBOX) serve to enable the functionality of the metaphors. A BBOX is only a simplified approximation to the shape of the real object, but it is sufficient for our study. More important is a correct positioning of the BBOX within the VE. We implemented a calibration procedure to estimate the relative position and orientation of the Kinect camera within the Lighthouse tracking space, which allows us to align both coordinate spaces. The calibration only needs to be repeated when the camera poses or the tracking area has changed.

5 Preliminary Study

With our prototypical implementation, we carried out a study in which the participants should try out four different signalling metaphors. We hoped to understand which metaphor is better or worse to identify obstacles for users while minimizing the perceived disturbance of immersion.

20 participants (13 male, 7 female) aging from 21 to 57 (mean: 34.16, sd: 8.38) took part in the study. Simulator sickness symptoms remained slight for all participants before (mean value: 12.67; sd: 10.58) and after the experiment (mean value: 12.25, sd: 12.5). Hence, we kept all the experimental data collected from the 20 subjects.

The space for the experiment consisted of an empty area of 7.8 × 5.7 m (tracking area 5 × 5 m). To use the room size effectively and due to the maximum cable length of the HTC Vive of 5 m (headset) + 1 m (breakout box) the computer was positioned relatively centrally (see Fig. 1). Part of the available space was thus allocated to the computer desk, where the software was controlled, and the participants were able to answer surveys on a separate notebook.

A Microsoft Kinect v2 was used to detect obstacles placed within a designated area. To prevent the participant from being accidentally recognized as an obstacle, we let the participants trigger the detection while they were not in front of the camera. For this reason, too, a larger tracking area for the VR application was beneficial.

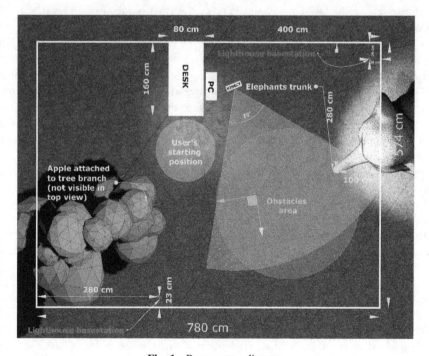

Fig. 1. Room setup diagram

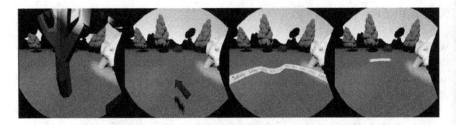

Fig. 2. The four used metaphors: placeholder (tree); arrow; rubber band; color indicator.

For the experiment, we defined four metaphors to signal the position and location of real obstacles in the virtual world (see Fig. 2).

The *Placeholder metaphor* uses a stationary virtual object to visualize the position and location of an obstacle. For our experiment, we wanted to increase the immersion and chose a tree as a placeholder object. The size and shape of the tree trunk correspond to those of the obstacles used, although the height is of course not the same.

The *Arrow metaphor* is based on the principle of the compass needle. The direction in which the arrow points is the position of an obstacle in space. The same principle is used in many computer games, to guide players to a certain position in the game. The arrow metaphor is not always visible, but only appears when the participant has reached a minimum distance of one arm length from the obstacle. The arrow floats unobtrusively in the lower peripheral FOV of the HMD when a user normally looks forward. If several obstacles are nearby, the direction of the nearer obstacle is indicated.

The *Rubber Band* metaphor is inspired by the work of Cirio et al. [15], but solely serves as an obstacle indicator alone and does not allow navigation. The Rubber Band floats constantly visible to the user around the virtual center of the body, in the lower FOV and approximately one arm's length away. Its appearance is based on a yellow-black colored safety band. If a user moves close to the position of an obstacle, the band deforms accordingly towards the user. The deformation of the band signals the position and direction of obstacles in space relative to the user. Several obstacles can be visualized simultaneously.

The *Color Indicator* metaphor visualizes the distance and direction to an obstacle by means of a spatially placed color-coded fade-in. The metaphor floats invisible to the user around the virtual center of the body, in the lower FOV and approximately one arm's length away. When approaching an obstacle, the area of the metaphor that comes into contact with the obstacle is first colored green. As the distance decreases further, the ribbon turns yellow, orange and finally red. Several obstacles can be visualized simultaneously.

An outdoor scene was chosen as the setting for our VE. An apple tree and an elephant flank an empty central area. Further away trees, bushes and rocks can be seen as decoration. Apple and elephant are the only interactive and thus essential elements of the scene. A highlighted illumination emphasizes them, and simple text boards give additional information. In addition, a recording of woods noises gives the static scenery more liveliness.

The task was to pick the apple from the tree with one of the two HTC Vive controllers. To do this, the controller must be held against the apple. As soon as the contours of the controller light up yellow, the trigger button at the back must be pressed briefly. The user must carry the apple to the elephant and hold it to the tip of his trunk. This attaches the apple to the elephant and completes the task.

6 Experiment

All study participants underwent the same procedure. The process was explained, and possible dangers were pointed out. A few demographical details, their experience with computers, video games and VR applications were recorded. In order to determine possible effects of simulator sickness, the participants were asked to fill-in a Simulator Sickness Questionnaire [31] before and after the experiments. Subsequently, the Vive controllers were explained to the participants and the HMD was adapted. The task to perform had been explained. All participants were given a few minutes to test (without obstacles) everything and ask questions. As soon as all questions were answered, the first metaphor was explained to the participant. The participant was led into the starting area and the VR headset was put on. Two obstacles were positioned arbitrarily in the detection zone and the test was started. While the task was being completed, attention was paid to whether or not there was a collision with one of the obstacles in the way. Directly after the task and before the glasses were taken off, we asked the participants to indicate the suspected position of the two obstacles with a controller. The result was evaluated by the 2 experimenters with a rating system of 0–3. Bad results were rated with a 0. Low results (a wrong direction to the object was suspected, but the position was still relatively close to the obstacle) were rated 1. With a 2, the result was rated "good" if the direction was correct but the position was suspected to be too close or too far away. Finally, the 3 corresponds to a high precision: the participant has determined the direction exactly and also indicated the position very precisely (Fig. 3).

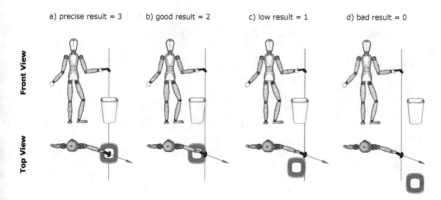

Fig. 3. Illustration of our evaluation scheme for the spatial understanding of the different metaphors.

The task was the same for each run. To avoid any ordering effect, we used a latin-square distribution for the presentation order of the metaphors. The position of the placed obstacles was determined randomly. After each metaphor, the participants were asked to give their subjective assessment on the metaphor. We also evaluated potential negative effect of the metaphor on presence. This was done through 3 questions assessing the actual level of presence. The questions and methodology were taken from Slater et al. [32]. It led to the calculation of a presence score that was calculated by counting the number of answers equal to 6 and 7 for each participant, leading to a score from 0 (very high negative effect on presence) to 3 (no effect on presence). To finish, the participants were given the opportunity to add any comments they desired.

7   Results and Discussion

We performed a Friedman rank sum test to evaluate the influence of the metaphor on spatial understanding. The results indicate a significant influence of the metaphor ($x^2(3) = 39.88$, $p < 0.001$). Pairwise comparisons show that Placeholder leads to a better spatial understanding of the position and direction of the obstacles given by the metaphor (median = 3) when compared to Color Indicator (median = 2, $p < 0.001$), Rubber Band (median = 2, $p < 0.001$) and Arrow (median = 2, $p < 0.001$) (see Fig. 4).

We performed a Friedman rank sum test to evaluate the influence of the metaphor on the presence score we calculated (see Sect. 6). The result indicates a significant influence of the metaphor on presence ($x^2(3) = 13.237$, $p = 0.0041$). The pairwise comparison show that Placeholder leads to a higher presence when compared to Color Indicator ($p = 0.011$), Arrow ($p = 0.022$) and Rubber Band ($p = 0.021$).

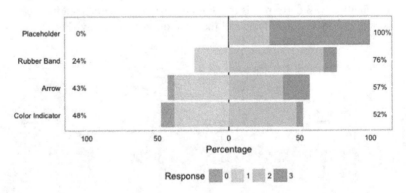

Fig. 4. Distribution of the users spatial understanding with each technique (0 = bad, 1 = low, 2 = medium, 3 = high).

After the last run, the participants were asked to sort the metaphors from best to worst. There is a significant influence of the metaphor ($x^2(3) = 29.28$, $p < 0.001$) on this ranking. Pairwise comparison show that the participants rated Placeholder significantly higher than Color Indicator (p = 0.043), Rubber Band (p = 0.009) and Arrow (p = 0.0013) (see Fig. 5).

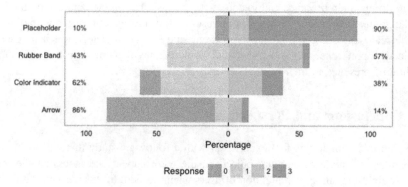

Fig. 5. Distribution of the subjective preference ranking (0 = lowest-rated, 3 = highest rated).

These results indicate that, in all extent, the Placeholder metaphor is the most efficient leading to higher spatial comprehension of the information. Subjective evaluation also indicates a strong preference of the subjects for this metaphor. Not only was it ranked as the preferred technique. It was also perceived as the most intuitive. It is not surprising that it is the technique with the lowest impact on presence.

Informal debriefing confirms these results. The Placeholder metaphor has been described by many as very easy to understand, as they are used to avoiding such obstacles from the real world. For many, the placeholder trees fit into the scenery "naturally", "realistically" and without being disturbing. Surprisingly, no one complained that the trees were standing in their way. Some participants regarded the Rubber Band as a "feel-good zone" or safety zone and saw it as an advantage. Although the Color Indicator metaphor is based on the same basic principles and differs only in the visual representation of the information, some participants reported that the Rubber Band metaphor was more intuitive than the Color Indicator metaphor (however the comparison was not significant). In particular, the spatial relation of the deformation made it easier for many to deduce the position and dimensions of the obstacles.

A striking number of participants were irritated by the Arrow metaphor. First of all, some people remarked that the displayed arrow reminded them of route and navigation systems. Many found it difficult to rethink and look at the arrow as an indication of the position of an invisible target in space.

These results indicate that the participants strongly favour the dimension of presence over the capacity of the metaphor to ensure safety. Indeed, Placeholder was strongly integrated in the VE, leading to lower impact on presence while the other metaphors promoted the strategy of foresight and cautious advancement on the chosen path. The participants had to devote part of their attention to the metaphors, which was probably felt as disturbing for the accomplishment of the task. In general, most of the participants found the not always visible metaphors less pleasant, and the "sudden" appearance of the metaphors was often criticized.

However, if the purpose of the metaphor is to alert the user of a possible danger in its immediate surroundings, it is mandatory that it can capture the user's attention. The effectiveness of the alert metaphor should therefore not be assessed on the basis of the ability to maintain presence and to cope with the task, but on the contrary on the basis of the ability to draw the attention of the user and to inform them of a possible danger in the real world. To that extent, the present study is incomplete, since it focuses on subjective preferences of the users, and demonstrates that they strongly prefer the feeling of presence in their evaluations.

8 Conclusions and Perspective

We plan to complete this first experiment with a follow-up study that will include the possibility of capturing moving objects in the environment and assessing the metaphor's ability to attract the attention of users when needed. In this respect, the Rubber Band, while it received lower rates from the participants in our study, could be a good metaphor. Using non-visual metaphors, e.g. spatialized sound, will also be considered. Finally, another important aspect we will evaluate in our upcoming study is the impact of such alert metaphors on cybersickness, and the effects on the stress felt by the users during and after immersion.

Acknowledgments. The authors would like to thank the participants of the study for their time and precious feedback.

References

1. Slater, M., Usoh, M., Steed, A.: Taking steps: the influence of a walking technique on presence in virtual reality. ACM Trans. Comput.-Hum. Interact. **2**(3), 201–219 (1995) https://doi.org/10.1145/210079.210084. Accessed Sept 1995
2. Waller, D., Loomis, J.M., Haun, D.B.M.: Body-based senses enhance knowledge c directions in large-scale environments. Psychon. Bull. Rev. **11**(1), 157–163 (2004) https:// doi.org/10.3758/BF03206476. Accessed 01 Feb 2004
3. Waller, d., Bachmann, E., Hodgson, E., Beall, A.C.: The hive: a huge immersive virtua environment for research in spatial cognition. Behav. Res. Methods **39**(4), 835–843 (2007 https://doi.org/10.3758/BF03192976. Accessed 01 Nov 2007
4. Usoh, M., Arthur, K., Whitton, M.C., Bastos, R., Steed, A., Slater, M., Brooks, Jr, F.P Walking> Walking-in-place> Flying, in Virtual Environments. In: Proceedings of the 26t Annual Conference on Computer Graphics and Interactive Techniques (SIGGRAPH 1999 pp. 359–364. ACM Press/Addison-Wesley Publishing Co., New York (1999). https://do org/10.1145/311535.311589
5. Ruddle, R.A., Lessels, S.: The benefits of using a walking interface to navigate virtua environments. ACM Trans. Comput. Hum. Interact. **16**(1), 5 (2009). https://doi.org/10.114: 1502800.1502805
6. Chance, S.S., Gaunet, F., Beall, A.C., Loomis, J.M.: Locomotion mode affects the updatin of objects encountered during travel: the contribution of vestibular and proprioceptive inpu to path integration. Presence **7**(2), 168178 (1998). https://doi.org/10.1162/1054746985656 9(1998)

7. Peck, T.C., Fuchs, H., Whitton, M.C.: The design and evaluation of a large-scale real-walking locomotion interface. IEEE Trans. Vis. Comput. Graph. **18**(7), 1053–1067 (2012). https://doi.org/10.1109/TVCG.2011.289

8. Matthew, G.: Roomscale 101 - An Introduction to Roomscale VR (2017). https://blog.vive.com/us/2017/10/25/roomscale-101/. Accessed 19 June 2018

9. Oculus guardian system (2018). https://developer.oculus.com/documentation/pcsdk/latest/concepts/dg-guardiansystem/. Accessed 19 June 2019

10. Aaron, P., Zeller, M., Wojciakowski, M.: Mixed reality: enthusiast guide - FAQ (2017). https://docs.microsoft.com/enus/windows/mixed-reality/enthusiast-guide/before-you-buy-faqs. Accessed 19 June 2019

11. Razzaque, S.: Redirected walking. Ph.D. dissertation. Chapel Hill, NC, USA. Advisor(s) Brooks, Jr., Fredrick, P. AAI3190299 (2005)

12. Bachmann, E.R., Holm, J., Zmuda, M.A., Hodgson, E.: Collision prediction and prevention in a simultaneous two-user immersive virtual environment. In: 2013 IEEE Virtual Reality (VR), pp. 89–90 (2013). https://doi.org/10.1109/VR.2013.6549377

13. Langbehn, E., Steinicke, F.: Redirected walking in virtual reality (2018). http://basilic.informatik.uni-hamburg.de/Publications/2018/LS18

14. Scavarelli, A., Teather, R.J.: VR collide! comparing collision avoidance methods between co-located virtual reality users. In: Proceedings of the 2017 CHI Conference Extended Abstracts on Human Factors in Computing Systems (CHI EA 2017), pp. 2915–2921. ACM, New York (2017). https://doi.org/10.1145/3027063.3053180

15. Cirio, G., Marchal, M., Regia-Corte, T., Lcuyer, A.: The magic barrier tape: a novel metaphor for infinite navigation in virtual worlds with a restricted walking workspace. In: Proceedings of the 16th ACM Symposium on Virtual Reality Software and Technology (VRST 2009), pp. 155–162. ACM, New York (2009). https://doi.org/10.1145/1643928.1643965

16. Cirio, G., Vangorp, P., Chapoulie, E., Marchal, M., Lecuyer, A., Drettakis, G.: Walking in a cube: novel metaphors for safely navigating large virtual environments in restricted real workspaces. IEEE Trans. Vis. Comput. Graph. **18**(4), 546554 (2012). https://doi.org/10.1109/TVCG.2012.60

17. Hoffman, H.G.: Physically touching virtual objects using tactile augmentation enhances the realism of virtual environments. In: Proceedings of the IEEE 1998 Virtual Reality Annual International Symposium (Cat. No. 98CB36180), pp. 59–63 (1998). https://doi.org/10.1109/VRAIS.1998.658423

18. Simeone, A.L., Velloso, E., Gellersen, H.: Substitutional reality: using the physical environment to design virtual reality experiences. In: Proceedings of the 33rd Annual ACM Conference on Human Factors in Computing Systems (CHI 2015), pp. 3307–3316. ACM, New York (2015). https://doi.org/10.1145/2702123.2702389

19. Simeone, A.L.: The VR motion tracker: visualising movement of non-participants in desktop virtual reality experiences. In: 2016 IEEE 2nd Workshop on Everyday Virtual Reality (WEVR), vol. 00, p. 14 (2016). https://doi.org/10.1109/WEVR.2016.7859535

20. Nescher, T., Zank, M., Kunz, A.: Simultaneous mapping and redirected walking for ad hoc free walking in virtual environments. In: 2016 IEEE Virtual Reality (VR), pp. 239–240 (2016). https://doi.org/10.1109/VR.2016.7504742

21. Sra, M., Garrido-Jurado, S., Schmandt, C., Maes, P.: Procedurally generated virtual reality from 3D reconstructed physical space. In: Proceedings of the 22nd ACM Conference on Virtual Reality Software and Technology (VRST 2016), pp. 191–200. ACM, New York (2016). https://doi.org/10.1145/2993369.2993372

22. Ghosh, S., et al.: NotifiVR: exploring interruptions and notifications in virtual reality. IEEE Trans. Vis. Computer Graph. **24**(4), 1447–1456 (2018). https://doi.org/10.1109/TVCG.2018.2793698(2018)
23. OpenVR API Documentation (2017). https://github.com/ValveSoftware/openvr/wiki/API-Documentation. Accessed 3 June 2018
24. Unity download archive (2018). https://unity3d.com/getunity/download/archive. Accessed 19 June 2018
25. HTC Vive Tracker (2018). https://www.vive.com/de/vive-tracker/. Accessed 3 June 2018
26. Steward, J., Lichti, D., Chow, J., Ferber, R., Osis, S.: Performance assessment and calibration of the Kinect 2.0 time-of-flight range camera for use in motion capture applications (7692). FIG Working week 2015 From the Wisdom of the Ages to the Challenges of the Modern World Sofia, Bulgaria, 17–21 May 2015 (2015)
27. Breuer, T., Bodensteiner, C., Arens, M.: Low-cost commodity depth sensor comparison and accuracy analysis. In: Proceedings of SPIE 9250, Electro-Optical Remote Sensing, Photonic Technologies, and Applications VIII; and Military Applications in Hyperspectral Imaging and High Spatial Resolution Sensing II, 13 October 2014, p. 92500G (2014)
28. Point Cloud Library 2018. Point Cloud Library (PCL) (2018). http://docs.pointclouds.org/trunk. Accessed 20 June 2018
29. Munaro, M., Basso, F., Menegatti, E.: Tracking people within groups with RGBD data. In: 2012 IEEE/RSJ International Conference on Intelligent Robots and Systems, pp. 2101–2107 (2012). https://doi.org/10.1109/IROS.2012.6385772
30. Fischler, M.A., Bolles, R.C.: Random sample consensus: a paradigm for model fitting with applications to image analysis and automated cartography. In: Readings in Computer Vision, pp. 726–740. Elsevier (1987)
31. Kennedy, Robert S., Lane, Norman E., Berbaum, Kevin S., Lilienthal, Michael G.: Simulator sickness questionnaire: an enhanced method for quantifying simulator sickness. Int. J. Aviat. Psychol. **3**(3), 203–220 (1993). https://doi.org/10.1207/s15327108ijap03033(1993)
32. Slater, M., Usoh, M., Steed, A.: Depth of presence in virtual environments. Presence: Teleoperators Virtual Environ. **3**(2), 130–144 (1994)

Digital Cultural Strategies Within the Context of Digital Humanities Economics

Evangelos Markopoulos[1](✉), Panagiotis Markopoulos[2],
Mika Liumila[3], Younus Almufti[1], and Chiara Romano[1]

[1] HULT International Business School, 35 Commercial Road, Whitechapel,
London E1 1LD, UK
evangelos.markopoulos@faculty.hult.edu,
{yalmufti2015, cromano2015}@student.hult.edu
[2] University of the Arts London, Elephant and Castle, London SE1 6SB, UK
p.markopoulos1@arts.ac.uk
[3] Turku University of Applied Science,
Joukahaisenkatu 3, 20520 Turku, Finland
mika.luimula@turkuamk.fi

Abstract. Staying sustainable in the world of cultural heritage is a major organizational challenge nationally and internationally. Due to the global financial crisis the funds available to sustain the operations of museums and libraries become difficult not only to obtain them but also to utilized them effectively and efficiently. The operational costs of museums increase over the time due to exhibit maintenance and acquisition costs. This cost is inversely proportional to the revenues that can be generated. Virtual reality, avatar technologies, virtual worlds, holograms, gaming and gamification can offer creative interactivity and unique experiences with low or no cost to the global visitor and introduce new revenue streams. This paper practically integrates the realization of digital cultural strategies and operations within the context of digital humanities economics that can turn museums and libraries from cost centres to profit centres for the benefit of the humanity and the society.

Keywords: Digital museum · Digital library · Gamification · Humanities ·
Technology · Virtual Reality · Avatar · Economics

1 Introduction

The sustainability of the world's cultural heritage is a major issue in national and international level. The global financial crisis had also an impact on the available funds museums and libraries operations and maintenance. The limited funds that can possibly be gathered from various sources need to be spend effectively and efficiently. This situation becomes worse as the global financial crisis affects the people's ability to travel for pleasure, and even more to spend on museum visits. At the same time the operational costs of museums and libraries, of any kind, continuously increases due to the cost of the exhibit's maintenance and acquisition. This cost is inversely proportional to the revenues that can be generated. The fact that the history and cultural

© Springer Nature Switzerland AG 2020
T. Ahram (Ed.): AHFE 2019, AISC 973, pp. 283–295, 2020.
https://doi.org/10.1007/978-3-030-20476-1_29

heritage of all nations is spread out in multiple museums around the world, reduces the interest of various target groups to visit a specific museum or a library. On the other hand, it is not possible to have dedicated and exclusive physical museums on specific civilizations, on all chronological periods or themes which can cover the whole spectrum of the existing exhibits globally. The Greek and the Egyptian history, for example, has been spread to hundreds of museums around the world. This impacts the Greek and Egyptian museums in terms of visitors and revenue as they operate with a small fraction of their history. The same is applied to other museums that operate with small bouquets of exbibits from different countries and civilizations, not enough to provide value to the visitor.

Under these circumstances, museums and libraries have turned out to be cost centres, living on the generosity of the donors, sponsors, corporate social responsibility programs, national and international cultural funds, etc. Museums and libraries must turn into profitable organizations not only for their sustainability but also for the secured preservation of their exhibits that contribute to the cultural heritage of the society and humanity. An inflection point on this cultural heritage preservation, dissemination and utilization equation can be achieved via the transformation of the physical museums and libraries into digital ones that can be supported by state-of-the-art technologies from other types of disciplines, industries and markets.

Virtual reality, avatar technologies, virtual worlds, holograms, gaming and gamification are only a few of the many technologies that can offer creative interactivity and unique experiences with low or no cost, to the global visitor. Based on our previous experiences virtual and augmented reality offers new ways to experience cultural context through the tools provided to create pre- and post-experiences to visitors in museums and libraries. Augmented reality in turn can be used to increase cultural experiences on-site. Even though there have been many museums already on digitization route, the actual digitization of their exhibits and artifacts is not enough. Digital operations strategy and business operations modeling must be in place to drive a museum into the digital era, and beyond.

2 Sizing the World Cultural Heritage

Cultural heritage refers to the customs, practices, places, objects, artistic expressions and values that are passed down from generation to generation. These values can be both tangible or intangible and often refer to the way previous generations lived [1]. Ancient cultural heritage mainly manifests itself through civilizations, as they had the tendency to record their history in written form revealing a greater heritage insight. The first records of history mainly occurred in late human development period five thousand years ago in ancient civilizations such the Sumerian the Greek and the Egyptian [2]. Great civilizations are in fact difficult to track, size or value primarily due to the unfortunate impact they had from during colonization period where most of such heritage and artefacts ended up fragmented across the colonizers. The British Museum for example, holds over one hundred thousand artefacts, while the Louvre holds more than seventy thousand. Thus, it is hard to track and set a certain value around these artefacts as they are associated with ambiguity when it comes to their volume, size and origin/ownership [3]. This artefact scattering needs to be carefully examined as it creates a challenge on sizing its value and restricts the ability to unify not only a culture, but the global cultural heritage.

3 Digital Cultural Heritage

Digitalization on the surface level is described as the transformation and conversion of information into a digital format, organized into a discrete form of binary data that modern devices process [4]. Digital Cultural heritage is defined as the opportunity of taking classical aspects of heritage, such as any unique resource of human knowledge and expression and transforming them into a digital material such as texts, databases, images and other digitalization aspects [5]. Accessibility, transferability and preservability are only a few of the digitize information benefits.

Digitized cultural heritage information allows people to access it without the need of traveling around the world to experience renowned artefacts and cultural landmarks. It can further more allow access to cultural heritage from inaccessible areas. Such an example can be the rich heritage of Iraq. The country is considered today as one of the most unsafe countries to travel due to armed conflict. This situation withholds from the public and the world as well, access to some of the most valuable heritage centres [6]. Digitization in this case can help to safely share, experience and transfer the wealth of knowledge from such heritage centres to all those who want to experience and learn from it, after all the digitized content can be a source for further and simpler education. Research indicates that the integration of technology in education can increase in learning by 12% [7]. Regarding the preservability of cultural heritage, most physical objects such as buildings, cities and even countries evolve over the time. Digitization can offer a gateway which allows access to information of such evolutions. Significant cultural heritage digitization projects are currently implemented in Egypt as part of national goals or under the UN Social Development Goals aiming to achieve digital transformation and industrial innovation that can increase the productivity in the information technology and communication sector [8].

Organizations like CULTNAT (Center for Documentation of Cultural and Natural Heritage) and the Library of Alexandria implement high impact strategic projects such towards the digitization of the Egyptian cultural heritage. CULTNAT aims to apply the latest technological innovations in documenting Egypt's tangible and intangible cultural heritage and Egypt's natural heritage, which includes information about natural areas and their biological components [9].

The project Eternal Egypt (shown in Fig. 1), brings to light over five thousand years of Egyptian civilization. The project is a living record of a land rich in art and history, people and places, myths and religions. The stories of Eternal Egypt are told using the latest interactive technologies, high-resolution imagery, animations, virtual environments, remote cameras, three-dimensional models and more [10].

Culturama a technology of CULTNAT (shown in Fig. 2), is another high impact innovation of Egypt on digital cultural heritage. It is a cultural panoramic show portraying the history of Egypt and is considered the first interactive nine-screen panorama worldwide. Culturama allows the presentation of a wealth of data layers, where the

Fig. 1. Eternal Egypt.

presenter can click on an item and go to a new level of details. The technology is a remarkably informative multi-media presentation of Egypt's heritage over 5,000 years of history up to modern times [11].

In the same context, cultural heritage projects are developed under Virtual Reality and Augmented Reality in various ways to increase sensory and cognitive experiences for tourism. The Medieval Gastro Box developed at the Turku Game Lab in Finland (shown in Fig. 3) is such a case were virtual reality has been applied successfully [12]. The Turku Castle and Turku Exhibition Centre is another similar example that implemented augmented reality gamified applications for visitors to enable rich content production as a combination of digital and physical artifacts [13].

Fig. 2. Cultrama technology

Predictions today indicate a rapid grow and rise on both virtual and augmented reality market. Augmented reality has reached today the maturity needed to be practically applied in many sectors. The Pokemon Go case indicated players engagement immersion and experiences [14]. Even that Pokemon GO was presented a few of year ago, the technology readiness level is appropriate for mass markets. However augmented reality in frequent use on cultural context has not been recorded yet. Learning from the Pokemon GO case, AR and VR can be applied similarly on archeological sites and cultural heritage areas to create a sense of living in places and spaces people cant walk or go.

Fig. 3. Enriching sensory and cognitive experiences for tourism and hospitality industry in Medieval Gastro Box

4 From Physical to Digital Museums and Libraries

Part of the cultural digitization process is the transition to digital museums and libraries as the world moves towards a digitally focused environment. Being part of cultural digitization, the digital museums and libraries can provide preservability and accessibility of highly important artefacts and books of knowledge. Digitalization of artefacts assures long term and secured preservation in the case artefacts get looted or damaged. In 2003 The Iraq National Museum was looted, and 8,000 important artefacts are still lost to this day. Therefore, digitization of museums also helps prevent and track objects from such instances [15]. Digital museums can also be a contribution to experts to study artefacts in a more accessible and interactive way provided that high-end technology is used to digitize them.

Utilizing the augmented reality megatrend, museums such as the Smithsonian National Museum of Natural History decided to currently combine the digitized information and the tangible artefacts, allowing viewers to use their phones to further inspect the artefacts [16]. Connectivity and convergence can be considered as another trend being utilized by digital museums and libraries. The British Museum allows now users to view museum areas through virtual reality, creating a fully immersive and digital experience without the need for the user to be there [17]. On the digital libraries end, in 2016 Egypt launched the Egyptian Knowledge Bank, one of the world biggest digital library hoping to increase productivity in research through better accessibility [18]. The Egyptian Knowledge Bank will be further expanded through national effort from Saudi Arabia, Egypt and the United Arab Emirates creating the Arab Digital Union which will merge all existing digital libraries from these countries [19].

5   Digital Museums and Libraries Operations

In order to create a cultural sustainable future, museums need to take care of their assets making sure that their galleries remain valuable to future generations. Innovation has a fundamental task in making cultural heritage information accessible to everyone. However, it is the overall digital operations framework that needs to be defined first in that will allow the development of a relationship with the audience.

Museums need to meet and exceed the visitors' expectations by understanding their interests and identify and offer their preferred channels to interact with. According to the Museums Report by Axiell in 2016, there has been an increase of 50% of visitor numbers due to museums' websites and a significant increase of 82% in activity on their social media platforms. Museums will have to constantly adapt to new changes in order to engage with the visitors through online channels [20] and to do that they need to do more than web-sites and on-line sales.

Today the term Digital Museums mostly refer to the development of a mobile device which gives the visitors access to audio files and images throughout the tour. Users can browse the entire list of the museum through the device in order to choose the desired option; such as the number of the stop, location or artist name, brief descriptions, etc. The primary reason in developing digital museums in this sense, was to create a more sustainable and reliable approach on dealing with the entire galleries. Such an approach was considered more efficient than developing online apps which are more costly, take long time to be implemented and only a small percent of the visitors downloads them; therefore, by having a platform which empowers associates in the organisation to offer mobile tours is relatively low cost and quick, perusing a cost-effective strategy. In such applications mobile devices create and improve experiences by combining both digital and non-digital libraries.

It is essential to note that digitized collections can reach large populations and increase the interest of the users in the long run leading to a physical visit. Despite the fact that digitized collections can't replace the genuine experience of visiting a museum, their prime purpose is to make them accessible to everyone. The main operations strategies for digital museums are presented in Table 1.

Table 1. Digital museum operations strategies.

Digital museum strategy	Description
Mobile technology	The use of mobile devices throughout a tour improves knowledge to non-specialist individuals of the public
Digital content exchange	Reduces the expenditures of digitized collections by sharing the information with other museums. Visual assets are costly yet can be effectively and efficiently shared between galleries, benefiting the consumer once the data is suitably collected and packaged
Social media	Maintain updated the social media platforms, since they are the modern channel for communication and networking. Social media platforms

(continued)

Table 1. (*continued*)

Digital museum strategy	Description
	allow organizations to interact with clients and provide a real one to one communication, giving them the freedom to share their opinions, experiences or anything they want
On-line monetization	Museums using online monetisation of small services and micro-sales, in order to increase their profit
Protection	Protection of valuable artefacts by having them digitized under various techniques

Despite the opportunities digital museums can offer, the universal development of digital museums solutions is still challenging due to the lack of standardized digitalization methods for museums. There is an uncertainty between big strategic visions and daily operations in the transformation of museums and libraries [21]. Han et al. [22] have stated that a quality model is needed to implement AR applications in Urban Heritage applications based on their findings in Dublin AR project. An EU H2020 project called ViMM (Virtual Multimodal Museum) has published a manifesto for all stakeholders from industry to end-users who are interested in the cultural heritage sector. They have emphasized audience participation, the use of harnessing technologies (e.g. artificial intelligence, virtual reality and augmented reality), openness to partnerships, digitalization based on globally used standards, and sufficient contextual information [23].

Similar to digital museums, digital libraries refer to an affordable media-management framework that allows the institutions to conserve their digital assets. Users are able to see, download, and use the assets uploaded in the system. These digital libraries give access to shared shelves, providing image collections from the world's driving exhibition halls, photograph documents, scholars, and artists in an effective repository. They can be a very complete asset accessible for instructive and academic use across disciplines with numerous rare and valuable accumulations that cannot be seen somewhere else [24]. The digital libraries can also satisfy the qualified majority of the users' requirements for digital images. They can offer access to its consistently developing collection and supply useful tools to enable users to organize and present.

The main advantage of digital libraries is that provide an easy and quick access to books, files and pictures of different types to everyone and everywhere. Digital libraries reduce the library's personnel costs and increase the potential to store much more data than a physical library and have all that accessible around the clock.

6 Advance Technologies for the Operations of Digital Museums and Libraries

The growth of technology in science and applications seems to be followed and adapted in the digital culture industry. The British Museum uses virtual reality to transport visitors to the bronze age. This project consists in engaging visitors with the past through virtual reality, using 3D headsets, tablets, and a projection of a dome structure sufficiently expansive to fit groups of five to enter. Although older generations find the VR confusing, younger audiences embrace it positively, encouraging visitors to move from the reproduced past to the genuine objects. Recent studies have shown that more than 60,000 young people have visited the museum due to the launch of virtual reality; therefore, it has expanded the enthusiasm to visit the exhibition hall [25].

Virtual reality is just one of the several digital strategies that museums can adopt to see the cultural world. The introduction of augmented reality allowed galleries to overlay instructive or entertaining content without adjusting the physical space, since real-world objects are augmented by PC produced perceptual data. It contains different sensory modalities, including visual, sound-related, haptic, somatosensory, and olfactory [26]. According to The New York Times, the Newseum has launched a new project using VR and AR, attracting a diverse audience including gamers.

Artificial intelligence is progressively being utilized today by historical centres of all sizes around the world, developing everything from robots, chatbots and sites, to devices that help analyse guest information and their collections. In 2018 it has been produced for the first-time humanoid robots to answer guests' inquiries and recount stories, utilizing voice, signals and an interactive touch screen. Also, they are able to dance, play games and pose for pictures. Artificial intelligence has been introduced in museums in order to engage the visitors by creating a playful and joyful experience [27].

However the use of advanced interactive technologies in digital museums and libraries open various research questions to be tackled in terms of cost-effectiveness (digital new value creation), methodology (for pre-defined preferred digital content), research methods (user centric design, rapid prototyping etc.), 3D modelling process (e.g. interoperability issues for various software platforms), commercial viability, test procedures, cultural differences, scalable business models (including business ecosystems).

7   Examples of a Gamified Digital Museums and Libraries

Due to digitalization era, museums and libraries have already digital content for new value creation through various types of applications. Gaming and gamification of cultural heritage is a growing sector in the digital culture industry. Digital property can be seen as a raw material for innovative gamified solutions which can be used in any possible way. An example is the Turku Castle application which represents renaissance era paintings in a form which can serve visitors who are interested in stories behind paintings but also those who are more interested in gamified features. Both of tourist and game-mode engage a physical visit to Turku Castle (shown in Fig. 4) [14].

Fig. 4. Digital content enables rich cultural experience.

The Turku Castle application as an example of AR applications has been extended and further developed to be suited for nearly any kind of tourism-based activity, event or a venue. Moreover, as the Medieval Gastro Box is located within two containers and a mobile unit, it can bring tourism related activities as a pop-up VR experience into various real-world locations, providing a new concept for many fields of tourism and hospitality industries, e.g. marketing, promotion, trade fairs and expositions.

Besides the museums VR and AR technology can also be used to promote cultural heritage in events such as exhibitions. Fair game platforms can be extended to cover for example cultural heritage topics such as Mikael Agricola (shown in Fig. 5) who is the author of the first book published in the Finnish language. In Mikael Agricola fair game AR was used to launch small mini games which were physically located around the exhibition center. Mikael Agricola post cards were used as visual markers for triggering mini games which were designed for pupils visiting in the exhibition as a part of their Finnish language studies.

Fig. 5. Mikael Agricola fair game designed for education in cultural heritage.

VR Ships for Forum Marinum (Fig. 6) is in another example of VR applications developed for cultural heritage. This application gives for the museum visitor an experience of boarding the famous Finnish ships and airplanes from wartime Finland in

Fig. 6. VR Forum Marinum demonstration Finnish ships & airplanes from wartime.

a realistic environment with VR glasses. In this experiment, focus was on providing historically accurate ships and other relevant content. In addition, also audio and sounds have been designed based on the location of the player (submarine, airplane etc.).

8   Cultural Heritage Digitization Funding

Digital libraries and digital museums compose the key areas of applications of the wider cultural heritage digitization strategies adapted by many nations globally over the last decade at least. The European Commission Directorate General for Communications Networks, Content & Technology has conducted extensive policy, coordination as well as funding actions to supplement Member States' cultural policy in areas of digitization and online access to cultural material and digital preservation.

Programs such as the Horizon 2020, the Interreg and others fund projects related to accessing cultural heritage through digital infrastructures, emotive virtual cultural experiences and other related initiatives [28].

On the same route the United Kingdom offers numerous funding opportunities as well. The Joseph Rowntree Charitable Trust provides funding to support cultural heritage digitization projects. The Heritage Lottery Fund (HLF) uses revenue generated from The National Lottery to provide funding for a variety of UK heritage projects including digitization and digital archiving projects. The Wellcome Trust offer funding for the digitization of UK based archive and library collections through it Research Resources Award grant scheme. The Grant Fund supports the digitization of collections relating to the history of medicine and health, the medical humanities or social sciences. The Arts & Humanities Research Council (AHRC) offers a number of funding streams that will support the creation of digital resources, with a variety of projects funded, including their Standard Research Grants and Follow on funding scheme [29].

Similar strategies can be found in Egypt, Mexico, China, and almost everywhere in the world. The new economy that rises around the implementation of cultural heritage projects primarily on digital museums and library, but also on managing and maintaining such digital operations cannot be ignored. Digital humanities via the digitization of the global cultural heritage reshapes the way art and knowledge is provided to the new generations, by creating new economies and new societies.

9 Areas of Further Research

The research conducted for this paper forms the base for a more techno-economic analysis on the design, implementation, management and maintenance of digital museums, libraries and other cultural heritage and digital humanities areas of applications. The effectiveness of a digital project, especially in culture and humanities, is related to the usage it achieves and the community it creates around it. Techniques, methods and practices such as gamification, immersive education via virtual reality, augmented reality, artificial intelligences and other technologies plays a key role to the success or such initiatives. If applied properly they also assure profitability which is essential for their maintenance and continuous enhancements needed to keep on with the user trends and needs. This work will be extended on studying, in depth, the cost-benefit analysis, user engagement, return on investments, revenue streams and other techno-economic elements in order to understand furthermore the cost, time, effort and investment (in human and financial capital) needed towards making the transition from the physical to the digital culture and humanities.

10 Conclusions

Digital cultural heritage and digital humanities, in a wide sense, is not anymore, a future trend but a present need. It is part of national and international digitate strategies designed and executed by nations and groups of nations over the last decade.

The increase of globalization creates the need to unify, share and experience the global cultural heritage. This need becomes bigger in our days due to the global financial crisis which makes access to arts, humanities and culture restricted for many. Digitization seems to be a solution to these challenges, but the cost, time and effort needed to move into a digital period must be studied against the world's readiness and willingness for such a transition.

The paper indicated with examples, operations and technologies the digitization movement that takes place on museums and libraries. However, this movement is at an early stage with low maturity on both technology adaptations, awareness and dissemination. Going digital on cultural heritage is certainly the way to proceed not only for the preservation, usage and utilization of such a wealth but mostly for the evolution of the society by understanding better and deeper that we are all connected in this planet and we shall always be.

References

1. ICOMOS: International Cultural Tourism Charter. Principles and Guidelines for Managing Tourism at Places of Cultural and Heritage Significance. ICOMOS International Cultural Tourism Committee (2002)
2. Kramer, S.: History Begins at Sumer, Thirty-Nine Firsts in Man's Recorded History. The University of Pennysylvania, Philadelphia (1990)

3. History of the collection. https://www.britishmuseum.org/about_us/departments/ancient_egypt_and_sudan/history_of_the_collection.aspx
4. WhatIs.com. What is digitization? https://whatis.techtarget.com/definition/digitization
5. Concept of Digital Heritage | United Nations Educational, Scientific and Cultural Organization. http://www.unesco.org/new/en/communication-and-information/access-to-knowledge/preservation-of-documentary-heritage/digital-heritage/concept-of-digital-heritage/
6. Iraq Travel Advisory. travel.state.gov/content/travel/en/traveladvisories/traveladvisories/iraq-travel-advisory.html
7. Tamim, R.M., Bernard, R.M., Borokhovski, E., Abrami, P.C., Schmid, R.F.: What forty years of research says about the impact of technology on learning. Rev. Educ. Res. **81**(1), 4–28 (2011)
8. Egypt Digitization 2020 – A New National Digital Transformation Initiative Planned to for Egypt by Resilience. http://resilienceand.co.uk/egypt-digitization-2020-a-new-national-digital-transformation-initiative-planned-to-for-egypt-by-resilience/
9. Bibliotheca Alexandrina. Center for Documentation of Cultural and Natural Heritage (CultNat). https://www.bibalex.org/en/center/details/thecenterfordocumentationofculturalandnaturalheritagecultnat
10. Eternal Egypt. http://www.eternalegypt.org/EternalEgyptWebsiteWeb/HomeServlet?ee_website_action_key=action.display.home&language_id=1
11. Bibliotheca Alexandrina Cultrama. https://www.bibalex.org/en/project/details?documentid=295
12. Qvist, P., Bulatovic Trygg, N., Luimula, M., Peltola, A., Suominen, T., Heikkinen, V., Tuominen, P., Tuusvuori, O.: Demo: medieval gastro box – utilizing VR technologies in immersive tourism experience. In: Proceedings of the 5th IEEE Conference on Cognitive Infocommunications, pp. 77–78 (2016)
13. Bulatovic Trygg, N., Luimula, M.: Cultural heritage in a pocket - case study "Turku castle in your hands". In: Proceedings of the 5th IEEE Conference on Cognitive Infocommunications, pp. 55–58 (2016)
14. Pyae, A., Luimula, M., Smed, J.: Investigating players' engagement, immersion, and experiences in playing Pokemon go. In: Proceedings of the 2017 ACM SIGCHI Conference on Creativity and Cognition, 27–30 June 2017, pp. 247–251 (2017)
15. Fifteen years after looting, thousands of artefacts are still missing from Iraq's national museum. http://theconversation.com/fifteen-years-after-looting-thousands-of-artefacts-are-still-missing-from-iraqs-national-museum-93949
16. Five Augmented Reality Experiences That Bring Museum Exhibits to Life. https://www.smithsonianmag.com/travel/expanding-exhibits-augmented-reality-180963810/
17. Explore ancient Egypt in our Virtual Reality tour (2017). https://blog.britishmuseum.org/explore-ancient-egypt-in-our-virtual-reality-tour/
18. Egypt Launches the 'World's Biggest Digital Library' (2016). http://cairoscene.com/ArtsAndCulture/Egypt-Launches-the-World-s-Biggest-Digital-Librar
19. El-Bakry, F.: Egypt, KSA, UAE to share knowledge via Arab Digital union (2018). http://www.egypttoday.com/Article/1/61565/Egypt-KSA-UAE-to-share-knowledge-via-Arab-Digital-union
20. Digital Transformation in the Museum Industry. Axiell ALM (2016). alm.axiell.com/wp-content/uploads/2016/07/Axiell-ALM-Digitise-Museums-Report.pdf
21. Leorke, D., Wyatt, D., McQuire, S.: More than just a library: public libraries in the 'smart city'. City Cult. Soc. **15**, 37–44 (2018)
22. Han, D.I., Jung, T., Gibson, A.: Dublin AR: implementing augmented reality (AR) in tourism. In: Xiang, Z., Tussyadiah, I. (eds.) Information and Communication Technologies in Tourism, pp. 511–523 (2013)

23. The ViMM Manifesto for Digital Cultural Heritage. https://www.vi-mm.eu/wp-content/uploads/2018/09/ViMM-Manifesto-Revised-Final-4-Sept.pdf
24. Artstor Digital Library. Collections from the World's Leading Museums and Archives. www.biblioteche.unical.it/banchedati/artstor_r_intro_1604.pdf
25. The Guardian, Guardian News and Media. British Museum Uses Virtual Reality to Transport Visitors to the Bronze Age. www.theguardian.com/culture/2015/aug/04/british-museum-virtual-reality-weekend-bronze-age
26. Museums Heritage Advisor. Technology in Museums – Introducing New Ways to See the Cultural World. https://advisor.museumsandheritage.com/features/technology-museums-introducing-new-ways-see-cultural-world/
27. The New York Times. Artificial Intelligence, Like a Robot, Enhances Museum Experience. www.nytimes.com/2018/10/25/arts/artificial-intelligence-museums.html
28. European Commission. Digital Cultural Heritage. https://ec.europa.eu/digital-single-market/en/digital-cultural-heritage
29. Towns Web Archiving. Guide: Sources of Funding for heritage digitisation projects. https://www.townswebarchiving.com/2018/02/sources-funding-for-heritage-digitisation-projects/

Understanding User Interface Preferences for XR Environments When Exploring Physics and Engineering Principles

Brian Sanders[1]([✉]), Yuzhong Shen[2], and Dennis Vincenzi[3]

[1] Department of Engineering and Technology, Embry-Riddle Aeronautical University, Worldwide, Daytona Beach, USA
sanderb7@erau.edu
[2] Department of Modeling, Simulation, and Visualization Engineering, Old Dominion University, Norfolk, VA, USA
vincenzd@erau.edu
[3] Department of Aeronautics, Graduate Studies, Embry-Riddle Aeronautical University, Worldwide, Daytona Beach, USA
YShen@odu.edu

Abstract. In this investigation we seek to understand user technology interface preferences and simulation features for exploring physics-based concepts found in engineering educational programs. A 3D magnetic field emanating from a bar magnet was used as the representative physics concept. The investigation developed three different virtual environments using three different technologies and exploiting the capabilities available through the use of those technologies. This provides a suite of platforms to compare and identify the types of interactions and control capabilities a user prefers when exploring this type of phenomena. These environments will now enable the behaviors and preferences to be identified through observation of the user as they navigate the event and manipulate various options available to them depending upon the technology used.

Keywords: Augment Reality · Virtual Reality · XR · Physics · Engineering

1 Introduction

Technology is readily available and affordable for creating visual and interactive simulated experiences across the spectrum from screen displays to fully immersive experiences using head mounted displays. This simulation capability can be used to create visual experiences that facilitate learning and understanding for engineering students as they explore physics concepts, particularly 3D events, those with steep gradients in the field response, and time dependent events. These simulations can help students connect basic principles and event geometries to the mathematical representations that explain the phenomena being presented to them.

There are multiple platforms for which to create visual and interactive virtual environments. For example, a mobile device can be used to create an augmented reality (AR) experience based on a predefined target. The user can then change their

© Springer Nature Switzerland AG 2020
T. Ahram (Ed.): AHFE 2019, AISC 973, pp. 296–304, 2020.
https://doi.org/10.1007/978-3-030-20476-1_30

perspective by controlling the position of the device camera. Interactive features can be controlled via touch screen capability. On the computer display this same interaction can be accomplished with the use of a mouse or gesture-based system. Likewise, interaction in the full Virtual Reality (VR) experience can be accomplished with handheld wands or a gesture-based control system. The challenge for the educator is to decide which to select for a given learning experience and the design of the overall environment. It is a natural touch point between education and human factors.

Effective computer interface design has been a primary objective of human factors and engineering professionals since the advent of digital displays and technologies began to be widely available in the late 1970s and early 1980s. Beginning with simple monochrome cathode ray tube (CRT) screens for a visual display and a keyboard for user input, the computer interface has evolved to include many different aspects, characteristics, qualities, and features that may or may not contribute to the functionality, goals, and objectives of the overall system. Enhanced keyboards, color displays, split keyboards, mice, trackballs, joysticks, and other such devices all began to flood the markets in search of the ideal ergonomically designed and user preferred configuration. Color, for example, can be used to the user's advantage to help highlight important information that will increase understanding, but it can also be used ineffectively and inconsistently in a way that will add little to the user experience or to achieving the goals of the system in general. When colors are chosen at random for use in similar applications, their effect can be one that promotes confusion and misinterpretation of information instead of enhancement of knowledge and understanding. Meteorology is a good example of an application that has applied color in a consistent and effective manner to make reporting of weather conditions clear and understandable to almost any viewer. Colors are effectively applied to depict temperature and other weather phenomena, and are used consistently to convey information in a way that enhances understanding and situational awareness. And, the weather industry in general has implemented the use of color and symbology in a consistent manner across applications, so regardless of what weather report is being viewed, the information is displayed in a consistent manner to convey understanding of weather conditions and phenomena in a clear and concise manner.

In the same way, interfaces whose objective is to impart information about a particular topic, in this case with the goal of producing enhanced learning and comprehension of physics and engineering based educational concepts, must be designed in specific and intentional manner to be able to exploit the capabilities available in different interfaces and technologies in order to enhance understanding and information transfer about those topics. All interfaces can impart some information just by virtue of presenting the information for viewing to the user, but interfaces that are specifically designed to exploit the capabilities of the technology being used may present certain advantages that will enhance learning in a measurable way. These specific capabilities, desirable to the user, can then be incorporated and developed further to benefit the learner.

State of the art technology began revolutionizing the Human-Machine Interface with the movement away from legacy control devices such as joysticks and other physical controls, toward more innovative interface technologies such as touchscreens, virtual reality displays (VR), augmented reality displays (AR), and mixed reality displays (MR). VR displays or VR-like displays are now affordable and commonplace, and are regularly used as the display of choice when immersion into a 3D environment is preferred [1]. Additionally, when used in combination with technologies such as head tracking, the ability to "move around the environment" and to "become immersed in the environment" is also possible, thereby affording the learner the opportunity to view new perspectives that could only previously be presented in two dimensional (2D) format.

A major reason for using more advanced type of displays is to enhance the user's ability to learn the concepts being presented. Several encouraging examples of applying virtual environments in the classroom have been documented (see for examples, [2–4]). More recently studies based in the augmented reality technology have emerged (see for examples, [5, 6]). These, and other, examples, show the promise of using this technology but they point out that the technology is just that-technology. For the educational environment the main focus is on learning, and the optimal implementation of technology needs to be carefully considered. Otherwise the effectiveness can quickly disappear after the initial excitement begins to fade. If the application or the interface is one that is not preferred by the user and does not facilitate learning beyond the level produced by 2D displays, then there would be little justification for using that technology. The ultimate goal in this research is to better identify the types of interactions and control capabilities a user prefers when exploring this type of phenomena, and to determine if the use of different interactions and control capabilities enhances learning and comprehension. The use of advanced display and control technologies is trendy and state-of-the-art, but for our purposes, it still needs to facilitate learning and enhance the learning experience for the user.

In summary, technology is available and documented results of its potential have shown the promise of integrating the capability 3D simulation offers with innovative instructional design. The challenge is beyond just developing a simulation though; that is doable. The technology is just another tool in the toolbox. The challenge is identifying those learning objectives, particularly those that are the best match of the given technology, and the design of the overall learning environment to include basic informational content, training, assessment components, and ease of navigating through the environment; that is the long-term objective of this investigation. This paper focuses on the design of the physics simulation itself and those desirable interactive features.

2   Learning Environment Design and Development

2.1   Learning Environment

Designing a simulation of a physics or engineering concept is certainly achievable. The challenge is more in the design of the overall learning experience. It needs to guide the student to the desired learning outcomes (instructor controlled) but also enable the student to control the learning experience. In this case the instructor can layout sequence of learning events for which the student can proceed at their own pace and have the ability to check back with certain components. It also needs to contain training on the use of the simulation otherwise the student can quickly get frustrated. The overall goal of this research is the establishment of the complete learning environment (Fig. 1).

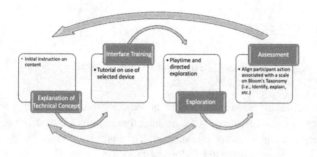

Fig. 1. Simulation environment

2.2   Platforms

To get started with this investigation three different technology options were selected to create the Exploration component of the learning environment shown in the figure above. This will provide the future participant with different sets of capabilities for exploring the learning environment. The first technology platform was a tablet-based platform consisting of a Samsung Galaxy Tab 8 with a 10.1-inch color display and a screen resolution of 1920×1200. The tablet platform is controlled using a touch screen to manipulate the visuals and navigate around the environment. The second technology platform was Screen Space utilizing an Alienware 19-inch laptop with a screen resolution of 1920×1080. Actually, any computer screen would do, the point is that it is a 3D representation in a 2D space. The Screen Space interaction was limited to use of a typical mouse interface. The third technology platform was the Oculus Rift which is an immersive VR headset with OLED color display and screen resolution of 2160×1200. The Oculus Rift platform can be used with either a Leap Motion Controller (LMC) or the use of handheld wands provided by the manufacturer.

2.3   Physics Model

A rectangular, magnetic bar shown in Fig. 2 was selected as the model concept for this investigation. It is a good concept to start with since it has several representative features and principles that cut across engineering concepts. For example, there is a source and a sink model (i.e., north and south poles of the magnetic). This concept is found in other engineering principles such as fluid mechanics and aerodynamics. There is a direction element as the magnetic flux flows from the north to the south pole and there is a gradient associated with the field strength as the distance from the source and sink increases. Some of the mathematical details to support these points are discussed below.

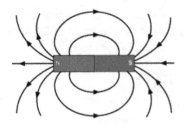

Fig. 2. Bar magnetic field model [7]

The model contains a magnetic source at the north-pole and a magnetic sink at the south-pole. It is a low order model that sufficiently captures the basic physics principles and visualizes field gradients and streamlines. The magnetic induction, B, at a point due to a monopole is governed by the following equation:

$$B = \frac{\mu_0 m}{4\pi \vec{d}} \tag{1}$$

Where

μ_0: permeability of free space $= 4\pi * 10^{-7} \frac{Newtons}{ampere^2}$

m: is the magnetization $\left[\frac{ampere}{meter}\right]$

d: a vector from monopole center to point of interest [meter]

The two-bar magnet is modeled as a source and sink combination or a dipole. The resulting governing equation is shown in Eq. 2. The effect of a magnetic dipole on an arbitrary point P(x, y) is shown in the schematic in Fig. 3 below. This model is easily extended to 3D in the simulation.

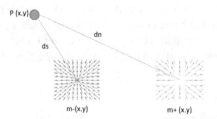

Fig. 3. Dipole model

$$B(x, y) = \frac{\mu_0 m}{4\pi} \left(\frac{1}{\vec{d_s}} + \frac{1}{\vec{d_n}}\right) \tag{2}$$

2.4   Simulation Development

The simulations were developed using Unity3D®. An AR simulation was created for the tablet in Unity with the aid of Vurforia, which can be used to create marker-based AR applications. In this case the user simply points the camera at an image or object and then the simulation is superimposed over it. The simulation is shown below in Fig. 4. The marker image is that of the bar magnet. Once that is detected, the directional arrows are displayed along with their magnitude. Careful consideration had to be given to the placement of the touch screen buttons on the right side. These buttons controlled the view options of grid point display, direction indicators, and streamlines. In their current position they are easily activated without moving the user's hand (i.e., can reach with their thumb). This approach allows the users to view the 3D field representation from several angles and even dive down into the field itself. The only drawback is that the marker image needs to be always in the camera's view.

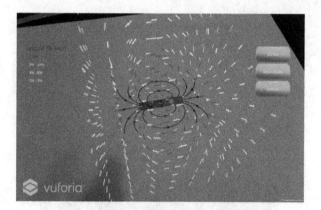

Fig. 4. AR application using samsung tablet

Figure 5 shows the visual display using the laptop and Fig. 6 shows the streamlines. The arrows provide good information on the strength and the streamlines provide good information on the field gradient and flow direction. The latter feature is illustrated through the use of the moving red spheres. In this case the interaction is controlled via the mouse, which typically has sub-millimeter accuracy. It controls the orientation and position of the camera as well as the selection of display options. It is interesting to note that the size of the interactive display selections can be small in this case, which may reduce cognitive loading. The fully immersive VR environment is similar to this one, but the user navigates around the visuals by physically moving their head and body.

Fig. 5. Simulation using laptop

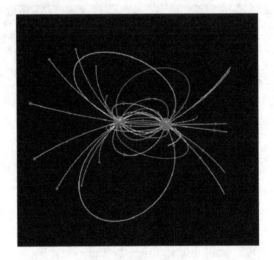

Fig. 6. Streamlines

3   Test Methodology and Procedure

The objective of this study is to identify what technologies would be better suited as an educational medium that would facilitate enhanced learning and comprehension of physics and engineering based educational concepts. Now that the environments are setup we will proceed into the assessment phase. This assessment needs to be based on (1) user preference for the technology being employed in a typical online learning environment, and (2) assessment by the participant regarding the capabilities of the specific technology being employed. These assessments will be based on observations of the user as they navigate the event and manipulate and employ the various capabilities available in each different technology platform. Post scenario discussions and interviews with the user after exposure to the various technology platforms will also be utilized in this study. This information would then be used to determine the most suitable platform for use in this application based on user preference and the platforms ability to enhance learning and comprehension of physics and engineering based educational concepts.

As a first step each participant will play in an environment that contains the same physics-based model and concepts regarding visual illustration of a rectangular bar magnet and the associated magnet field generated by that magnet. This allows participants to become familiar with the user environment produced by a combination of the model developed for the rectangular bar magnet and associated magnetic field technology being employed. As stated previously, the magnetic field model was visually represented by building the required models in Unity and then using that model as the basis for the learning application presented in each of three different technology platforms.

This free-flow approach of allowing the participant to examine and explore the concepts to be learned, and allowing the participant to utilize the different capabilities available in each different type of technology platform will provide some insight into the future selection of the most efficient and desirable platform for these types of educational opportunities. Data gathered in this study will include live observations of the participant while using the different platforms, and time spent in the exploratory environment. Posttest interviews will be used to collect structured data using Likert type questions, and free-flow open ended interview questions at the conclusion of the exploratory period for each platform. It is expected that all participants will be exposed to each type of technology platform in a randomized order of presentation to minimize learning effects produced by exposure to each platform in the same order every time.

Summary

This paper described the development of a representative physics-based simulation. The intent was to provide platforms for users to communicate their preferences for how they would like to interact with the simulation and what visual representations they thought are the most helpful for learning the proposed concept. Three types of simulations were developed. One was an AR using a typical tablet, the other was the 3D representation on the 2D computer screen and the third was a fully immersive VR capability. Finally, a general methodology was presented for how to measure participant's responses. It is anticipated that the information obtained will be seminal in the design and development of future interactive physics and engineering based educational applications that place high importance on user preference on use of technology at the fore front of any design initiatives to produce enjoyable and effective applications. This information will also be extremely important in designing applications that focus on enhanced learning and comprehension.

References

1. Sanders, B., Vincenzi, D., Holley, S., Shen, Y.: Traditional vs gesture based UAV control. In: Chen, J. (ed.) The Proceedings of the Applied Human Factors and Ergonomics Conference. AHFE 2018, AISC 784, pp. 15–23 (2019). https://doi.org/10.1007/978-3-319-94346-6_2
2. Pan, Z., Cheok, A.D., Yang, H., Zhu, J., Shi, J.: Virtual reality and mixed reality for virtual learning environments. Comput. Graph. **30,** 20–28 (2006)
3. Hashemipour, M., Manesh, H.F., Bal, M.: A modular virtual reality system for engineering laboratory education. Comput. Appl. Eng. Educ. **19,** 305–314 (2011)
4. Ibanez, M.B., Rueda, J.J., Maroto, D., Kloss, C.D.: Collaborative learning in multi-user environments. J. Netw. Comput. Appl. **36,** 1566–1576 (2013)
5. Akçayır, G., Akçayır, M.: Advantages and challenges associated with augmented reality for education: a systematic review of the literature. Educ. Res. Rev. **20,** 1–11 (2017)
6. Akçayır, G., et al.: Augmented reality in science laboratories: the effects of augmented reality on university students' laboratory skills and attitudes toward science laboratories. Comput Hum. Behav. **57,** 334–342 (2016)
7. Retrieved from http://igcse-physics-edexcel.blogspot.com/2015/12/64-understand-term magnetic-field-line.html

Appreciation of Proportion in Architecture: A Comparison Between Facades Primed in Virtual Reality and on Paper

Stijn Verwulgen[1(✉)], Sander Van Goethem[1], Gustaaf Cornelis[1], Jouke Verlinden[1], and Tom Coppens[2]

[1] Department Product Development, Faculty of Design Sciences,
University of Antwerp, Antwerp, Belgium
{stijn.verwulgen, sander.vangoethem, gustaaf.cornelis,
jouke.verlinden}@uantwerpen.be
[2] Department Architecture Interior Architecture and Urban Planning,
Faculty of Design Sciences, University of Antwerp, Antwerp, Belgium
tom.coppens@uantwerpen.be

Abstract. Virtual reality (VR) is a rapidly emerging tool in design and architecture. VR allows to define and evaluate design equations at full scale providing the user a sense of depth, volume and distances, as opposed to drawings, plans and CAD designs displayed on screen. Currently, VR enables to provide subjects with a preliminary experience of their own body and relative position towards the virtually designed space. That real time virtual body experience is expected to become more realistic the forthcoming decade, given the uprising of motion tracking and posture generating algorithms, and the expected breakthrough of augmented reality. These evolutions will make VR a tool that is complementary to design and simulation of body near products with 3D anthropometry and rapid prototyping such as 3D printing. Whereas the latter allows simulation of interaction with artifacts at scale of the human body or smaller, VR will enable to simulate and evaluate structures that surrounds the human body: rooms, spaces, buildings and urban implants. In particular, evaluation of scale with relation to human proportion is expected to benefit from VR. The present research is a first step in comparing classical drawing techniques with VR. This paper is confined to evaluate the aesthetic preference of subjects when exposed to facades in a virtual urban planning environment. The preference for a certain proportion is evaluated by exposing subjects to the same facades in both mediums. In this first pilot study, we isolated one variable: the length to width ratio of windows in the facade of a house with a flat roof. Subjects were primed with three length to width ratios: the square root of 2, the golden (in approximation 1.618) section and a value that was intuitively designed to be most pleasing, turned out to be 1.80. A total of n = 30 subjects were enrolled in this study. Only descriptive statistics were used due to small sample size. Results were interpreted as tendencies. There were no differences found in preferences when exposed to drawings compared to when exposed to the same models in VR. These results indicate that-restricted to our simple and controlled setting- evaluation in VR coincides with regular evaluation techniques, and thus could extend the latter, not only to relative proportion, but also in relation to subject's body dimensions as a sense of absolute scale. The test

© Springer Nature Switzerland AG 2020
T. Ahram (Ed.): AHFE 2019, AISC 973, pp. 305–314, 2020.
https://doi.org/10.1007/978-3-030-20476-1_31

setting presented in this pilot can be used to plan large scale-controlled studies to further investigate this hypothesis.

Keywords: Virtual reality · Proportion · Facade · Paper image

1 Introduction

Virtual reality (VR) is a technique to emerge the user in a non-existing but highly realistic looking world and/or intensive experience. Concepts and techniques for VR emerged almost three decades ago, mainly driven by the uprising of computer graphics and increasing computing power. VR immerses the user in an interactive simulated environment in which virtual objects have spatial presence [1, 2]. Currently, VR is commonly implemented by coupling a real time tracking system to a head-mounted stereographic display, projecting images in function of the user's orientation and position, at >30 Hz. The immersive experience can even be enhanced by adding stereo sound e.g. through headphone or earphone and by adding tactile feedback e.g. force-exerting gloves. Traditionally, VR has been assigned many promising applications including entertainment, education, tourism, art and design [3]. To date, a reliable VR setting is available as a plug-and-play system, which makes it feasible for use in research and education.

Highly immersive interactive VR environments are omnipresent in gaming, entertainment and animations. Underlying data can be displayed and handled on a VR platform, e.g. through Unity 3D [4]. These gaming/entertainment environments follow a "what you see is what you get" (WYSIWYG) approach [5]. The sole purpose of mimicking real world behavior of virtual objects is enhancing entertainment value. For the development of WYSIWYG VR environment, a major requirement is balance between ease of use and functionality that can be incorporated [6]. By no means, it is ensured that resulting VR is an accurate, scientifically validated model of the real world. This WYSIWYG approach limits interaction of end users in a VR objects to the intended, pre-programmed use.

Computer aided design (CAD) takes a radically different approach to real world modeling, compared to game development and entertainment. Shape and structures in CAD files are pinpointed to represent real world functions with maximal accuracy allowing verification and optimization of physical properties in design, for example thermal, structural, mechanical and kinematical characteristics. In architecture and urbanism, CAD is mainly used for simulating and evaluating aesthetics and acceptance CAD models are complemented with physical representations for visualization and decision-making. Virtual reality is a promising tool both in architecture and design to complement existing design, prototyping, visualization and verification techniques since it works at different scales. Objects that are smaller than the human body are experienced by the user as full 3D objects with fidelity and the user is embodied in architectural objects: buildings, streets, rooms, as if there. The potential of VR in both industrial design and architecture is huge, to improve the design process for workload and efficiency, and to support the emerging practice of collaborative design and co-creation in architectural design. A necessary step in the deployment of VR in design sciences, is an extensive validation of simulated characteristics. As said above, for industrial design simulations are merely of thermal, structural, mechanical and kinematical nature. For architecture and urbanism, a first question is how design artifacts are experienced in VR compared to classical representations and, finally, to the actual object.

In particular, to support the transition from physical and CAD based design to a workflow that is complemented with VR, the question arises how screen and paper representations of design artifacts are experienced in VR. A sub-question is whether and to what extend aesthetics is perceived in VR compared to flat representation, in design artifacts. The latter 2 dimensional projections are commonly used in early stage design verifications. In this paper we take a first step to compare aesthetic preference. We present a method to assess preference of ratio. The method was tested on facades, both in VR and on paper, and outcomes of both media were compared.

The research question is thus confined to perception of facades since (1) facades can be represented easily and truthfully on a flat surface as 2D projections, since they are inherently flat, (2) they are a basic component in architecture subject to easy to implement and some features that are relevant for aesthetics are relatively easy to isolate and implement, omitting potentially co-founding factors.

2 Materials and Methods

A facade digital model was designed with one door and eight windows. Each window had the same length-to-height ratio as the facade but scaled with a fixed factor 1/7, as shown in Fig. 1.

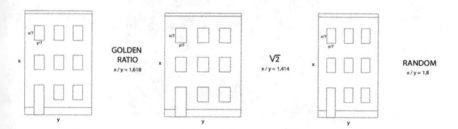

Fig. 1. Design of the facade with controlled proportions.

The geometry and volume are as designed with the intention to be as simple as possible without affecting the fidelity of the design. The design represents a facade of a flat roof house in which all other elements that might contribute to aesthetic preference were reduced to a minimum. Fractal features are needed to create ordered diversity in many figurative levels of architecture, particularly in texture, colors, ornaments, light and shade. This can lead to a composition that is meaningful and appealing. Consequently, for a more realistic embedding in the environment, we added detail in the facade finishes like: brick texture, window finishes, shadow, etc. The windows have windowsills that follow the length of the windows with a fixed offset. Also, windows and doors have a fixed offset inwards the house to simulate regular architectural models. The setting was placed in a street with side-by-side constructed houses that all exhibit the same fidelity as the facade model. Moreover all houses have the same height. The entire model was thus parameterized with one single parameter: the length-to-height ratio of the facade model and windows inside. As the height is fixed, the model is tunable by a single real value; the length of the facade can be adjusted according to the length-to-height ratio. In our tunable model, the ratio of facade-to-door length is fixed by definition/construction. Consequently, the height of the door is the same in each model, set to 1/3 of the height of the facade.

Three representations were generated from the parameterized design, by setting the length-to-height ratio to the golden ratio $(\sqrt{5} + 1)/2 = 1.618...$, the square root of two $\sqrt{2} = 1.414...$ and a ratio that was constructed by master students without considering particular numerical values or geometric relations, denoted 'Random' in Fig. 1. This value turned out to be 1.80.

The values were chosen for their particular geometrical properties. The square root of two is well known as a length-to-height ratio in rectangles, since it results in the same proportions when the rectangle is halved: substituting length and height in the original rectangle by height/2 and length for a new smaller ratio, yields the same ratio. This halving can easily be achieved by folding, so the square root of two format is default in a series of A1, A2, A3,... and B1, B2, B3,... paper formats, see Fig. 2. Since it is encountered in daily professional practice and it exhibits a straightforward geometrical property, it is considered worth to evaluate.

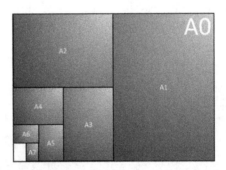

Fig. 2. Square root of two emerging in Ax paper formats not on scale [7].

The golden ratio is characterized by dividing a line in a small and a big part, such that the ratio of big part to the small is the same as the ratio of the total length to the big part. This leads to a second order polynomial equation with $(\sqrt{5} + 1)/2$ as a positive solution. The golden section has some fascinating and useful mathematical properties and also occurs in good approximation in biology and physics [7]. It has been hypothesized that the golden section forms a basis for universal aesthetics in art, design and aesthetics and there is historical evidence that from 1850 and up, some artists and designers incorporate the golden section deliberately in their work. However, there is no evidence that the golden section automatically emerges as a ratio for natural beauty.

nor is this hypothesis falsified universally [7, 8]. Occurrence of the golden ratio in a rectangle means that the rectangle can be reduced by taking of a square and then the remaining small rectangle also exhibits the same length-to-height ratio, as shown in Fig. 3.

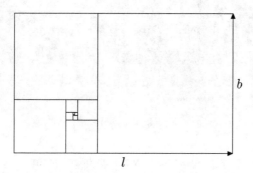

Fig. 3. Characteristic property of a rectangle that is shaped along the golden ratio.

Because of its occurrence in mathematics, nature, deliberately use in free and applied arts, many speculations and interesting geometrical properties; it is worth valuating the perception of the golden ratio proportion in architecture and in particular to plug it in our single parameter model.

The third ratio 1.80 was achieved by setting an arbitrary value in the model, assessed by master students in architecture as being a correct and intuitively pleasing proportion for the given facade model, without any mathematical arguments or underlying reasoning. In the original and comparable study by Gustav Fechner back in the 19th century, 75.6% of the respondents choose rectangles with a ratio between 1.5 and 1.77–35% went for the golden ratio rectangle [9].

The three facades are designed with BIM-software (Autodesk – Revit). The three facade models were paired and arranged horizontally in a street. Each pair was shown to a subject, as in Fig. 4. Subjects were asked to select one facade in a pair along their esthetic preference, by picking either left or right selection option, corresponding to their facade preference. Left-Right permutations where considered as different pairs. Thus each subject was exposed to six pair of facades. Each sequence of six pairs was exposed in two groups with varying medium: on paper and in VR. The six pairs are implemented in a street view and surrounded by photographic facades. Trees and people are added to this street view to improve the perception of a realistic environment. For the experiment with VR as a medium, the Revit model was converted to a virtual-reality readable file by Iris VR - Prospect Pro. For the part where paper acts as the medium, screenshots are taken from a frontal point of view in the virtual reality environment, for maximal fidelity.

Golden section   √2

Fig. 4. Facade pairs.

A total number of n = 30 subjects were enrolled in this study. The study wa
conducted as an exercise in research for master students in architecture. Subjects were
master students too. They are expected to have developed a 'solid' opinion abou
aesthetics in architecture. Eleven of them are female and nineteen are male. They are
all aged between twenty-one and twenty-six. Subjects were randomly assigned to
two groups of fifteen. One groups was exposed to the pair of facades on paper (group
P) and the other group to the same facades in VR (group V).

At the start of the experiment, each subjects completed a questionnaire for gender
age and grade. The V group was asked to fill in an additional set of control questions
called ITQ-Immersive tendency Questionnaire test meant, to account for tendencies in
individual VR experience [10]. Participant were assigned to non-overlapping time slot
to avoid mutual influencing aesthetic preferences. The VR environment was displaye
with Oculus Rift headset. Subjects were given the necessary time to get used to VR
experience and to the controls of the Oculus Rift system. In order to navigate around
two controllers are handed over to the subject, as shown in Fig. 5.

Fig. 5. VR test setting: Oculus rift VR-goggles and controls (left) and VR test environme
(right).

Subjects were immersed in the six virtual streets, with only the facades altered. The pairs are shown one by one to the subjects. After a few seconds are given to look and examine the facade pair, subjects are asked to give their 'aesthetic preference' by saying or pointing left or right. Subjects were asked to explain their choice. The P group was exposed to the same pairs as in the VR, printed on A4 sized paper. Similar to the experiment in VR, subjects were asked to give their aesthetic preference for each pair. Each subject was asked to motivate the choice. In both groups, we categorized the subjects' arguments for their aesthetic preference in different options called: ratio, mass, feeling, width, size doors and size windows.

3 Results

Results are summarized in Table 1, that lists the subjects' aesthetic preference between facade pairs. Choice frequencies are displayed for VR (V) and paper (P) groups. Each comparison of ratios is represented in three columns, with the place also representing the facade configuration.

Table 1. Frequency of selected preference when ratios in facades are shon in VR (V) and on paper (P). In each entry, the denominator or maximal number is 15. The C symbol indicates the consequence by which the preference is selected when the facade pair reverses order.

	Golden section (G) vs. √2			Golden section (G) vs. 1.8			1.8 vs. √2						
	G	√2	√2	G	C	G	1.8	1.8	G	C	√2	1.8	1.8	√2	C
V	11	4	6	9	*11*	10	5	6	9	6	6	9	8	7	*10*
P	11	4	5	10	*10*	6	9	3	12	7	6	9	11	4	*13*

When the facade designed by the golden section is shown on the left side, this ratio is chosen equally 73% (11/15) in the two different media, namely paper and VR. When the same facade is shown on the right side the golden section is chosen 66% (10/15) when paper acts as a medium and 60% (9/15) when represented in VR. Adding up the results of the two media, the facade designed by the golden section is chosen 73% (22/30) when it was shown on the left and 63% (19/30) when it was shown on the right. The consistency of the answers of the subjects, when the facades were showed a second time in a different order is 66% (10/15) on paper and 73% (11/15) when shown in VR. Adding up the two media an average is found of 70% (21/30). In the second main column the results of the subjects' aesthetic preference between the golden ratio and 1.80 ratio is listed. When the facade designed by the golden ratio is shown on the left side, this ratio is chosen 40% (6/15) when paper acts as a medium and 66% (10/15) when represented in VR. When the same facade is shown on the right side the golden section is chosen 80% (12/15) on paper and 60% (9/15) when represented in VR. Adding up the results of the two media, the facade designed by the golden section is chosen 53% (16/30) when it was shown on the left and 70% (21/30) when it was shown on the right. The consistency of the answers of the subjects, when the facades were showed a second time in a different order is 50% (7/15) on paper and 40% (6/15) when shown in VR. Adding up the two media an average is found of 43% (13/30). In the third main column, the results of the subjects' aesthetic preference between 1.80 and square root of two, are listed. When the facade designed by $\sqrt{2}$ is shown on the left side, this ratio is chosen 40% (6/15) when paper acts as a medium and 27% (4/15) when shown in VR. When the same facade is shown on the right side $\sqrt{2}$ is chosen 40% (6/15) on paper and 50% (7/15) when represented in VR. Adding up the results of the two media, the facade designed by $\sqrt{2}$ is chosen 40% (12/30) when it was shown on the left and 37% (11/30) when it was shown on the right. The consistency of the answers of the subjects, when the facades were showed a second time in a different order is 87% (13/15) on paper and 66% (10/15) when shown in VR. Adding up the two media an average is found of 77% (23/30).

4 Discussion

In the results we can see an inconsistency in the answers of the subjects when the facades were shown a second time but in a different order. One reason for this can be that subjects often got more focused on the consistency of their choices, than on their aesthetic preference, as some of them mentioned during the experiment. Another reason can be that in the first part of each experiment, the subjects are more focused than at the end, which causes fatigue and might also affects their choices. A third explanation could be that the present ratios were not perceived as fully relevant for aesthetics, or that the difference was not clear to subjects. This proposition is supported by the tendencies that consistency increased when the difference in ratio increases.

The golden section is chosen more in comparison with $\sqrt{2}$, which is consistent with studies like Fechner's. The choice of the aesthetic preference is independent on whether the golden section is shown on the left or on the right side. Results are of the same magnitude when the facades are presented on paper and in VR.

The second comparison is the golden section against the proportion based on intuitive design. The golden section is chosen more than the 1.80 design but only with a small lead. We can also see that the subjects are less consistent in their choices. In this comparison, it is notable that when the golden section is shown on the right side, it is chosen more than when it is shown on the left side. The results are slightly different when the facades are represented with the different media: paper and VR. We can see that the subjects are less consistent in their choices when presented in VR.

The last comparison is $\sqrt{2}$ against the proportion based on intuitive design. The proportion based on intuitive design is chosen more in this comparison. Here it is also independent for the consistency of the choices whether $\sqrt{2}$ is shown on the left or on the right side. The results are almost the same when the facades are represented with the different media: paper and VR.

We stress that this paper only reports tendencies. These observations confirm that behavior and decision making in virtual environments seems to be close to what people do in the real world [11].

For fully significant interpretation, larger scale observations and inductive statistical analysis are required.

5 Conclusion

The present research is a first step in comparing classical design and drawing techniques with VR. This paper is confined to evaluate the aesthetic preference of subjects when exposed to facades in a virtual urban planning environment. The preference for a proportion in facades is evaluated by pinning proportion down to a single parameter. In this first pilot study, we isolated one variable: the length to width ratio of windows in the facade of a house with a flat roof. Then, subjects are exposed to facades with different parameter values in different in both mediums: virtual reality and on paper. Subjects were primed with three length to width ratios: the square root of 2, the golden ratio and a value that was intuitively designed to be most pleasing, turned out to be 1.80. A total of n = 30 subjects were enrolled in this study. Only descriptive statistics were used due to small sample size. Results were interpreted as tendencies. There were no differences found in preferences when subjects were exposed to drawings compared to when exposed to the same models in VR. These results indicate that-restricted to our simple and controlled setting- evaluation in VR coincides with regular evaluation techniques, and thus could extend the latter, not only to relative proportion, but also in relation to subject's body dimensions as a sense of absolute scale. The test setting presented in this pilot can be used to plan large scale-controlled studies to further investigate the hypotheses that arise from observed tendencies.

References

1. Steuer, J.: Defining virtual reality: dimensions determining telepresence. J. Commun. **42**, 73–93 (1992)
2. Bryson, S.: Approaches to the successful design and implementation of VR applications. Virtual Real. Appl. 3–15 (1995)
3. Wexelblat, A.: Virtual Reality: Applications and Explorations. Academic Press (2014)
4. Kim, S.L., Suk, H.J., Kang, J.H., Jung, J.M., Laine, T.H., Westlin, J.: Using unity 3D to facilitate mobile augmented reality game development. In: 2014 IEEE World Forum on Internet of Things (WF-IoT), pp. 21–26 (2014)
5. Ahearn, L.: 3D Game Textures: Create Professional Game Art Using Photoshop. CRC Press (2014)
6. Cowan, B., Kapralos, B.: An overview of serious game engines and frameworks. In: Recent Advances in Technologies for Inclusive Well-Being, pp. 15–38. Springer, Cham (2017)
7. Verwulgen, S., Cornelis, G.: Universele esthetiek! De gulden snede? Over zin en onzin omtrent een merkwaardig getal. Academic and Scientific Publishers, Antwerp (2018)
8. Livio, M.: The Golden Ratio: The Story of Phi, The World's Most Astonishing Number. Broadway Books (2008)
9. Green, C.D.: All that glitters: a review of psychological research on the aesthetics of the golden section. Perception **24**, 937–968 (1995)

10. Witmer, B.G., Singer, M.J.: Measuring presence in virtual environments: A presence questionnaire. Presence **7**, 225–240 (1998)
11. Foreman, N.: Virtual reality in psychology. Themes Sci. Technol. Educ. **2**, 225–252 (2009)

Immersive Virtual Reality Beyond Available Physical Space

Nicholas Caporusso[✉], Gordon Carlson, Meng Ding,
and Peng Zhang

Fort Hays State University, 600 Park Street, 67601 Hays, USA
{n_caporusso, gscarlson, m_ding6.se,
p_zhang15_sia.se}@fhsu.edu

Abstract. Technology provides Virtual Reality (VR) with increasing levels of realism: high-performance head-mounted displays enable delivering immersive experiences that provide users with higher levels of engagement. Moreover, VR platforms, treadmills, and motion tracking systems add a physical dimension to interaction that aims at increasing realism by enabling users to use their body to control characters' movements in a virtual scenario. However, current systems suffer from one main limitation: the physical simulation space is confined, whereas VR supports rendering infinitely large scenarios. In this paper, we investigate the human factors involved in the design of physically-immersive VR environments, with specific regard to the perception of virtual and physical space in locomotion tasks. Finally, we discuss strategies for designing experiences that enable optimizing the use of the available physical space and support larger virtual scenarios without impacting realism.

Keywords: Virtual reality · Spatial awareness · Motion tracking · Wearable

Introduction

In the recent years, the increasing quality and efficiency of Virtual Reality (VR), combined with the development of high-resolution Head-Mounted Displays (HMD), resulted in novel paradigms such as Immersive Virtual Reality, which are pushing the boundaries of VR experiences. Several studies demonstrated the importance of loco-motion in VR [1] and suggested that actual movement results in better immersion and realism than traditional screen-based or immersive VR [2]. Physically-Immersive Virtual Reality (PIVR) enables individuals to walk on omnidirectional platforms and treadmills [3]. Alternatively, higher levels of realism can be achieved using systems that enable users to move in a contained physical space where a low-latency motion-capture infrastructure acquires their position, orientation, and movement, and represents them in the simulated environment, in real-time. As a result, this type of wearable technology for immersive virtual reality detaches the infrastructure from the user [4] who can explore a sophisticated virtual world by walking in an empty room while experiencing the VR scene using a headset; also, they can interact with the environment thanks to specific controllers such as wands, wearable devices [5, 6], and objects equipped with appropriate markers. Consequently, PIVR is especially suitable for

Springer Nature Switzerland AG 2020
Ahram (Ed.): AHFE 2019, AISC 973, pp. 315–324, 2020.
https://doi.org/10.1007/978-3-030-20476-1_32

applications that require higher levels of engagement. For instance, previous studies proposed its use for training law enforcement officers and emergency responders [7] and for simulating of safety- or mission-critical tasks [4, 8]. Nevertheless, it can enhance gameplay and experiences involving digital art.

Nevertheless, one of the main limitations of PIVR based on motion tracking is the physical space in which users can move, which is constrained by the area of the motion-capture infrastructure and by the size of the room where the system is installed although virtual environments are potentially infinite, boundaries are set in the VR scene to prevent users from walking out of the motion-sensing area or into a wall Several devices (e.g., concave omnidirectional platforms that keep users walking on the spot) attempt to address this discrepancy, though they reduce realism because they introduce restraints that prevent free movement and physical interaction between multiple players.

In this paper, we focus on the relationship between physical and virtual spaces in PIVR. Specifically, we analyze the main challenges in the design of scalable VR simulations that are congruent with the limitation of their fixed-size motion-capture infrastructure. Furthermore, we study the human factors involved in the perception of physical and virtual areas, and we detail several methods, such as, folding, layering and masking, that can be utilized for reducing the perceived size discrepancy. Finally we discuss the findings of an experiment in which we demonstrate how multiple techniques and geometries can be combined to generate the illusion of a much larger physical area that matches the size of the virtual scene. As a result, small-scale motion capture systems can support large PIVR simulations without affecting the user experience.

2   Related Work

Physically-Immersive Virtual Reality based on low-latency motion tracking has the potential of taking VR beyond the barriers and limitations of current immersive omnidirectional locomotion platforms. The authors of [7] introduce a modular system based on PIVR for training emergency responders and law enforcement officers with higher levels of realism. In addition, the advantage of the system is two-fold: scenarios can be created and loaded dynamically without requiring any modification to the physical space; moreover, as the system is portable, it represents an affordable alternative to travelling to disaster cities and training grounds. Similarly, [9] describes the concept of a museum that uses PIVR to create a walkable virtual space for artwork Given its depth and realism, physically immersive VR can open new opportunities for currently available software: in [10], the authors evaluate users' reaction to alternative input methods and dynamics, such as, motion sickness, in the context of a porting of Minecraft as a tool for promoting user-generated virtual environments [11]. Novel applications include sports, where physically-immersive VR can be utilized as a simulation platform as well as a data acquisition system for evaluating and improving the performance of athletes [12]. Similarly, embodied experiences [13] can be utilized to increase physical engagement of eSports practitioners [14].

Unfortunately, one of the limitations of PIVR is that the size of the virtual scenario is limited by the physical space where the simulation takes place. Although current motion tracking technology enables covering larger areas, this type of infrastructure is associated with high assembly, running, and maintenance costs. Conversely, more affordable commercially-available technology supports tracking users over smaller areas (typically 5 × 5 m). Ultimately, regardless of the potential size of a virtual world, the VR scenario is limited by the infrastructure, which, in turn, defines the boundaries of the physical simulation area. Traditional approaches based on virtual locomotion have been studied in the context of PIVR and primarily consist in techniques for enabling users to navigate large VR environments using controllers [14]. Although they are especially useful when the physical space of the simulation is limited, they affect realism and engagement. Potential solutions based on the use of GPS for tracking the user over open spaces [15] are not suitable for real-time applications. Alternative techniques confine users in the simulation space by creating virtual boundaries that they are not supposed to cross. For example, they position players on a platform where they can move and interact with elements located beyond edges. Appropriate skybox and scenario design give the illusion of being in a larger space, though the walkable area is limited.

Conversely, the aim of our research is to leverage the same physical space of the simulation area in ways that support designing infinitely-large virtual scenarios that conveniently reuse the same physical space without having the users realize it. By doing this, a small simulation area can support a much larger virtual scenario. Different strategies might be suitable for dynamically reconfiguring the virtual scenario in order to reuse the same walkable physical space without the user realizing it, or without affecting user experience. For instance, the virtual space could be organized over multiple levels stacked vertically or horizontally and elements in the scenario (e.g., elevators, corridors, portals, and doors) or narrative components could be utilized to transport the user from one level another.

3   Experimental Study

We developed a preliminary pilot study aimed at evaluating whether it is possible to implement techniques that programmatically change the virtual scenario in a way that triggers users into thinking that they are moving in a physical space that is much larger than the actual simulation area. To this end, we designed an experiment that compares a static scenario with a dynamic environment that programmatically changes as the user walks in the simulation area. Our purpose was to test the concept and study individuals' awareness of the relation between virtual space and physical space, and to evaluate their reaction and user experience. The goal of this pilot study was to determine efficacy of the experimental design and applicability of the instrument.

The experimental software consisted of a set of virtual scenarios implemented using Unity3D, one of the most popular VR engines. For the purpose of this study, two scenarios were utilized, each representing a simple, square maze surrounded by walls (see Fig. 2), so that subject could explore them.

3.1   Participants

We recruited 21 participants (8 females and 13 males) to realize a preliminary study and test our hypothesis. All participants were aged 18–24 and healthy, drawn from a student population at a medium-sized university in the American Midwest. Most of them were gamers but none had any significant experience with immersive headsets or PIVR before the experiment, other than testing the device for a very short time at exhibitions. Internal Review Board (IRB) approval was granted for this protocol.

3.2   Hardware and Software Setup

The experiment was hosted in an empty space in which we created a dedicated simulation area of 6 x 6 meters. Two base stations were located at two opposite corners facing one another so that they covered the entire area. We utilized an HTC Vive Pro head-mounted display (HMD) equipped with a wireless adapter. This allowed subjects to move freely minimizing spatial cues or safety issues related to corded setups. The wireless setup makes it easier for subjects to be immersed into the virtual environment. The equipment was connected to a desktop PC supporting the requirements of the VR setup. Figure 1 demonstrates the infrastructure of the experiment and its configuration.

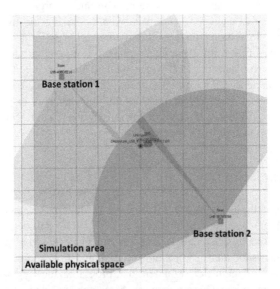

Fig. 1. Configuration of the simulation area as defined by the two low-latency sensors (base stations), head-mounted display, and DisplayLink wireless transmitter.

The mazes were the same size (approximately 4 × 4 m) and their structure was similar, because they were built using the same type of components. We built three reusable building blocks: each module consists of a square structure measuring 2 × 2 m surrounded by walls and had two openings located on two adjacent sides. The inne

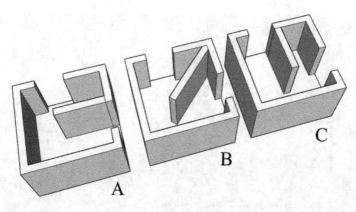

Fig. 2. The building blocks of the experimental mazes. Two modules consist of perpendicular walls, whereas a diagonal wall was introduced in block B to provide users with a visual clue for recognizing it.

configuration of each module was different, in that it contained walls varying in shape and position.

Each module was designed so that they could be attached next to one another by appropriately rotating them and connecting their openings, creating the illusion of a never-ending path. Four modules are utilized to create a walkable 4 × 4 maze enclosed in walls. The height of all the walls, including those surrounding the maze, was the same and it was set at 2.75 m so that subjects could not see over them. The VR scenario was built using 1:1 scale and was placed in the middle of the simulation area. The two mazes utilized in the experiment (see Fig. 3) were different: the static maze was obtained by assembling four blocks that did not change during the experiment. The dynamic maze was implemented by adding a software component that, upon certain events, programmatically replaced the opposite module of the maze with a building block selected at random and appropriately rotated to be consistent with the rest of the structure of the maze. Each module is designed so that the participant cannot see the "exit" opening while standing near the "entrance" opening. This allows for manipulating two modules out of sight of the participant which makes it possible to dynamically generate changes to the maze layout. Colliders (a particular type of trigger in the Unity3D software) were placed in each of the openings: walking through an opening resulted in intersecting a collider which, in turn, triggered the event for replacing the module opposite to the position of the user. This was to prevent the user from witnessing changes happening in the structure of the scenario. As a result, by dynamically changing no more than two pieces at a time, the dynamic maze resulted in a never-ending sequence of randomly-selected modules.

The experimental software was designed to track the subject in real time and collect data from the simulation. Specifically, the position of the VR camera (representing the location of the subject's head) and its orientation (representing the rotation and line-of-sight of the subject's head) over three axes were recorded at 1 Hz interval, which was enough for the purpose of the experiment. This, in turn, was utilized to calculate the approximate distance travelled by the subject. Moreover, the simulation software

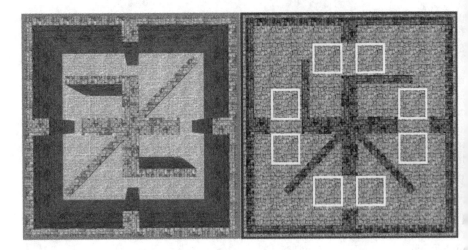

Fig. 3. The static maze and an example configuration of the dynamic maze showing the location of colliders.

recorded data associated to events (e.g., hitting a collider) when they occurred, including time of the event, subject's position, and configuration of the maze. In addition to acquiring experimental data, a debriefing mode was developed to enable importing the data points from the experiment and reviewing the path walked by the subject, their orientation, and the events triggered during the test.

3.3   Experimental Protocol and Tasks

Participants entered the experiment room blindfolded (so that they could not see the size of the simulation area) and they were seated on a chair positioned in the simulation area, where they were equipped with the VR headset. Their ears were covered by the earphones of the HMD to avoid auditory cues that could help them estimate the size of the room or their position and orientation. Two assistants closely supervised subjects during the experiment to avoid any incidents and to prevent them from accidentally walking out of the simulation area. The experiment consisted in two different tasks, one involving the static maze (Task 1) and one realized in the programmatically generated maze (Task 2).

Task 1 - static maze. In this task, subjects were asked to keep walking in a direction until they passed a total of six slanted walls (see Fig. 2, item B), which corresponded to 3 laps of the maze.

Task 2 - dynamic maze. This task involved a dynamically-generated maze and asked subjects to walk past six slanted walls as in task 1. However, as the software was programmed to change one module in the scenario and replace it with a random one, the number of lapses was a factor of the dynamic configuration of the maze.

Tasks were divided into trials, each consisting in walking past two slanted walls. After they completed a trial, subjects were interrupted, and they were asked to solve a simple arithmetic problem (i.e., calculate a sum) before continuing. Then, they were

ısked to indicate the spot in which they started the task by pointing it with their finger
ts direction. In the third trial, subjects were asked to close their eyes and they were
aken to a different location of the maze, before asking them to point the starting spot.

At the beginning of each tasks, subjects were placed in the same spot, located in an
nset between two walls, so they could not see the full width of the maze. At the end of
:ach task, subjects were asked to point (with their finger) to the spot in which they
started, and to estimate the distance that they walked. At the end of the experiment,
subjects were asked to estimate whether the physical space in which they moved was
arger or smaller than the virtual space, using a Likert scale, and to report the perceived
lifferences between the two mazes. The experiment had no time limit, but duration was
ecorded.

↓ Results and Discussion

All the subjects were able to successfully complete the experiment and commented that
he experience was very realistic, despite the simple structure of the scenario. None of
hem deviated from the path even if they did not receive any specific instructions, and
hey avoided inner and outer walls. Table 1 shows a summary of the experimental
esults. Regardless of individuals' accuracy in the perception of space, 15 participants
71%) reported that they walked further in the dynamic maze than in the static maze.
This is consistent with the structure of the dynamic maze, in which slanted walls were
urther apart. As a result, participants walked a longer distance, as measured by the
xperiment software.

able 1. Summary of experimental results, showing orientation dispersion (OD), measured in
egrees, and Dispersion Index (DI) in the three trials, and the total perceived distance walked in
sk 1 and 2.

Task	OD Trial 1	OD Trial 2	OD Trial 3	DI Trial 1	DI Trial 2	DI Trial 3	Distance walked
1 - Static maze	31.43 ±35.23	41.43 ±40.86	88.57 ±57.43	1.05 ±1.17	1.38 ±1.36	2.95 ±1.91	82.57 ±101.86
2 - Dynamic maze	67.14 ±45.27	105.71 ±56.78	100 ±57.32	2.24 ±1.51	2.24 ±1.51	3.52 ±1.89	84.48 ±58.88

Data show that subjects were able to preserve their spatial awareness in the static
aze, as the difference between the actual starting point and the area they indicated was
)proximately 30°, on average. Conversely, changing the configuration of the maze
sults in a shift of more than 60°, in the first trial. Indeed, subjects' spatial awareness
·graded in the second trial, in which orientation dispersion increased. Nevertheless,
·e static maze resulted in 10° increase, whereas the programmatically-generated
enario had the most impact (50° increase). Both trial 1 and 2 show that the dynamic
aze is twice as effective in disorienting subjects compared to the static maze. Figure 4
nows the increase in orientation dispersion. Specifically, changing the virtual structure

of the maze produced similar results to the effect obtained by physical disorientation obtained by blindfolding subjects and physically moving them in a different location of the maze.

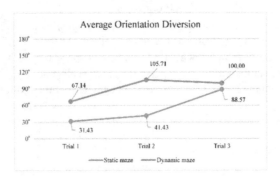

Fig. 4. Average orientation diversion (measured in angle) in the task 1 (static maze) and task 2 (dynamic maze) in each of the three trials.

Figure 5 represents the angle distribution of orientation diversion. Data show that static structure had minor impact on spatial awareness, whereas changing the configuration of the maze disoriented individuals. A one-way between-subjects ANOVA was conducted to compare the differences between task 1 and task 2 (first two trials): results show that changing the structure of the maze has a statistically significant effect at $p < 0.05$ level [$F(1,82) = 4.11$, $p = 0.046$)].

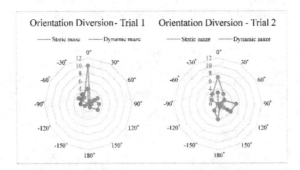

Fig. 5. Comparison between orientation diversion in the static and dynamic mazes, in the first (left) and in the second (right) trial. Data are projected as a cumulative distribution with respect the diversion angle.

Moreover, when they were asked if they noticed any difference between the first and second maze, only 3 subjects (14%) were able to identify the difference between the two mazes, whereas most of the subjects perceived the dynamic maze as larger. All participants associated the static maze to a square shape. In contrast, the dynamic maze was perceived as having a polygonal shape with more than 4 sides by 23% of the subjects.

5   Conclusion and Future Work

PIVR provides incredible opportunities for improving the user experience of virtual environments and opens new possibilities for designing innovative applications studying interaction dynamics, despite the infrastructure constraints limit the available physical space, which, in turns, prevents creating infinitely large, walkable virtual environments.

In this paper, we introduced a new technique for overcoming some of the issues that affect locomotion PIVR especially in small simulation areas. Specifically, we explored the possibility of designing dynamic scenarios that programmatically reconfigure the environment without the users realizing it, so that a new piece of the virtual world unfolds as they walk over the same physical space. Although the small size of the sample might limit the applicability of this study, our findings demonstrate the feasibility of the proposed technique and suggest it as an interesting direction for improving the design of immersive spaces. Our experiment showed that the proposed method can be utilized to modify individuals' spatial awareness and, consequently, their ability to correctly perceive the dimensional relationship between physical and virtual space without disrupting the user experience. This, in turn, enables to seamlessly generate the feeling of a much larger walkable physical space that matches the size of the virtual world. Additionally, the technique described in this paper can be combined with other methods to further modify users' perception of the size of the simulation area. In our future work, we will explore factors, such as, size and configuration of the environment, lighting, visual and auditory cues, and cognitive aspects. Furthermore, we will incorporate data from spatial ability tests (e.g., visuospatial imagery, mental rotations test, topographical memory, etc.) to evaluate the relationship between spatial ability in physical and in simulated worlds, to be able to compare our data with previous research [16]. Finally, in a follow-up paper, we will report data regarding to users' orientation while exploring the maze: as suggested by previous studies on immersive content [17], users might show similar navigation patterns, which, in turn, can be leveraged to introduce changes in the environment that contribute to modifying perception of space.

References

1. Nabiyouni, M., Saktheeswaran, A., Bowman, D.A., Karanth, A.: Comparing the performance of natural, semi-natural, and non-natural locomotion techniques in virtual reality. In: 2015 IEEE Symposium on 3D User Interfaces (3DUI), pp. 3–10, March 2015. IEEE (2015)
2. Huang, H., Lin, N.C., Barrett, L., Springer, D., Wang, H.C., Pomplun, M., Yu, L.F.: Analyzing visual attention via virtual environments. In: SIGGRAPH ASIA 2016 Virtual Reality Meets Physical Reality: Modelling and Simulating Virtual Humans and Environments, p. 8, November 2016. ACM (2016)
3. Warren, L.E., Bowman, D.A.: User experience with semi-natural locomotion techniques in virtual reality: the case of the Virtuix Omni. In: Proceedings of the 5th Symposium on Spatial User Interaction, p. 163, October 2017. ACM (2017)

4. Grabowski, M., Rowen, A., Rancy, J.P.: Evaluation of wearable immersive augmented reality technology in safety-critical systems. Saf. Sci. **103**, 23–32 (2018)
5. Scheggi, S., Meli, L., Pacchierotti, C., Prattichizzo, D.: Touch the virtual reality: using the leap motion controller for hand tracking and wearable tactile devices for immersive haptic rendering. In: ACM SIGGRAPH 2015 Posters, p. 31, July 2015. ACM (2015)
6. Caporusso, N., Biasi, L., Cinquepalmi, G., Trotta, G.F., Brunetti, A., Bevilacqua, V.: A wearable device supporting multiple touch-and gesture-based languages for the deaf-blind. In: International Conference on Applied Human Factors and Ergonomics, pp. 32–41, July 2017. Springer, Cham (2017). https://doi.org/10.1007/978-3-319-60639-2_4
7. Carlson, G., Caporusso, N.: A physically immersive platform for training emergency responders and law enforcement officers. In: International Conference on Applied Human Factors and Ergonomics, pp. 108–116, July 2018. Springer, Cham (2018). https://doi.org/10.1007/978-3-319-93882-0_11
8. Caporusso, N., Biasi, L., Cinquepalmi, G., Bevilacqua, V.: An immersive environment for experiential training and remote control in hazardous industrial tasks. In: International Conference on Applied Human Factors and Ergonomics, pp. 88–97, July 2018. Springer, Cham (2018). https://doi.org/10.1007/978-3-319-94619-1_9
9. Hernández, L., Taibo, J., Seoane, A., López, R., López, R.: The empty museum. Multi-user interaction in an immersive and physically walkable VR space. In: Proceedings of 2003 International Conference on Cyberworlds, pp. 446–452, December 2003. IEEE (2003)
10. Porter III, J., Boyer, M., Robb, A.: Guidelines on successfully porting non-immersive games to virtual reality: a case study in minecraft. In: Proceedings of the 2018 Annual Symposium on Computer-Human Interaction in Play, pp. 405–415, October 2018. ACM (2018)
11. Lenig, S., Caporusso, N.: Minecrafting virtual education. In: International Conference on Applied Human Factors and Ergonomics, pp. 275–282, July 2018. Springer, Cham (2018). https://doi.org/10.1007/978-3-319-94619-1_27
12. Cannavò, A., Pratticò, F.G., Ministeri, G., Lamberti, F.: A movement analysis system based on immersive virtual reality and wearable technology for sport training. In: Proceedings of the 4th International Conference on Virtual Reality, pp. 26–31, February 2018. ACM (2018)
13. Ekdahl, D., Ravn, S.: Embodied involvement in virtual worlds: the case of eSports practitioners. Sport Ethics Philos. 1–13 (2018)
14. Bozgeyikli, E., Raij, A., Katkoori, S., Dubey, R.: Locomotion in virtual reality for room scale tracked areas. Int. J. Hum.-Comput. Stud. **122**, 38–49 (2019)
15. Hodgson, E., Bachmann, E.R., Vincent, D., Zmuda, M., Waller, D., Calusdian, J.: WeaVR: a self-contained and wearable immersive virtual environment simulation system. Behav. Res. Methods **47**(1), 296–307 (2015)
16. Coxon, M., Kelly, N., Page, S.: Individual differences in virtual reality: Are spatial presence and spatial ability linked? Virtual Real. **20**(4), 203–212 (2016)
17. Caporusso, N., Ding, M., Clarke, M., Carlson, G., Bevilacqua, V., Trotta, G.F.: Analysis of the relationship between content and interaction in the usability design of 360° videos. In: International Conference on Applied Human Factors and Ergonomics, pp. 593–602, July 2018. Springer, Cham (2018). https://doi.org/10.1007/978-3-319-94947-5_60

Gazing Pattern While Using AR Route-Navigation on Smartphone

Asami Tanabe[(⊠)] and Yohsuke Yoshioka

Graduate School of Engineering, Chiba University, 1-33 Yayoi-Cho, Inage-Ku,
Chiba-Shi, Chiba, Japan
rrhxg755@gmail.com,
yoshioka.yohsuke@faculty.chiba-u.jp

Abstract. Recently, many route-navigation applications that combines AR (augmented reality technology) and a map have appeared, and it is possible to understand the route more intuitively.

However, when using the navigation with AR, the vision of the user has been considerably limited because the route needs to be checked by superimposing the smartphone screen on the real street scape. In this research, the user's gazing pattern while using the conventional smartphone navigation and the AR navigation were compared and analyzed with eye movement tracking technology, for considering an efficient route guidance method that makes comprehensibility and safety compatible.

Keywords: Augmented reality · Gazing pattern · Navigation wayfinding · Mobile systems

1 Introduction

AR stands for "Augmented Reality". By virtually displaying visual information on real landscapes, it is "virtually expanding" the world in front of us. The impact of this technology is huge, and in particular, it has been easily implemented as a smartphone service, attracting attention as an innovative technology capable of improving the convenience of everyday life and creating new experiences. Recently, a high number of route navigation applications that combine AR and maps have emerged that enable a more intuitive understanding of the route.

However, when using AR navigation, the vision of the user is considerably limited because the route needs to be checked by superimposing the smartphone screen on the real streetscape. Therefore, it is necessary to develop a route guidance system that can understand the route in a short time.

Many studies have focused on the methods of presenting information on navigation tools taking into account the users' spatial cognition.

Fukukawa [1] proposed a route planning algorithm, which weighs the users' difficulty (or ease) of locating their own current position as well as determining the total physical distance to the target location. They conducted an experimental study in a real situation using a prototype system to examine and refine the model for a safe and easy route planning.

© Springer Nature Switzerland AG 2020
T. Ahram (Ed.): AHFE 2019, AISC 973, pp. 325–331, 2020.
https://doi.org/10.1007/978-3-030-20476-1_33

Ishikawa [2] examined the effect of different methods for presenting navigational information on the wayfinding behavior and spatial memory of the users.

Goldiez [3] presented the experimental results of various AR effects regarding display strategies on human performance in a simulation based analog of a "search and rescue" navigation task, demonstrating the promising benefits of mobile AR usage in specific navigation tasks.

In this research, the user's gazing pattern while using the conventional smartphone navigation and the AR navigation were compared and analyzed with an eye movement tracking technology, for considering an efficient route guidance method that makes the system comprehensible and safe.

2   Method

2.1   Participants

Ten college students (5 men and 5 women) participated in this experiment, with ages from 22 to 24 years and a mean age of 22.9 years. They had previously used conventional navigation applications, but not the AR navigation application. In addition, they were unfamiliar with the routes we selected.

2.2   Study Area

For the experiment, two routes were selected from a quiet residential area in Chiba city, Japan (Fig. 1). The reasons for this selection were as following. ① There was no building that could become a clear landmark, ② the distance was 5 min on foot (approx. 400 m), and ③ the paths had the same number of corners.

2.3   Navigation Tools

In route 1, the AR navigation application (Yahoo map!/Yahoo) was used, while in route 2, the conventional navigation application (Google Map/Google) was used for guiding the subjects.

The AR navigation application (Fig. 2a): Route guidance by superimposing the map information such as the distance to the route line, the goal. and the signboard showing the turning on the actual landscape shown on the smartphone camera.

The conventional navigation application (Fig. 2b): On the device screen (diagonal 4.7 in), a map was always displayed so that the traveling direction was upward. Due to the limited size of the screen, the start and goal positions were not simultaneously shown on the screen, and only part of the whole route was presented, so the subjects could change the map display range by operating the smartphones.

2.4   Design and Procedure

The subjects were asked to head to the destination as guided by the navigation applications. They wore a 120 Hz precise eye tracking device (EMR-9/NAC) so tha

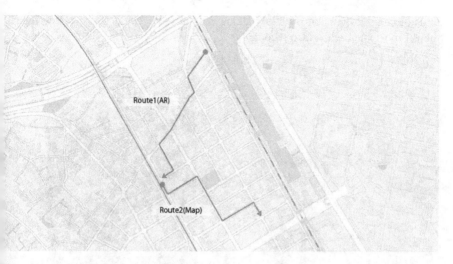

Fig. 1. Map of the study area in Chiba city, Japan. Two routes were selected from this area. In route 1, the AR navigation application was used, and in route 2, the conventional navigation application was used for guiding the subjects.

(A)   (B)

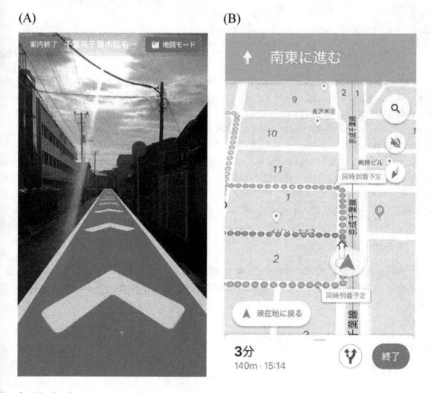

Fig. 2. Navigation tools. (a) AR navigation application (Yahoo map!/Yahoo), (b) conventional navigation application (Google Map/Google)

the subject's gaze position while walking on the route could be detected (Fig. 3)
Subject movement was video recorded.

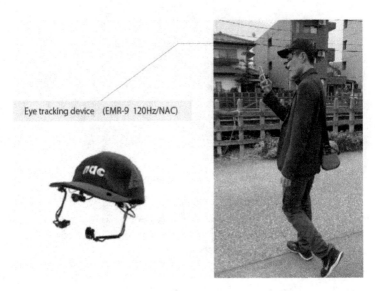

Fig. 3. Experimental equipment. The subjects wore the 120 Hz eye tracking device (EMF
9/NAC) so that the subject's gaze position while walking on the route could be detected.

The experimenter described each tool and ensured that subject knew how to us
them. When they indicated that they were ready, subjects started walking toward th
first goal. When the subjects reached the first goal, they were told that they would hav
to engage in the second route and started going towards the second goal using
different application.

Subjects had to fill out the Sense of Direction Question paper, a simplified versic
(SDQ - S), which is a method that evaluates according to 5 grades the efficiency of th
sense of direction using a score [4]. We used this scale to possibly corelate it with th
subjects' wayfinding behavior, but there was no significant association.

2.5   Measured Variables

Concerning subjects' behavior, we measured (a) travel speed, (b) percentage of gazir
at the screen time relative to the walking time, (c) average gazing at the screen duratic
in each condition.

3 Results and Discussion

The gazing at the smartphone screen duration and frequency of the subjects when using the conventional navigation and the AR navigation were compared by analyzing the gazing data detected by the eye tracking system. The movements were verified, and the following results were obtained from this experiment.

(a) Traveling speed (Fig. 4a): For each subject, we computed the distance traveled and the mean travel speed and examined the differences between the AR navigation and the conventional navigation using a mixed analysis of variance, with sex (male or female) as a between-subject variable. There were no significant main and interaction effects in terms of the traveling speed.

(b) The percentage of gazing at the screen time relative to the walking time (Fig. 4b): For each subject, we computed the walking time and the gazing at the screen duration. In AR navigation, the percentage of gazing time relative to the walking time was larger than that for conventional navigation.

(c) Average gazing at the screen duration in each condition (Fig. 4c): A multiple comparison performed on the profiles revealed a significant main effect of the application used ($p < 0.01$). In AR navigation, the average gazing at the screen duration per trial was longer than that of conventional navigation.

Using conventional navigation, the subjects tended to look at the screen more frequently around the starting point and the destination for checking the traveling direction. The subjects tended to gaze at the smartphone screen before turning a corner in order to check the next route.

When employing AR navigation, there was a tendency to look at the screen of the smartphone before the subject turned the corner, or when they arrived at the corner. Some subjects walked past the correct corner. The comparison between the gazing patterns while using conventional navigation and that while using AR navigation revealed that the percentage of the gazing at the smartphone screen duration relative to the walking time was larger, and that the average gazing at the screen duration in each condition was longer when using AR navigation. These results suggest that the users of the AR route-navigation felt uneasy because it is difficult to understand and imagine the whole route till the destination due to the lack of a bird's eye view of the route.

In other words, it is suggested that the gazing at the screen duration is longer when examining the entire route than that while checking for direction because the route needs to be checked by superimposing the screen on the real streetscape when using AR navigation.

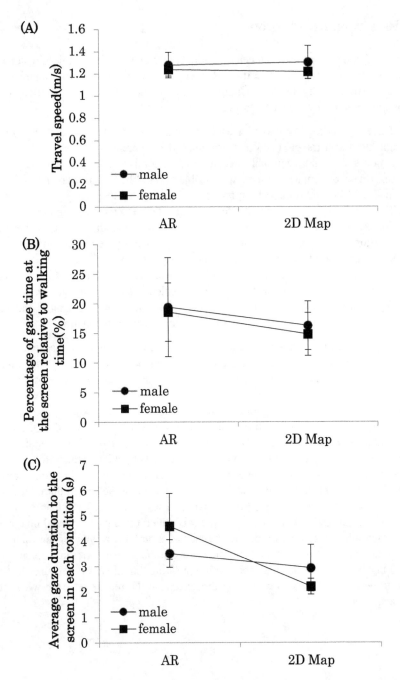

Fig. 4. Comparison of wayfinding measure for different tools. (a) travel speed, (b) percentage c gaze time to the screen relative to walking time, (c) average gaze duration to the screen in eac condition.

4 Summary and Conclusion

Several significant findings emerged from the present study. According to a comparison between the gazing patterns while using the conventional navigation and that while using the AR navigation, it was found that both the percentage of the gazing at the smartphone screen duration relative to the walking time and the gazing frequency were larger when using AR navigation. In addition, the average gazing at the screen duration in each condition was longer while using AR navigation.

These results suggest that the users of the AR route-navigation felt uneasy because it is difficult to understand and imagine the whole route until the destination due to the lack of a bird's eye view of the route.

We are planning further experiments to verify the format of AR navigation that will present navigation information in a way to help the users understand the whole route.

Acknowledgements. This work was supported by JSPS KAKENHI Grant Number JP17H03359.

References

1. Fukukawa, H., Nakayama, Y.: A pedestrian navigation method for user's safe and easy wayfinding. In: Kurosu, M., (Eds.) Human-Computer Interaction. Users and Contexts of Use. HCI 2013. Lecture Notes in Computer Science, vol. 8006. Springer, Berlin, Heidelberg
2. Ishikawa, T., Takahashi, K..: Relationships between methods for presenting information on navigation tools and users' Wayfinding behavior. Cartogr. Perspect. **75** (2013)
3. Goldiez, B.F., Ahmad, A.M., Hancock, P.A..: Effects of augmented reality display setting on human wayfinding performance. IEEE Trans. Syst. Man Cybern. Part C (Appl. Rev.) **37**(5), 839–845 (2007)
4. Stark, A., Riebeck, M., Kawalek, J.: How to Design an Advanced Pedestrian Navigation System: Field Trial Results (2007)

Global Virtual Teams Course at Arnhem Business School

Florentin Popescu and Robert Warmenhoven[✉]

HAN University of Applied Sciences, Ruitenberglaan 31,
6826 CC Arnhem, The Netherlands
{Florentin.Popescu,Rob.Warmenhoven}@han.nl

Abstract. Based on previous research at Arnhem Business School (ABS), the Netherlands regarding Global Virtual Teams (GVT), students consider virtual teams as an added value to the existing modules and extracurricular activities offered by the school [1]. This paper assesses the influence of Global Virtual Teams in Business Education at ABS regarding the development and design of new curriculum modules based on the experiential learning. It seeks to determine the significant challenges GVT confront and existing frames for the success of GVT at ABS and confirm the usefulness of Global Virtual Teams as teaching approaches by improving the understanding of the learning opportunities by utilizing experiential learning. We look at the skills and competences that students need to gain from participating in GVT Module to be able benefit of prospective job opportunities provided by companies.

Keywords: Global Virtual Teams Course · X-Culture · Higher education ·
Virtual communication introduction

1 Introduction

Currently, the new generation claim for personalized learning is higher and universities such as HAN must respond to this demand. The industry and employers have redefined the educational results and this is obvious in the recent literature. Both industry and employers request to the universities to train students in order to be prepared for the flexibility that describes the societies of the 21st century. From employers perspective this flexibility means to think creatively, to improve intercultural competence, to think critically, to be able to work across systems, to come up with new ideas, to develop new skills to support the "green" economy, to be able to adapt to changes and face new circumstances and the prove "moral compass" [2–9].

According to the companies in the region of Arnhem we interviewed regarding the advantages of virtual teams academic experiences intercultural and global in nature offer to the students the opportunity to have a preview regarding the globally inter connected workplace that is nowadays part of the real world [10]. The need of inte grated and imbedded global virtual communication skills and competences in the curriculum is then at most importance.

This study is framing the creation of a new course "Introduction to working in Global Virtual Teams", which is designed to teach students how to work better in a

T. Ahram (Ed.): AHFE 2019, AISC 973, pp. 332–342, 2020.
https://doi.org/10.1007/978-3-030-20476-1_34

global virtual team. The course can be taught in a joint project between Arnhem Business School and another university or only done by ABS students alone (the course is designed to be taught synchronously with team members who are not physically in Arnhem to obtain the full value of instruction).

First, students are taught global team competencies to promote successful interactions on a global virtual team. Second, students are taught and provided the opportunity to use virtual communication skills to promote successful communication on global virtual teams. Along with specific instruction in these areas, students are provided the opportunity to improve their cross-cultural knowledge and experiment with virtual communication technologies during the class instruction, assignments and project tasks.

The authors suggest expanding the GVT literature by using a case study of the Arnhem Business School students' experiences with GVT regards the requirements and opportunities that this kind of learning environment can be suitable in the curriculum of higher education programmes. We look at the skills and competences that students need to gain from taking part in GVT Module to be able to take proper advantage of prospective job opportunities offered by the companies.

Problem Description and Research Design

Student participation in the X-Culture project (www.X-Culture.org) was the research framework for this paper. X-Culture project represents a continuous educational activity that is focusing to train university students to develop a business proposal by participating in GVTs. In this project, teams of five to seven individuals were formed and each student was distributed to a team. The condition was that each member of a team is from another country. The task for the students was to develop a business plan for a real company chosen from a given list in order to help that company to expand into foreign markets with new services or products. By participating into this GVT-based project students were able to get in touch with real challenges related to international and intercultural virtual team cooperation. Moreover, students were capable to form more realistic expectations regarding international cooperation through virtual teams [10].

The impact of Global Virtual Team Course regarding the experiential learning in business education at ABS is evaluated in this paper. In order to develop this research, the large amount of data obtained across 2 years from the X-culture project participants was used. This data was obtained between the last semester of 2015 and the final semester of 2017, representing 5 seasons and roughly 4000 students per season. These 4000 students were grouped in about 800 teams of students representing more than 40 countries in each season. 2 instructors and 66 students from Arnhem Business School were participating in this project together with other 14 SME's who were interested in GVTs in the region of Arnhem and Nijmegen.

Meaningful learning related to the challenges of using virtual communication and the use of technology was revealed by the surveys applied after the project implementation to different participating groups and also to a control group of Arnhem Business School. These results are demonstrated by the longitudinal waves of responses obtained through the surveys from undergraduate students prior to the X-Culture project, during this project and after the implementation of it. These surveys were applied to the instructors and students in the pre-project phase, every week during the implementation phase and as post-project interviews, evaluations and surveys.

Different measures were adopted based on the individual and team results and performance. These measures include the capacity to meet deadlines, peer assessments, multi-dimensional evaluation of team report quality, satisfaction and also different records useful for content and qualitative study.

Elementary descriptive coding and statistics were used in order to analyze and present the data obtained through the X-Culture project surveys. This research relies on two sources of data in order to assess the impact of participating in the X-Culture international project: (i) Pre-, mid and post-project managed surveys and (ii) a study specific survey created locally at Arnhem Business School and operated as post-project survey. This paper assesses the influence of Global Virtual Teams in Business Education at ABS regarding the development and design of new curriculum modules based on the experiential learning.

3 International and National Frameworks

Research performed among alumni and various companies in the Netherlands has shown that the profession-al field requires International Business graduates with a solid grip of generic or so-called 21st century competences, like communication skills, collaboration, critical thinking, intercultural sensitivity, innovative thinking, information management, flexibility and personal and professional development [11].

Also Sent [12] concluded that job life cycles are becoming progressively shorter. Students will therefore change jobs regularly throughout working career. For this reason, International Business orientated students will need to be able to cope with flexible labor market, along with the concomitant requisite lifelong learning implied (See Kan [13] as well as Knottnerus [14]. These documents emphasize the importance of 21st century skills which are characterized by transferability from job context to job context [15].

The professional field does not only advocate a strong focus on generic skills in education, but due to the globalization of the workplace, employees are also expected to perform successfully in an international context, stressing the importance of intercultural communication and cultural sensitivity [11].

In their report 'Culture at Work', The British Council summarizes this as follows:

The workplace in modern society is more and more competitive and globalized. Consequently, communicating all over the world and across international borders with partners, colleagues and customers is a daily activity for numerous workers around the globe. Accordingly, employers should find employees that are both technically competent and culturally skillful in order to be able to provide efficiency in the international workplace. This situation puts employers under a strong pressure to find the best employees [16].

An in-depth analysis of the 21st century skills [17, 18], the generic skills [19] and the transferable skills [20] resulted in the adoption of the KSAVE model. This model ensures the substance of the International Business Degree Programmes, the Dublin descriptors and the HEO Standard [11] (Table 1).

Table 1. Relation between the 14 themes of the new IB Framework, HEO Standard and Dublin Descriptors

Domains in KSAVE model	Themes in IB new framework	HEO standard	Dublin descriptors
Ways of Thinking	Critical Thinking	2	3
	Innovation & Creativity	3	2, 3
	International Business Awareness	3	2, 3
Ways of Working	International Business Communication	3	4
	Collaboration	3	4
	Management of information as digital citizen	2	3
Living in the World	Personal & Professional Development	4	3,5
	Ethical & Social Responsibility	4	3,5
	Intercultural Proficiency	4	3,5
Tools for Working & Management	Marketing & Sales	1	1,2
	Finance & Accounting	1	1,2
	Operations & Supply chain management	1	1,2
	Organization & People	1	1,2
	Business Research	2	1,2,3

The KSAVE model defines generic 21st century skills in terms of knowledge, skills, attitudes, values and ethics that are learned and acquired through core subjects. These elements of KSAVE (knowledge, skills, attitude, values and ethics) are incorporated in the Programme Learning Outcomes (PLO's). In the IB Framework these elements are not elaborated any further as the framework assumes that institutes will elaborate these for every PLO themselves; this then serves as what was formerly known as the 'Body of Knowledge and Skills (BOKS)' [11].

The IB Framework for Arnhem Business School has been defined as follows Table 2):

Next to these workplace requirements, several standards are further outlined in national and international higher educational degree expectations. The national expectations are embedded in the Dutch Higher Economic Education Standard "HEO standard" and this new IB PLO framework reflects this national perspective. By this means, the study programme ensures that IB graduates possess the following attributes:

Table 2. Programme Learning Outcomes new IB Framework

Ways of thinking		
Critical Thinking	WT1	Use the process of thoughtful evaluation to formulate a reasonable conclusion deliberately
Innovation & Creativity	WT2	Create innovative ideas in a changing business environment systematically
International Business Awareness	WT3	Analyze patterns in global macro-economic factors and policies that drive international trade and business development
Ways of Working		
International Business Communication	WW4	Communicate (business) messages effectively persuasively using advanced English to an (un) informed audience
	WW5	Optional: Use one or two additional languages to facilitate international business
Collaboration	WW6	Collaborate effectively with different kinds of stakeholders, in different cultural, organizational and political landscapes to contribute to achieving agreed goals
Management of Information as digital citizen	WW7	Produce management information from various data sources in an international business environment
Living in the world		
Personal & Professional Development	LW8	Express reflections on his personal development with the aim of personal growth
	LW9	Respond appropriately to an unfamiliar, or unexpectedly changing, business environment
Ethical & Social Responsibility	LW10	Formulate his own position concerning ethical and social responsibility in a professional environment
Intercultural Proficiency	LW11	Mitigate the pitfalls of cultural differences in business and social contexts
	LW12	Display willingness to work with people from other cultures and to work in countries with different cultural backgrounds
	LW13	Use appropriate verbal and non-verbal communication in an intercultural setting
	LW14	Assess the effect of cultural differences upon organizational behavior and strategic choices
Tools for working and management		
Marketing & Sales	TWM15	Develop a well-founded marketing plan to support the creation of value for international customers
	TWM16	Use appropriate sales techniques in support of durable customer relationships
	TWM17	Incorporate developments of the digital landscape in a marketing strategy

(*continued*)

Table 2. (*continued*)

Ways of thinking		
Finance & Accounting	TWM18	Evaluate financial performance of the organization from different stakeholders' perspectives
	TWM19	Recommend financing possibilities in a dynamic international environment
Operations & Supply chain management	TWM20	Evaluate the operations processes within and between organizations
	TWM21	Manage the operations processes within and between organizations
Organization & People	TWM22	Draft the strategic cycle of part(s) of the organization (process and content)
	TWM23	Assess the impact of change on the organization
Business Research	TWM24	Analyze a complex business problem in an international business setting with use of an adequate research design, resulting in an evidence based feasible solution

- A solid theoretical basis.
- Research skills that will enable them to contribute to the development of their chosen profession.
- A sufficient set of professional skills.
- A professional, ethical, and social orientation.

In April 2016 within the domain of the Dutch Higher Economic Education (HEO) the decision was made that the programmes IBMS (34936), IBL (34407), IB (30009) and Trade Management for Asia (34041) are to be united in a programme named International Business (IB) as from 2018–2019, with one Central Register of Higher Education Programs (Centraal Register Opleidingen Hoger Onderwijs) - CROHO number 30029. Therefore, Arnhem Business School had to comply with the national IB framework by developing a new IB programme. Arnhem Business School offers the International Business Programme, where students can choose between four differentiations in the main phase: Finance, Supply Chain Management, Marketing & Sales and Management.

4 GVT Course Outline

Based on the new IB development at ABS, a new set of Elective modules that meet the IB requirements had to be designed. Matching with the new IB requirements and the IB profile at ABS, the Global Virtual Teams Course (Module) has been designed taking into account above mentioned IB profile, the IB Programme Learning Outcomes, the Module Learning Outcomes and specific competencies and skills.

After following this module, the students will able to:

- Describe the availability and points out adequate use of cooperation technologies in a Global Virtual Team.
- Proves the capacity to work in a global team or group toward a shared aim by utilizing different approaches that involve the GVT's cultural diversity.
- Utilize technology as a mechanism to organize, to research, assess and disseminate information.
- Utilize information technologies like communication or networking tools and social networks in order to be able to access, operate, integrate, assess and create information that will function well in the knowledge economy.
- Demonstrates expertise and knowledge to follow into an appropriate manner the rules of cultural communication when communicating with people across borders.
- Characterize the influence of culture on building trust among virtual team members.
- The alignment with the PLO's is also set as follows:
- WW6: Collaborate effectively with different kinds of stakeholders in different cultural, organizational and political landscapes to contribute to achieving agreed goals.
- WW7 - Management of information as digital citizen (Produce management information from various data sources in an international business environment)
- LW11: Mitigate the pitfalls of cultural differences in business and social contexts
- LW12: Display willingness to work with people from other cultures and to work in countries with different cultural backgrounds.
- LW13: Use appropriate verbal and non-verbal communication in an intercultural setting.

The course, "Introduction to working in Global Virtual Teams", is designed to teach students how to work better in a global virtual team. The course can be taught in a joint project between Arnhem Business School and another university or only done by student of the Arnhem Business School; the course was designed to be taught synchronously with team members who are not physically in Arnhem. As suggested by the name the course contains two major objectives. First, students are taught global team competencies to promote successful interactions in a global virtual team. Second, students are taught and provided the opportunity to use virtual communication skills to promote successful communication in global virtual teams. Along with specific instruction in these areas, students are provided the opportunity to improve their cross-cultural knowledge and experiment with virtual communication technologies during the class instruction, assignments and project tasks.

- Lecture 1: Introduction to Virtual Communication Technology, Globalization and Global Competences, GVT Team Processes
- Lecture 2: Building and Maintaining Trust, Conflict Resolution, Cultural Dimensions
- Lecture 3: Global Virtual Team Leadership, Virtual Communication, Cross-Cultural Communication

The second part of the course repeats coverage of these topics but in a virtual seminar format. Students are provided with readings associated with each topic

Students are teamed with international partners to present the topic and relate it to their global virtual team experience or the experience of classmates. A list of current readings is provided where it will be continuously updated as needed to reflect the current research and understanding of global virtual teams. It should be noted that this type of instruction and interaction adds several levels of complexity to normal instruction. Those teaching the course should become familiar with the types of virtual communication technologies to be used in the course.

It is strongly recommended that advanced trial use of each technology be undertaken with the technology to be used, in the location where it is to be used and at the time it is to be used, prior to classroom use. This would ensure proper connectivity, adequate bandwidth and troubleshooting of any unanticipated problems. If technical support in each location is available, use it.

Students from all locations should have a procedure in place to receive local credit for participating and successfully completing both this course and the project. Ideally, there would be an instructor or assistant in each location to hand out, proctor examinations and assist in instruction. If none is available, adjustments to instruction, dissemination, and collection of course materials need to be modified. Each student should also have access to a computer with Internet access as interaction during and outside lectures is planned and important to this course.

One person should be designated as the course instructor. This person should maintain regular contact with all members. This will help to clarify and specify due dates, assignment questions and promote interaction between the instructor and student. Typically, messages should be sent out immediately after a lecture to specify due assignments. This message may also be used to reinforce key principles, but it should be kept brief.

A second message should be sent a day or so prior to the course giving class presentation assignments and reminders. In between, the instructor should specify and be available through virtual communication technologies.

While classrooms may vary, the following is a suggested arrangement as course instruction. In this model, all students are taught by their own local university lecturer (see Fig. 1). The lecturer has direct connections to the local students and, through virtual technology to the two international universities. Students at these universities must communicate to the class through the virtual technology media as represented through the thick blue lines. However, as the students are asked to discuss or work together in small groups, they would communicate directly with one another as indicated by the red lines. Although this is not shown on the diagram students in university A could pair up with students in University B.

Still, it is encouraged to have teammates pair up in discussions. This allows them to increase contact with each other and practice using virtual communication technology. Depending on the virtual communication technology used and class size, the teacher may assign pairs within the communication technology or simply allow students to connect directly.

This model allows for group presentation and discussion similar to a traditional classroom. It also allows students to interact virtually with one another. If the interaction is limited to project group, it provides for the opportunity to build trust and a working relationship to promote team unity while gaining greater understanding of

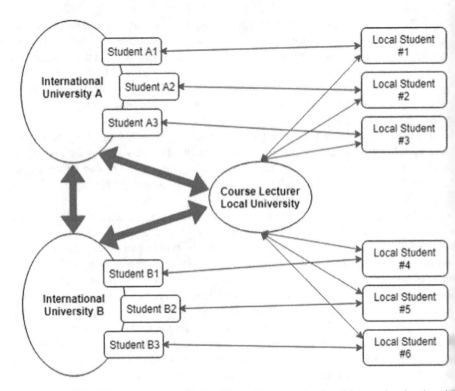

Fig. 1. Model of Synchronous instruction for students at a local and international universi
Source: Lecture Plan ME495r, Holt Zaugg, 2010 [21].

cultural concepts. If a greater emphasis on cultural interactions is desired, then studen
may be paired up with international classmates outside of their project group. Th
format allows students to interact in large, small and paired international groups.

Depending on the time zone differences, 2 different class times may be schedule
This will allow students to alternate meeting at early and late times. This furth
simulates real world virtual teams and prevents one group of students from burnout
meeting times that are too early or too late in the day. The instructor would also need
be willing to adjust his/her teaching schedule to accommodate these switches.

Working collaboratively in culturally diverse virtual teams, students select a top
to discuss for 3 GVT meetings and present the joint results at the end of the exchan
in the form of a report (2500 words). Besides the outcome of these meetings
individual reflection (500 words) is expected and a digital with a recording of a GV
meeting (15 min).

Students have to compare a product, service or aspect of business (e.g. manageri
styles, business innovations etc.) across at least two different cultures. Examples
topics that students will research on and report results:

- The influence of cultural differences on attitudes towards sleeping pills in Europe and Asia.
- Influences of cultural differences on Daimler's business cooperation with the Mitsubishi Motor Company.
- Key elements of McDonald's marketing mix adaptation – A country comparison between the USA and India within the framework of high and low-context culture according to Hall.
- McDonald's marketing strategy in China and the Netherlands – an analysis with Hofstede's cultural dimensions.
- A cross-cultural analysis of the German and Slovenian electric car market – Findings deduced from Hofstede's national cultural dimensions applied to the Marketing Mix.
- Lenovo's acquisition of IBM – a prime example of how to bridge cultural differences.
- A cross-cultural comparison of BMW's brand positioning in Germany and Egypt.
- Can different life insurance shares in Japan and Lithuania be explained by their respective cultures?
- An analysis of cultural influences on life insurance coverage compared to non-life insurance.
- A cross-cultural analysis of attitudes towards car sharing in Germany and China using the example of the Daimler AG.
- A cross-cultural analysis of gender quotas in Germany and Romania.
- The use of self-checkout terminals in grocery stores: A cultural comparison between the Netherlands and Hungary.

These projects, grounded in cross-cultural theory, are examined in terms of both content and communication skills in a GVT context. Cross-border cooperation within the multidisciplinary learning communities is on-going all participants experienced as very positive; it gives new impulses, leads to the stretching of the network considered relevant and to competence and vision development among the participants. The development of a multidisciplinary, sustainable professional learning community between education, teacher training and research, in which not only the participants learn, but also contribute to the innovation or transformation of the underlying organizations and usable knowledge and products, requires attention, time and effort, certainly from the organizations involved [22].

In this elective, students will learn how to communicate and collaborate effectively using the latest technology to drive collaboration from Global Virtual Teams, both as team members and as team leaders. When the course ends, students will have gained the knowledge and skills required to work effectively with their colleagues/peer-students around the world. They will understand the different social dynamics of remote work - how to build trust, manage conflict, ensure results, ideate, overcome cultural and linguistic gaps and use technology to mitigate social distance, all while driving innovation and creating value.

References

1. Popescu, F., Warmenhoven, R.: Use of technology and virtual communication via global virtual teams at arnhem business school. In: Kantola, J., Nazir, S., Barath, T. (Eds.) Advances in Human Factors, Business Management and Society. AHFE 2018. Advances in Intelligent Systems and Computing, vol. 783. Springer, Cham (2019)
2. ACS: Mind the skills gap: The skills we need for sustainable communities. Academy for Sustainable Communities. http://image.guardian.co.uk/sys-files/Society/documents/2007/10/08/Final_full.pdf (2015)
3. Barber, M., Donnelly, K., Rizvi, S.: An Avalanche is Coming: Higher Education and the Revolution Ahead. Institute of Public Policy Research, London (2013)
4. BITC: Leadership skills for a sustainable economy. Business in the Community/EDF Energy. pp. 1–24 (2010)
5. British Council: The Global Skills Gap: Preparing young people for the new global economy. http://tinyurl.com/globalskillsgap (2011)
6. IBM: People and Skills for a Sustainable Future: Report based on proceedings at IBM Summit at Start. The Bathwick Group (2010)
7. IPPR: The Futures Green: Jobs and the UK low carbon transition. Institute of Public Policy Research. http://www.ippr.org/images/media/files/publication/2011/05/futuregreen_1737.pdf (2009)
8. Ipsos-Mori: Skills for a Sustainable Economy: The Business Perspective. Ipsos-MORI Reputation Centre. http://www.ipsos-mori.com/Assets/Docs/Publications/skills-for-a-sustainable-economy-the-business-perspective_170610.pdf (2013)
9. SKY: The Sustainable Generation: The SKY Future Leaders Study. Isleworth, Middlesex (2011)
10. Warmenhoven, R., Popescu, F.: GVT challenges at arnhem business school. In: The European Proceedings of Social & Behavioural Sciences EpSBS, ISSN: 2357-1330 (2018)
11. Sijben, G., Stoelinga, B., Molenaar, A., Ubachs, M.: Framework International Business https://www.vereveniginghogescholen.nl/system/profiles/documents/000/000/224/original/international-business.framework.2018.pdf?1518520108 (2018)
12. Sent, E.M., et al.: Wendbaar in een Duurzame Economie. Vereniging Hogescholen, Den Haag (2015)
13. Kan, A.R., et al.: Adviesrapport: Flexibel Hoger Onderwijs Voor Volwassenen (2014)
14. Knottnerus, J.A., et al.: Naar Een Lerende Economie. Amsterdam University Press Amsterdam (2013)
15. Gullik van, P., Sijben, G., Stoelinga, B.: Outcomes Validation Research (2015)
16. The British Council, "Culture at Work: The Value of Intercultural Skills in the Workplace." https://www.britishcouncil.org/sites/default/files/culture-at-work-report-v2.pdf (2013)
17. Trilling, B., Fadel, C.: 21st Century skills: Learning for Life in Our Times. Wiley, San Francisco (2009)
18. Voogt, J., Pareja, N.: 21st Century skills: Discussienota. Universiteit van Twente, Twente (2010)
19. Goldsworthy, A.: Developing Generic Skills: Examples of Best Practice. http://www.bhert.com/documents/B-HERTNEWSNo.16_001.pdf (2003)
20. Andrews, J., Higson, H.: Graduate employability, 'Soft Skills' Versus 'Hard' business knowledge: a european study. High. Educ. Eur. 33(8). Routlegde, UK (2008)
21. Lecture Plan ME495r, Holt Zaugg, (2010)
22. Van Vijfeijken, M., Van Der, N., Uerz, D., Kral, M.: iXperium/Centre of Expertise Leren met ICT, Faculteit Educatie HAN Samen leren innoveren met ICT, Tijdschrift voor Lerarenopleiders 36(4) (2015)

Research on Virtual Simulation Evaluation System for Passenger Compartments Lighting of Subway Trains in China

Jian Xu[1], Ze-rui Xiang[1,2(✉)], Jin-yi Zhi[1,2], Xiao-fei Xu[1,2], Si-jun He[1], Jin Wang[1], Yang Du[1], and Gang Xu[3]

[1] School of Architecture and Design, Southwest Jiaotong University, Chengdu 610031, China
xiangzerui@163.com
[2] Institute of Design and Research for Man-Machine-Environment Engineering System, Southwest Jiaotong University, Chengdu, China
[3] CRRC QingDao SiFang CO., LTD, Qingdao 266111, China

Abstract. Lighting in the subway passenger compartments is one of the many factors influencing riding comfort. The research and design of lighting evaluation system for subway passenger compartments aim to establish evaluation standard and models for the design of light environments in this area. By using the virtual simulation technology of optical software Dialux and referring to the existing lighting standards in China, Lighting Evaluation System 1.0 (LES 1.0), a light environment evaluation system for subway passenger compartments, is developed. In the process of visual simulation, the lighting scenes of subway passenger compartments are simulated to obtain comfort evaluation index data for evaluating the comfort. On this basis, the evaluation process is constructed, including the establishment of the evaluation standard model for light environments. The research is helpful for designers and engineers to improve the design and evaluation of lighting environments in subway passenger compartments in China.

Keywords: Subway passenger compartment · Lighting design · Virtual simulation · Evaluation system

1 Introduction

Rail transit in China's major cities has been developing rapidly with the urbanization of the country. As of the end of 2017, urban rail transit in mainland China has included subway, light rail, monorail, inner-city rapid rail transit, modern tram, maglev transportation, and other systems. The subway system has a total mileage of 3,883.6 km, accounting for 77.2% of the total [1]. It can be seen that subway has become a favored means of transport in cities for most Chinese people. However, subways merely as a traffic tool can no longer satisfy passengers, given that more emphasis on the riding comfort is laid with the improving of Chinese people's living standard and education level. Lighting in the subway passenger compartments is one of the main factors influencing such comfort.

© Springer Nature Switzerland AG 2020
F. Ahram (Ed.): AHFE 2019, AISC 973, pp. 343–353, 2020.
https://doi.org/10.1007/978-3-030-20476-1_35

All countries and certain international organizations have proposed their technical regulations and guidelines for the passenger compartment lighting design of rail transit trains (as shown in Table 1). According to the *Railway Applications – Electrical Lighting for Rolling Stock in Public Transport Systems* (EN 13272-2012) [2], detailed provisions are stipulated for the illuminance of passenger compartments for rolling stock and the lighting requirements for passenger compartments of other trains are specified as follows: The average illuminance of the passenger compartment seat area which is 800 mm away from the floor surface shall be 150 lx, and that of the standing area and aisle which are 800 mm way from the floor shall be 50 lx. According to the *Illuminance for Railway Rolling Stock – Recommended Levels and Measuring Method* (JIS E 4016-2009) [3], the illuminance of passenger compartments for trail transit rolling stock is specified as follows: The average illuminance of the area which is 850 mm above the floor surface within passenger compartments in the normal lighting condition shall be more than 200 lx. According to the *General Technical Specification for Metro Vehicles* (GB/T 7928-2003) [4], the average illuminance of the area which is 800 mm above the floor surface shall be not less than 200 lx, with the minimum value not less than 150 lx. We can see from the standards that there is no united technical standard for lighting designs. In addition, there remains such a problem that, in these standards, the recommended illuminance values are determined by assuming predictability or technical feasibility of lighting instead of considering people's responses, especially the comfort of the lighting environment in subway passenger compartments. One of the reasons is that comfort is difficult to quantify. At present, Chinese technical standards for lighting designs are mainly based on various international standards. As for the lighting design of subway passenger compartments, China has not worked out a technical standard through its own independent basic research, let alone a standard developed according to the characteristics of Chinese users themselves. This situation leads to the lack of objective and reasonable bases for design in this respect.

Table 1. Recommended illuminance values in standards for passenger compartment lighting of rail transit trains.

Lighting area	Illuminance	EN 13272-2012 [2]	JIS E 4016-2009 [3]	GB/T 7928-2003 [4]
Seat area	Average illuminance	150 (lx)	–	–
	Illuminance uniformity	0.8–1.2 (lx)		
Standing area and aisle	Average illuminance	50 (lx)	–	–
	Illuminance uniformity	0.5–2.5 (lx)		
Normal lighting area	Average illuminance	–	> 200 (lx)	≥ 200 (lx)

The research of lighting evaluation system and design of subway passenger compartments in this project is divided into two parts. The first part mainly plans to design the working process of the light environment evaluation system for subway passenger compartments and study and sort out the *Evaluation Standard Model for Light Environments of subway passenger compartments* according to the existing standards. The second part mainly plans to modify the evaluation standard for light environments as specified in the first part by using a research method of Chinese users' preference for lighting comfort. At the same time, the evaluation standard model for light environments will be expanded with a large number of example tests for application experiments. The research of this paper is the first part. The research and design of lighting evaluation system for subway passenger compartments aim to establish evaluation standard and model for the design of light environments in this area with the characteristics of Chinese users. By using the optical virtual simulation technology and referring to the existing lighting standards in China, Lighting Evaluation System 1.0 (LES 1.0), a light environment evaluation system for subway passenger compartments, is developed. In the process of visual simulation, the lighting scenes of subway passenger compartments are simulated to obtain comfort evaluation index data for evaluating the comfort. On this basis, the evaluation process is constructed, including the establishment of the evaluation standard model for light environments. The research is helpful for designers and engineers to improve the design and evaluation of lighting environments in subway passenger compartments in China.

Selection of Lighting Evaluation Indexes for Subway Passenger Compartments

As passengers will be psychologically and physically influenced by lighting mainly in a visual way, lighting in subway passenger compartments is an important factor influencing the comfort level of subways. Therefore, visual comfort is used to show the comfort of subways on the aspect of lighting in subway passenger compartments. Although there is no widely agreed definition of visual comfort, the most widely accepted approach is the "NON-annoyance approach" based on the assumption that "comfort is not discomfort" [5]. Therefore, visual comfort can be construed as a state in which people can clearly and unobstructedly receive the visual information from the visual environment without any influence on their physiological and psychological health. In the whole visual environment of subway passenger compartments, passengers mainly receive the information from the subway information display system and that on the appearance of passenger compartments. The information display system is mainly used to show the train riding information of passengers. To find such type of information is the visual tasks of passengers. The factors influencing visual performance include illuminance, glare and glare control and veiling reflection. The information on the appearance of subway passenger compartments has an impact on the visual psychology of passengers, which is influenced by color temperature, color rendering, shade and other factors. These infuencing factors and their respective algorithms constitute the evaluation elements of LES 1.0 for subway passenger compartments lighting. Through data collation and expert analysis, LES 1.0 finally selecte illumination, color temperature, color rendering, and glare as the evaluation index of subway passenger compartments lighting.

2.1 Illuminance

Illuminance is a physical quantity that indicates the extent to which a certain area is exposure to light, and is expressed by the area density of the luminous flux at the illuminated site in lux (lx). The illuminance (E) at one point on the surface is defined as the ratio of the incident luminous flux $d\Phi$ to the unit area dA: $E = d\Phi/dA$. Illuminance is a direct index to determine the brightness of an object. In a certain range, visual ability can be improved by increasing illuminance. When the illumination is within the appropriate range, it is beneficial to protect people's vision. If the illumination is excessive or insufficient, it will damage human health and cause visual fatigue.

Passengers' behaviors in subway passenger compartments include walking within cars, looking for seats, reading, playing with mobile phones and checking road maps. There is a low requirement for vision as to such behaviors as walking within cars and looking for seats, so illuminance has a trivial effect on this kind of visual task. Visual behaviors such as reading, playing with mobile phones and checking road maps have high requirements for vision. The illuminance value, if insufficient, will affect passengers' vision, cause visual fatigue and undermine visual comfort. Therefore, the illuminance of the illuminated surface of the subway passenger compartment facilities (e.g. display system) is the main aspect of the lighting quality of passenger compartments.

2.2 Color Temperature

The color temperature is the temperature of the surface of the illuminator, which can be used to indicate the color appearance of the light source. In a light environment, the combination of color temperature and illumination of different light sources make people a different feeling. The research shows that the combination of color temperature and illuminance reflecting light properties can affect the comfortable feeling of people in different light environments.

However, the warm-cool feeling resulted from colors is not the only effect of color temperature. Color temperature is of great significance to control people's mood. According to the findings of Swedish scholar Küller R et al., illuminance and color temperature can affect people's mood fluctuation under the natural light condition [6]. Special public places are of higher requirements for color temperature. For example, the light environments of subway passenger compartments, light sources with low brightness and low color temperature should not be used for lighting, because the light from such resources will make the narrow space in the subway cars seem to be narrower and smaller, causing passengers to feel afraid and depressed. Conversely, high color temperature is used, passengers will psychologically feel that the car space expanded, and thus they will feel less depressed.

2.3   Color Rendering Index (Ra)

Color rendering index is defined as the *"effect of an illuminant on the color appearance of objects by consciousor subconscious comparison with their color appearance under a reference illuminant"* [7]. According to the definition of Ra, in the living and working environments, it is necessary to display the color characteristics of various facilities and implements, which is conducive to the protection of eye vision and people's psychological health. In some specific venues, higher requirements are put forward for the color reducibility of objects, so as to meet the needs of the user work, such as workspace with matching colors.

From the last section, which has explained that color temperature of light sources (light colors) is directly related to people's psychology, we can see that colors will affect people's visual psychology, and so will the material colors of subway passenger compartments. The inherent color reducibility of the objects in the passenger compartments determines the visual comfort of the passengers. Ra of the light sources can be used to indirectly predict the rendering quality of material colors of subway passenger compartments. Generally, the higher Ra is, the better the color rendering of light sources is, and the stronger the color reducibility of objects is.

2.4   Glare

Glare can be defined as *"the sensation produced by luminance within the visual field that is sufficiently greater than the luminance to which the eyes are adapted to cause annoyance, discomfort or loss in visual performance and visibility"* [8]. Generally, glare can be divided into disability glare and discomfort glare. Disability glare, which weakens the vision, mainly occurs in the indoor and outdoor lighting. For lighting in subway passenger compartments, the main problem is discomfort glare, and the effective control of such glare can help avoid disability glare.

3   Establishment of Visual Simulation Scenes for Lighting in Subway Pssenger Compartments

The simulation and data of lighting scenes for subway passenger compartments are mainly calculated with Dialux, professional software for lighting design and calculation developed by German DIAL GmbH. It can be added with parameter plug-ins of various lamp manufacturers, that is, the directories of electronic lamps. The software has been widely recognized and applied in the world. With strong functions, Dialux software can be widely used for lighting design and calculation in various scenes and types of indoor, outdoor and road lighting [9, 10]. Basically, the mainstream lamp manufacturers in the world have cooperated with Dialux to release the lamp database plug-ins, including Philips, OSRAM, BEGA and iGuzzini. The lamp manufacturers in China include NVC, Shanghai Yaming Lighting, Pak and Zhejiang Haixing Lighting. As Dialux uses an accurate photometric database and an advanced and professional algorithm, its calculation results are very close to the real measurement results after the construction. Dialux does very well in the output of computed results, and can provide complete written reports (including spot illuminance, lamp data and other information), as well as 3D simulated diagrams, and isophot diagrams. Dialux software can import the dimension information of subway car space model by loading DWG or DXF files, and interact with AutoCAD and other software in terms of design parameters, so as to avoid the repeated process of manually setting up subway car models and inputting parameters, as shown in Fig. 1. On this basis, designers and engineers can work out a unified glare rating, which is combined with illuminance, Ra and other parameters to evaluate the visual comfort of light environments in subway passenger compartments.

Fig. 1. Model for visual simulation scenes inside subway passenger compartments in DIAlux

4   Design of Evaluation Process for Lighting in Subway Passenger Compartments

The core of the light environment evaluation system for subway passenger compart ments is to establish evaluation standards, the model accuracy of which will affect the entire evaluation results. Such accuracy is related to specific groups. Some specia trains serve specific groups of people. The Disney-themed Trains in Hong Kong, Chin (see Fig. 2), for instance, are of specially designed passenger compartments for chil dren. In addition, China will face a severely aging population in the future, so it i necessary to provide trains customized for the elderly. We have taken children and th elderly into consideration and regarded subway cars as mobile buildings. On this basis in the study of the evaluation standard model for light environments in subway pas senger compartments, we have sorted out a series of convenient and feasible evaluatio standards, namely "a evaluation standard model for light environments" with the re erence to the international and Chinese national standards, such as *Lighting of Indoo Work Places* (CIE S 008/E-2001) [11] of Commission International de L'Eclairage *Standard for Lighting Design of Buildings* (GB 50034-2013) [12] and other relate standards and in combination with the requirements of different age groups of users fc lighting and the glare value calculation methods.

After the above said model is established, the lighting data of any plane and an point in the space of a lighting design plan can be calculated according to the lightin

Fig. 2. Hong Kong disney-themed train.

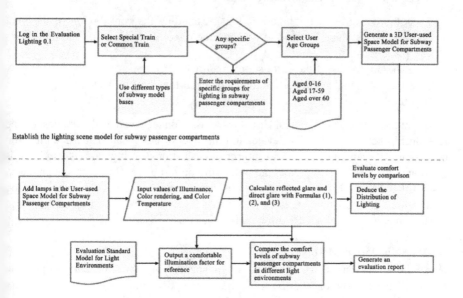

Fig. 3. Workflow chart of lighting evaluation system 1.0.

design and the design software Dialux. These data can be used as the input parameters for calculating a unified glare rating, and compared with the standard data in the evaluation model to obtain a comfort evaluation report for the lighting design plan, as shown in Fig. 3.

Among the four evaluation indexes, i.e. illuminance, color rendering, color temperature and glare, the former three are derived from the intrinsic parameters of the selected lamps, and the last is the later response after the subway passenger compartments lighting system is set up. Therefore, the first three parameters are relatively easy to quantify and evaluate, while the glare control is just on the contrary. To this end, LES1.0 is developed based on the simulation scenes provided by Dialux to design the evaluation methods that can quantify the lighting glare of subway passenger compartments.

4.1   Introduction and Calculation Method of Unified Glare Rating

As an important index in the lighting design for subway passenger compartments, discomfort glare is regarded as a significant lighting quality index in Chinese lighting standards (e.g. *Standard for Lighting Design of Buildings* (GB 50034-2013)). LES1.0 also uses the UGR (Unified Glare Rating) system proposed by Commission International de L'Eclairage (CIE) for glare calculations.

The UGR value is calculated as shown in Formula (1) [11]:

$$UGR = 8Log_{10}(0.25/L_b) \sum \left[L_s^2 \cdot \omega/p^2\right] \tag{1}$$

Where

UGR is the Unified Glare Rating value;

L_b is the background brightness (cd/m^2), excluding the contribution of the light source, and its value is equal to dividing the direct illuminance on the observer's eye plane by π;

L_s is the brightness of the glare source (cd/m^2);

ω is the solid angle (ster) of the glare source at the observer's eye;

p is the Guth position index of the glare source

From the literature [13], it can learn that Akashi conducted an evaluation test on the relationship between subjective evaluation and UGR, and found that the correlation coefficient between UGR value and subjective evaluation is as high as 0.89. so UGR is up to now the best way to predict the indoor discomfort glare. However, the predicted results of UGR are often too severe when the light source is small (< 0.005 m^2) and too loose when the light source is large (> 1.5 m^2). Therefore, in case of a small or large light source, a modified UGR formula [14] needs to be used.

4.1.1   Calculation Method of Small Light Source

When the projection area (A_p) of a small light source is $A_p < 0.005$ m^2 and its position deviates from the line of sight by > 5^0, the UGR is determined by the light intensity (I) reaching the eye. and The brightness (L_s) is as follows [14]:

$$L_s = \frac{I}{A_p} = 200 \cdot I.$$

Therefore, the UGR calculation formula of small light source is as follows [14]:

$$UGR = 8Log_{10}(0.25/L_b) \sum \left[200 \cdot I^2/r^2 \cdot p^2\right]. \tag{2}$$

where

r is the distance (m) from the glare source to the eye

4.1.2 Calculation Method of Great Light Source

Great light source glare needs to consider both "luminous ceiling and indirect lighting" and "transition lamp between the luminous ceiling and ordinary lamp". in case of a "luminous ceiling or indirect lighting", a simple evaluation method, i.e. if illuminance ≤ 1600 lx, UGR = 22, shall be selected. In case of a "transition lamp between luminous ceiling and ordinary lamp", the calculation method is as shown in Formula 3) [14]:

$$GGR = UGR + \{1 - (0.18/CC)\} \cdot 8 \cdot \log_{10}[2.55 \cdot \{1 + E_d/E_i\}] \qquad (3)$$

Where

GGR is the great source glare rating;
CC is the ceiling coverage;
E_d is the direct illuminance resulted from the light source at the eye;
E_i is the indirect illuminance at the eye

.2 Evaluation Standard Model for Light Environments in Subway Passenger Compartments

a the spaces within the subway passenger compartments regarded as mobile buildings, their lighting shall be analyzed and designed from the perspective of the indoor lighting f buildings, so there is a need to refer to relevant standards of building lighting. In the lighting of Indoor Work Places (CIE S 008/E-2001) of Commission International de 'Eclairage, and Standard for Lighting Design of Buildings (GB 50034-2013), the standard lighting values of various buildings are managed [11, 12]. LES1.0 extracts and organizes the data from the lighting in passenger compartments for special trains ustomized for specific groups, such as children and the elderly) and that for common ains, arranges the lighting parameters to be corresponding to each other in combination with the effect of glare on comfortable feeling of users, and classify the reference alues for lighting of public buildings and public places, so as to build a reference andard of the evaluation system, namely the *Evaluation Standard Model for Light nvironments of subway passenger compartments*. The main reference values in the aluation standards include standard value of illuminance (lx), color rendering index a), and unified glare rating (UGR).

The *Evaluation Standard Model for Light Environments of subway passenger mpartments of subway passenger compartments* developed by LES1.0 mainly consts of the lighting standards for passenger compartments of special and ordinary ains. For other lighting spaces of subway passenger compartments not included in the aluation Standard Model for Light Environments of subway passenger compart- ents, LES1.0 calculates illuminance, Ra and uniform glare rating based on actual put values, and displays the results without comparison with the Standard Model. As per different subway passenger compartments environments, behaviors and groups of users, the *Evaluation Standard Model for Light Environments of subway passenger compartments* arranges the lighting parameters to be corresponding to each other, and establishes a large database. Of course, the next step of the research is to conduct example texts and modify the accuracy of the data in this database to revise, so as to provide lighting designers and engineers with improvement plans and effective tools for verifying the lighting quality.

5 Conclusions

Based on the research on visual comfort in light environments, this paper presents an evaluation standard of non-glare comfortable lighting in subway passenger compartments and designs an evaluation model. Thanks to such an effort, LES1.0, a convenient and applicable evaluation system for light environments in subway passenger compartment, is designed. The system is developed on the basis of Dialux software. It can not only get high-fidelity simulation scenes, but also generate complete written report and analysis results of lighting environments in subway passenger compartments. It is suitable for all kinds of designers and engineers.

The main target of this research is to introduce the workflow design of LES1.0 and the study and arrangement of *Evaluation Standard Model for Light Environments of subway passenger compartments*. Next research will focus on two aspects. On one hand, we will modify the light environment evaluation standards by using a research method of Chinese users' preference for lighting comfort, so as to better combine the users' preference for comfort with lighting parameters. At the same time, the standard model of evaluation system will be expanded with a large number of example tests for application experiments. On the other hand, we will develop immersion scene experience with virtual reality headsets as carriers and an evaluation platform based on the existing virtual simulation systems, in order to facilitate users' evaluation.

Acknowledgments. The author would like to acknowledge the support from the National Key Research and Development Programs of China (No. 2017YFB1201103-9/ 2017YFB120110 12) and the Humanities and Social Sciences Research Youth Foundation of Ministry of Education of China (No. 17YJC7601021).

References

1. China Association of Metros: 2017 Statistics and Analysis Report in Urban Rail Transit. J. Urban Rail Transit. **26**, 8–27 (2018)
2. European Committee for Standardization: EN 13272. Railway Applications – Electric Lighting for Rolling Stock in Public Transport Systems (2012)
3. Japanese Industrial Standards Committee.: JIS E 4016. Illuminance for Railway Rolling Stock – Recommended Levels and Measuring Method (1992)
4. Chinese Standard: GB/T 7928. General Technical Specification for Metro Vehicles (2008)
5. Iacomussi, P., Radis, M., Rossi, G., et al.: Visual comfort with LED lighting. J. Energy Proc. **78**, 729–734 (2015)

6. Küller, R.: The impact of light and colour on psychological mood: a cross-cultural study of indoor work environments. J. Ergon. **49**, 1496–1507 (2006)
7. Commission Internationale de L'Eclairage: CIE 17.4. International lighting vocabulary, Vienna, Austria (1987)
8. Illuminating Engineering Society of North America: IESNA. The Lighting Handbook, 9th edn. New York (2000)
9. Gómez-Lorente, D., Rabaza, O., Espín Estrella, A., et al.: A new methodology for calculating roadway lighting design based on a multi-objective evolutionary algorithm. J. Expert Syst. Appl. **40**, 2156–2164 (2013)
10. Huang, X.: Lighting design of factory building based on dialux simulation analysis. J. China Illum. Eng. J. **24**, 120–124 (2013)
11. International Standard.: CIE S 008/E. Lighting of Indoor Work Places (2001)
12. Chinese Standard.: GB 50034. Standard for Lighting Design of Buildings (2013)
13. Yang, G., Yang, X.: Discomfort glare and discomfort glare rating. J. China Illum. Eng. J. **17** (2), 11–15 (2006)
14. Yang, G., Yang, X.: Discomfort glare and discomfort glare rating. J. China Illum. Eng. **17** (3), 9–12 (2006)

A Usability and Workload Investigation: Using Video Games to Compare Two Virtual Reality Systems

Crystal S. Maraj and Jonathan Hurter[✉]

Institute for Simulation and Training, University of Central Florida,
Orlando, FL, USA
{cmaraj, jhurter}@ist.ucf.edu

Abstract. Comparison of the effectiveness and user perception of different commercial Virtual Reality (VR) systems for similar locomotion tasks is lacking. This gap is filled in the present paper by comparing two VR systems: the Oculus Rift and the HTC Vive. Three game scenarios of increasing elaboration served as the basis for comparative measures: objective performance and subjective questionnaires (i.e., usability and workload). The between-subjects differences showed the Rift had higher subjective ease of use; an ergonomic recommendation is thus given. The within-subjects differences showed scenario workload differences within both the Rift and Vive. A trend of nonsignificant change in workload between scenarios 1 and scenarios 2 suggests using scenario 1 as a tutorial; a significant change in workload between scenarios 2 and 3 might relate to a lack of competency, signal certainty, scaffolding, and boredom. A higher-workload scenario may be added after scenario 3 in future studies.

Keywords: Virtual Reality · Research game · Usability · Workload · Head-Mounted Displays · Locomotion

1 Introduction

Virtual Reality (VR) involves the technological display of a simulated scenario, where users may interact with the simulation, or virtual environment. Virtual environments li on the extreme end of a real-virtual continuum, the latter ranging from completely rea to virtual (i.e., synthetic) object display systems [1]. One method to achieve VR is vi Head-Mounted Displays (HMDs), which provide visual information through screen based displays that block out real-world visuals. A controller or interface allowin interaction is also possible with HMDs. The appeal of HMD-VR is viewed in a myria of emerging applications; although not an exhaustive list, these applications includ experiential learning (i.e., learning-by-doing) [2], remote inspections of industri equipment [3], three-dimensional data visualization [4], rehabilitation [5], and vide games [6]. The present spawn in VR investigation is likely exacerbated by the con mercialization of accessible equipment, and the subsequent increase in consumer use

Along with this shift towards different applications of VR, research has accumu lated for the human experience of VR systems. One route has been to test how HMD

T. Ahram (Ed.): AHFE 2019, AISC 973, pp. 354–363, 2020.
https://doi.org/10.1007/978-3-030-20476-1_36

VR may differ from more traditional formats: for example, comparing humans' subjective reactions towards the same video game played through either the Microsoft Xbox gaming console or an HTC Vive VR system [7]. Another avenue is understanding variations of the HMD-VR experience within a sole system: investigating locomotion, or control, techniques [8–10], types of player perspectives (i.e., first-person and third-person) [11] and various drivers of subjective user experience and objective usability [12] illustrate this line of thinking.

Yet, it is unclear which VR systems (here an HMD and a controller), are suitable to certain tasks. Specifically, a comparison between two different Commercial Off-The-Shelf (COTS) HMD-VR systems, with respect to controller-based locomotion tasks, is lacking. Some comparison research has appeared, yet with a focus on largely passive tasks (i.e., watching immersive media) [13–15]. Further, a question is raised towards how to determine the best method of evaluating VR, in terms of user experience (e.g., [16]). By examining how users perform in and react to a VR video game (i.e., a research game), this paper (a) evaluates the usability and workload indicators of two separate HMD-VR systems, and (b) evaluates the method of evaluation for the emerging domain of HMD-VR systems. If VR is poised to replace or supplement traditional formats, our goal is to determine the drawbacks and advantages of different VR systems.

1.1 Usability

One definition of usability describes the product used in a timely manner to optimize productivity [17]. Another definition describes usability as multidimensional, consisting of five components: learnability, efficiency, memorability, errors, and satisfaction [18]. This definition focuses on the user's ability to complete a task without hindrance, hesitation, or frustration. The present research adopted the International Organization of Standardization (ISO) [19] definition of usability, which describes a user's interaction with a product through effectiveness, efficiency, and satisfaction. Usability may be divided into objective and subjective metrics. For example, usability may be quantified objectively (e.g., via performance aspects of effectiveness and efficiency) or subjectively (e.g., via satisfaction surveys). Past research conducted by [20] measured efficiency through the NASA-TLX survey, a measure of workload [21]. Specifically, mental workload was used to assess efficiency as it pertains to resource allocation.

1.2 Workload

Workload expends a user's task-completion resources and arises from the nexus of task and performer [22]. Workload is considered an emergent property, rather than an inherit property, and may be determined subjectively, such as by asking a user about the demands of the task itself, as well as the user's response to a task [21]. Relevant to workload is information-processing demands: in one model, the human is equated to a computer, where perceptual, cognitive, and motor subsystems work together during human-computer interaction [23]. Further, integral to information processing is attention and memory [24]; resources are also thought to have different pools [25]. Further, learning deals with different types of load [26], suggesting the benefit of a VR system in education and training might depend on the technology's injection of load.

1.3   Research Questions

The following Research Questions (RQs) were generated to aid in evaluating two HMD-based VR systems: the HTC Vive (i.e., Vive) and the Oculus Rift (i.e., Rift). RQs 1–3 focus on between-subjects' differences, whereas RQs 4 and 5 focus on within-subjects differences.

RQ1. Is there a statistically significant difference between the Vive and Rift for completion rate, per-scenario effectiveness, per-scenario time duration, total-time duration, and time-based efficiency for each of the three scenarios?

RQ2. Is there a statistically significant difference between the Vive and Rift for workload subscales and global workload for each of the three scenarios?

RQ3. Is there a significant difference between the Vive and Rift for usability survey subscales after all three scenarios?

RQ4. Is there a significant change in workload across all three scenarios for the Vive condition?

RQ5. Is there a significant change in workload across all three scenarios for the Rift condition?

2   Method

2.1   Participants

A sample of 40 participants were recruited through the University of Central Florida There were 26 males and 14 females, and the mean participant age was near 20 for both males ($M = 19.81$; $SD = 1.36$) and females ($M = 19.93$, $SD = 1.36$). Several inclusion criteria were mandated: U.S. citizenship, normal (or corrected-to-normal) vision, no history of seizures, and no color-blindness. A maximum of $10 USD was given to each participant as compensation.

2.2   Experimental Design

RQs 1 through 3 followed a between-subjects experimental design, with the independent variable as the type of VR system (i.e., the Vive or Rift), and the dependent variables as performance and survey responses. In contrast, RQs 4 and 5 followed a within-subjects approach, with each VR system's independent variable as the scenarios, and the dependent variable the workload survey.

The role of the participant was to complete three different VR game scenarios using either the Vive or Rift system. All participants ran the scenarios in the same order, and each successive scenario task became more elaborate. Besides the HMD, controllers differed between systems: the Vive used a trackpad to indicate direction, whereas the

Rift used a joystick (i.e., the Oculus Touch) to indicate direction. The Vive and Rift allowed both rotational and positional tracking, with six degrees of freedom. Both Vive and Rift HMDs presented the participant an egocentric (i.e., first-person) perspective of the virtual environment in all conditions.

Per scenario, the goal was to collect all available objects within a time limit. The first scenario included a flat, red-tiled floor, with thirty blue tiles appearing as the desired objects. Only one blue tile would light up at a time; once collected, a new blue tile would appear in a different location. The second scenario involved collecting thirty blue spheres in an outdoor landscape. Again, one item would require collecting before another appeared. Yet, unlike the first scenario, hills and other objects were introduced. The final, third scenario involved a village: various buildings and other barriers were introduced. Twenty collectable, blue spheres were shown simultaneously, peppered around the village. In all cases, the user was required to use the controller to move around the environment to complete the scenario.

.3   Testbed

A desktop computer ran the scenarios, with the latter developed using the Unity game engine. The desktop had a 64-bit Windows 10 operating system, an Intel Core i7-700 K CPU (at 4.20 GHz) processor, 32 GB of RAM, an Intel HD Graphics 630 graphics card, and an NVIDIA GeForce GTX 1080 Ti graphics card. Each HMD used its own positional trackers. The software used were the Oculus application, Steam VR application, and Unity application.

.4   Subjective and Demographic Surveys

Likert-based subjective surveys were used to cover usability and workload. The usability survey was created in-house, with subscales covering visual quality, comfort, ease of use, and effectiveness. At 0.73, the Cronbach's alpha for the usability survey was acceptable. Workload was taken via the NASA-TLX [21], the latter given after each scenario. The NASA-TLX consists of multiple subscales: the task's mental demand, physical demand, temporal demand, performance (i.e., how well participants thought they performed in the task), effort, and frustration. The survey also allowed creating a cumulative global workload score. Demographic measurements included general background questions, as well as specific questions related to interactive technology habits.

5   Objective Performance Measurements

The two main types of measurements are classified based on time and the number of objects collected. Further, a meta-score was created by combining time- and collection-based measurements. Time-based measurements included both per-scenario time duration (i.e., the time spent in each scenario) and total-time duration (i.e., a cumulative score of time durations). Object collection measurements included per-scenario effectiveness (i.e., total number of objects collected per scenario) and completion rate (i.e., a cumulative score of how many scenarios received perfect scores). The meta-score score integrating both time and objects collected was calculated as time-based efficiency. To create the time-based efficiency score, the completion code of each of the three scenarios was used (i.e., 1 for a completed scenario and 0 for an uncompleted scenario), each code was divided by the time participants spent in the code's respective scenario, leading to three separate scores. These three separate scores were then summed, and the sum divided by the total number of scenarios (i.e., three).

2.6   Procedure

The main procedure of the experiment is given in Table 1, below.

Table 1. Procedure in brief.

Experiment section	Participant actions completed
Pre-scenarios	Informed consent, color-blindness test, demographics, and VR interface training
Experiment scenarios	Scenarios, 1-minute breaks between scenarios, and NASA-TLX after each scenario
Post-scenarios	Final NASA-TLX, usability survey, receipt for compensation, and dismissal routine

3   Results

The data set was analyzed using the Statistical Package for the Social Sciences (SPSS) software version 24. For performance and survey data, the analysis included running descriptive statistics, as well as tests for assumptions of normality, homogeneity of variance, and outliers for the Vive and Rift groups. The analysis revealed that the data set violated the assumptions of normality, leading to non-parametric tests (i.e., Mann-Whitney U Tests) to evaluate differences between the Vive and Rift conditions for RQ 1, 2, and 3.

In regard to RQ 1, the findings showed no statistically significant differences between the Vive and Rift for completion rate, per-scenario effectiveness, per-scenario time duration, total-time duration, and time-based efficiency for each of the three scenarios. Similar results were also found for RQ 2: there were no statistically significant differences between the Vive and Rift for workload subscales and global workload, per each of the three scenarios. For RQ 3, there was no statistically significant differences between the Vive and Rift for subjective usability, except for ease of use. A Mann-Whitney U Test revealed a statistically significant difference between the Vive and Rift HMD for ease of use with Vive ($Md = 4$, $n = 19$) and Rift ($Md = n = 21$), $U = 129.5$, $z = -2.12$, $p = <0.05$, $r = 0.34$.

RQs 4 and 5 showed statistically significant changes in workload across all three scenarios using the Friedman Test. Tables 2 and 3 present the findings for the Vive and Rift condition, respectively.

Table 2. Vive condition ($n = 19$) workload findings over three scenarios.

Workload	Friedman test statistic	p value	Percentiles 50th (Median)		
			Scenario 1	Scenario 2	Scenario 3
Mental demand	$\chi2(2) = 24.28$	<.005	5	15	40
Physical demand	$\chi2(2) = 17.33$	<.005	5	5	25
Temporal demand	$\chi2(2) = 16.51$	<.005	15	5	45
Performance	$\chi2(2) = 23.84$	<.005	5	0	30
Effort	$\chi2(2) = 27.91$	<.005	5	15	45
Frustration	$\chi2(2) = 21.04$	<.005	0	5	20
Total	$\chi2(2) = 29.31$	<.005	10	8.33	35.83

Table 3. Rift condition ($n = 21$) workload findings over three scenarios.

Workload	Friedman test statistic	p value	Percentiles 50th (Median)		
			Scenario 1	Scenario 2	Scenario 3
Mental demand	$\chi2(2) = 29.49$	<.005	10	10	40
Physical demand	$\chi2(2) = 27.56$	<.005	5	15	40
Temporal demand	$\chi2(2) = 12.16$	<.005	25	10	40
Performance	$\chi2(2) = 29.77$	<.005	5	5	35
Effort	$\chi2(2) = 27.46$	<.005	15	20	50
Frustration	$\chi2(2) = 29.63$	<.005	5	10	30
Total	$\chi2(2) = 33.06$	<.005	12.5	15	40.83

3.1 Exploratory Data Analysis: Post-hoc Tests

Having established statistically significant differences in workload across all three scenarios for the Vive and Rift, exploratory analyses were conducted for the workload subscales of performance, effort, and frustration. The Wilcoxon Signed-Rank Test was used to examine statistically significant changes from one scenario to another. Further, the Bonferroni adjusted alpha value ($p = 0.025$) was applied to the research findings. The effect size listed was based on Cohen criteria [27]: a small effect = 0.1, a medium effect = 0.3, and a large effect = 0.5. All significant findings had a large effect size. Table 4 shows the outcomes of the exploratory analysis, where workload subscale score changes were tested for differences across select scenarios.

4 Discussion

4.1 Research Questions 1 Through 3

The main finding of RQs 1 through 3 was the significant difference of subjective usability between the Rift and Vive: The Rift was shown to be easier to use than the Vive, at least within the context of the walking-and-collecting task. All other aspects of usability were essentially equal, leading to a recommendation towards the Rift if task ease is relevant. Nevertheless, it is unclear whether the ease of use derived from the headset, the controller, or some mixture therein. Further inquiry may show if the Vive's joystick, which is common in many video games involving locomotion, is a major factor (e.g., via learnability) in ease of use.

Although ease of use, or efficiency, was measured in objective and subjective terms, only subjective ease of use differed; this disassociation highlights the need to measure various forms of similar concepts, with the purpose of gaining insight into both performance and reactions towards performance.

Table 4. Post-hoc tests. Empty cells indicate a nonsignificant difference between scenarios.

Device	Workload subscale	Scenario change	Direction	z score	p value	Effect size
Vive	Performance	1 to 2	–	–	–	–
Vive	Performance	1 to 3	Decrease	$z = -3.14$	<0.025	$r = 0.52$
Vive	Effort	1 to 2	–	–	–	–
Vive	Effort	2 to 3	Increase	$z = -3.56$	<0.025	$r = 0.59$
Vive	Frustration	1 to 2	–	–	–	–
Vive	Frustration	2 to 3	Increase	$z = -3.48$	<0.025	$r = 0.58$
Rift	Performance	1 to 2	–	–	–	–
Rift	Performance	2 to 3	Decrease	$z = -3.83$	<0.025	$r = 0.59$
Rift	Effort	1 to 2	–	–	–	–
Rift	Effort	2 to 3	Increase	$z = -3.86$	<0.025	$r = 0.60$
Rift	Frustration	1 to 2	–	–	–	–
Rift	Frustration	2 to 3	Increase	$z = -3.49$	<0.025	$r = 0.54$

4.2   Research Questions 4 and 5

The different workload scores between scenarios 2 and 3 may be discussed in terms of a lack of competency, certainty, scaffolding, and boredom. In terms of competency, the game controls may have yet been mastered by scenario 3: research has found significant positive correlations between the mastery of controls and the psychological needs of competency [28]. Hindering this competency, shown through a user's inability on task, leads to needs frustration.

If one takes the game controls as mastered, we then consider other distinctions between the tasks of scenario 2 and scenario 3. Scenario 3 included a larger, object-cluttered environment, which occluded or hid some objects. Although all objects were given at once, they could not be seen from a single vantage point and may require additional movement to search. Ultimately, the certainty of where the next collectable could be found was dampened in scenario 3, likely requiring a different search strategy than scenario 2. A similar finding occurred in [29]: decreasing the certainty of opposing team-player locations was found to increase mental workload (although not explicitly changing effort, frustration, or performance) in a video-game version of capture-the-flag

There were also physical impediments to avoid via locomotion in scenario 3. Although scenario 3 had less objects than both scenario 1 and 2, the strategy difference and impediments may have led to both the heightened effort and frustration between scenarios 2 and 3, as well as changes in perceived performance from scenarios 1 to 3 (for the Vive) and scenarios 2–3 (for the Rift). It might be difficult to disentangle all factors of workload from the data, but one corroborated explanation of the increased frustration arises with the lack of incremental scaffolding from the previous scenarios, and lack of boredom from the difficult task [30].

In terms of performance, effort, and frustration, the first two scenarios are identical within the Vive and Rift. Overall, these trends suggest scenarios 1 and 2 invoke a similar level of workload. Thus, the first scenario could be used to onboard the user with interface training.

If users may use VR in high-workload conditions, or extreme situations, this should be reflected in the spectrum of scenarios via workload distinctions. Thus, an additional condition may be added after scenario three, with higher workload. The same control scheme may be used, but information-processing twists (e.g., in attention and memory demands) could be leveraged to improve the difficulty of the current final scenario. For example, potential new scenario designs include forcing users to collect balls in a certain order (e.g., numbering balls and requiring participants to collect the balls by their increasing number), forcing the user to avoid enemies that are chasing them, making the balls less distinct in the environment (i.e., hiding the balls better), and layering a secondary change-detection task onto the primary collection task.

4.3   Limitations

Although the controls and scenarios were simple, perception of the system may have been influenced by a short time-period of testing. Giving more scenario time (e.g., to build muscle memory with the controllers), could eliminate short-term effects. Since future VR applications may be used for extended periods, the study should cater to such a need. In terms of ease of use, pinpointing the causal element(s) of each device's ease is arduous. Introducing finer scales for ease of use (e.g., controller ease and visual display ease) would allow pinpointing.

5   Conclusion

By examining the usability and workload of two different VR systems, both between differences (in ease of use) and within differences (for workload) were found: higher ease of use in the Oculus condition provides grounds for further use and research, while scenario workload differences informs a need to manipulate scenarios (i.e., use scenario 1 as interface training and add a more difficult scenario after scenario 3) in future research games. Overall, the present research elucidates valuable human-based concepts underneath the technology of VR systems.

Acknowledgments. This research was sponsored by Gino Fragomeni of the U.S. Army Research Laboratory Human Research Engineering Directorate Advanced Training and Simulation Division (ARL HRED ATSD), under contract W911QX-13-C-0052. However, the views, findings, and conclusions contained in this presentation are solely those of the author and should not be interpreted as representing the official policies, either expressed or implied, of ARL HRED ATSD or the U.S. Government. The U.S. Government is authorized to reproduce and distribute reprints for Government.

References

1. Milgram, P., Kishino, A.F.: Taxonomy of mixed reality visual displays. IEICE Trans. Inf. Syst. **E77-D**(12), 1321–1329 (1994)
2. Kwon, C.: Verification of the possibility and effectiveness of experiential learning using HMD-based immersive VR technologies. Virtual Real. 1–18 (2018)
3. Linn, C., Bender, S., Prosser, J., Schmitt, K., Werth, D.: Virtual remote inspection—a new concept for virtual reality enhanced real-time maintenance. In: 2017 23rd International Conference on Virtual System & Multimedia (VSMM), pp. 1–6. IEEE (2017)
4. Millais, P., Jones, S.L., Kelly, R.: Exploring data in virtual reality: comparisons with 2D data visualizations. In: Extended Abstracts of the 2018 CHI Conference on Human Factors in Computing Systems, p. LBW007. ACM (2018)
5. Rose, T., Nam, C.S., Chen, K.B.: Immersion of virtual reality for rehabilitation-review. Appl. Ergon. **69**, 153–161 (2018)
6. PSVR Games, https://xinreality.com/wiki/PSVR_Games
7. Lum, H.C., Greatbatch, R., Waldfogle, G., Benedict, J.: How immersion, presence, emotion, & workload differ in virtual reality and traditional game mediums. In: Proceedings of the Human Factors and Ergonomics Society Annual Meeting, vol. 62, no. 1, pp. 1474–1478. Sage Publications, Los Angeles (2018)
8. Martel, E., Su, F., Gerroir, J., Hassan, A., Girouard, A., Muldner, K.: Diving head-first into virtual reality: evaluating HMD control schemes for VR games. In: Proceedings of the 10th International Conference on the Foundations of Digital Games. (2015)
9. Frommel, J., Sonntag, S., Weber, M.: Effects of controller-based locomotion on player experience in a virtual reality exploration game. In: Proceedings of the 12th International Conference on the Foundations of Digital Games. ACM (2017)
10. Bozgeyikli, E., Raij, A., Katkoori, S., Dubey, R.: Point & teleport locomotion technique for virtual reality. In: Proceedings of the 2016 Annual Symposium on Computer-Human Interaction in Play, pp. 205–216. ACM (2016)
11. Monteiro, D., Liang, H.N., Xu, W., Brucker, M., Nanjappan, V., Yue, Y.: Evaluating enjoyment, presence, and emulator sickness in VR games based on first-and third-person viewing perspectives. Comput. Animat. Virtual. Worlds **29**(3–4), 1–12 (2018)
12. Tcha-Tokey, K., Loup-Escande, E., Christmann, O., Richir, S.: Effects of interaction level, framerate, field of view, 3D content feedback, previous experience on subjective user experience and objective usability in immersive virtual environment. Int. J. Virtual Real. **1**(3), 27–51 (2017)
13. Papachristos, N.M., Vrellis, I., Mikropoulos, T.A.: A comparison between Oculus Rift and low-cost smartphone VR headset: immersive user experience and learning. In: 2017 IEEE 17th International Conference on Advanced Learning Technologies (ICALT), pp. 477–481. IEEE (2017)

4. Rupp, M.A., Odette, K.L., Kozachuk, J., Michaelis, J.R., Smither, J.A., McConnell, D.S.: Investigating learning outcomes and subjective experiences in 360-degree videos. Comput. Educ. **128**, 256–268 (2019)
5. Somrak, A., Humar, I., Hossain, M.S., Alhamid, M.F., Hossain, M.A., Guna, J.: Estimating VR sickness and user experience using different HMD technologies: an evaluation study. Future Gener. Comput. Syst. **94**, 302–316 (2019)
6. Wienrich, C., Döllinger, N., Kock, S., Schindler, K., Traupe, O.: Assessing user experience in virtual reality–a comparison of different measurements. In: International Conference of Design, User Experience, and Usability, pp. 573–589. Springer, Cham (2018)
7. Dumas, J.S., Redish, J.: A Practical Guide to Usability Testing. Intellect books (1999)
8. Nielsen, J.: Usability Engineering. Elsevier (1994)
9. Organisation Internationale de Normalisation (ISO), https://www.iso.org/obp/ui/#iso:std:iso:9241:-11:ed-2:v1:en
10. Riccio, A., Leotta, F., Bianchi, L., Aloise, F., Zickler, C., Hoogerwerf, E.J., Kübler, A., Mattia, D., Cincotti, F.: Workload measurement in a communication application operated through a P300-based brain–computer interface. J. Neural Eng. **8**(2), 1–6 (2011)
11. Hart, S.G., Staveland, L.E.: Development of NASA-TLX (Task Load Index): results of empirical and theoretical research. In: Hancock, P.A., Meshkati, N. (eds.) Human Mental Workload, pp. 139–183. Elsevier Science, Amsterdam (1988)
12. Hart, S.G., Wickens, C.D.: Workload assessment and prediction. In: Booher, H.R. (ed.) Manprint, pp. 257–296. Springer, Dordrecht (1990)
13. Card, S.K., Newell, A., Moran, T.P.: The Psychology of Human-Computer Interaction. Erlbaum Associates, Hillsdale (1983)
14. Wickens, C.D., McCarley, J.S.: Applied Attention Theory. CRC, Boca Raton (2008)
15. Wickens, C.D.: Multiple resources and mental workload. Hum. Factors **50**(3), 449–455 (2008)
16. Sweller, J., Ayres, P., Kalyuga, S.: Cognitive Load Theory. Springer, New York (2011)
17. Cohen, J.W.: Statistical Power Analysis for the Behavioral Sciences, 2nd edn. Lawrence Erlbaum Associates, Hillsdale (1988)
18. Przybylski, A.K., Deci, E.L., Rigby, C.S., Ryan, R.M.: Competence-impeding electronic games and players' aggressive feelings, thoughts, and behaviors. J. Pers. Soc. Psychol. **106**(3), 441–499 (2014)
19. Parasuraman, R., Galster, S., Squire, P., Furukawa, H., Miller, C.: A flexible delegation-type interface enhances system performance in human supervision of multiple robots: empirical studies with RoboFlag. IEEE Trans. Syst. Man Cybern. Part A: Syst. Humans **35**(4), 481–493 (2005)
20. Sharek, D., Wiebe, E.: Using flow theory to design video games as experimental stimuli. In: Proceedings of the Human Factors and Ergonomics Society Annual Meeting, vol. 55, no. 1, pp. 1520–1524. SAGE Publications, Los Angeles (2011)

Virtual and Augmented Reality Gamification Technology on Reinventing the F1 Sponsorship Model not Purely Focused on the Team's and Car's Performance

Evangelos Markopoulos[1(✉)], Panagiotis Markopoulos[2],
Mika Liumila[3], Ya Chi Chang[1], Vasu Aggarwal[1],
and Jumoke Ademola[1]

[1] HULT International Business School, 35 Commercial Road, Whitechapel,
London E1 1LD, UK
evangelos.markopoulos@faculty.hult.edu,
{ychang2014,vaggarwal2016}@student.hult.edu,
ademola.jumoke@yahoo.com
[2] University of the Arts London, Elephant and Castle, London SE1 6SB, UK
p.markopoulosl@arts.ac.uk
[3] Turku University of Applied Science, Joukahaisenkatu 3,
20520 Turku, Finland
mika.luimula@turkuamk.fi

Abstract. Formula 1 economics is highly related with the investments on the F1 cars and their performance. Winning teams have multidimensional benefits that can assure sustainability, development, reputation and profitability. On the other hand, not everyone can be a winner and this reality generates high risk for F1 the F1 investors, the teams, and the F1 itself. This research work aims to identify new sponsorship methods for F1 which are not purely focused solely on the car's and team's performance. The paper presents a gamification approach on which an innovative and disruptive Virtual Reality game can be developed to contribute on gaining new fans from all financial and social levels globally. This new fan base can be transformed into the new target group for sponsorships not only on the physical F1ccars but also on the digital ones that can be owned and driven by anyone, anywhere at any time.

Keywords: Gaming · Gamification · Serious games · Esports ·
Entertainment games · Monetization · Sponsorship · F1 · Formula 1 ·
Virtual reality · Augmented reality · Blue ocean strategy

1 Introduction

The winner takes it idea applied in any business and Formula 1 could not be exception. The economics of this highly expensive sport are significantly related wi the investments made on the F1 cars based on their performance. The winners, either team or individual victories, have multidimensional benefits that can assure sustai ability, development, reputation and profitability to the manufacture and the spo

© Springer Nature Switzerland AG 2020
T. Ahram (Ed.): AHFE 2019, AISC 973, pp. 364–376, 2020.
https://doi.org/10.1007/978-3-030-20476-1_37

However, this is not the case for the teams that follow, which have also invested tremendous amounts of funds to be part of Formula 1 and race at such level.

The fact that not everyone can be a winner generates a very big risk for F1 investors, the teams and the F1 itself. This research work attempts to identify a new sponsorship model for Formula 1 which is not based only on the car's or team's performance. The suggested approach aims to achieve creative re-engagement of the F1 followers with the sport and the manufacturers. This gamification approach is based on an innovative and disruptive operations framework to contribute on gaining new customers (fans), recapture essential markets and targets groups, but also establish the base for effective strategy development. This approach is based on the development of a Virtual Reality brand-exclusive Formula 1 game addressing two key concepts which are the players experience on playing within an exclusive F1 manufacturer environment, and the utilization of the data collected from the players performance and behavior, in order to develop more effective sponsorship strategies through data analytics.

The proposed manufacturer brand exclusivity can be adjusted to any F1 manufacturer such as Ferrari, McLaren, Williams, Sauber, etc., and for any role of the F1 racing teams such as the driver, the mechanics, etc. The data collection can be used for marketing purposes, sponsor engagement, fans engagement, and other initiatives that can contribute to the financial and reputational development of the F1 manufacturer. Data gathering and management can be based on brands, advertisements, colours selection, preferences, logos, demographics, user behavior, and other options given to the player to customize its paying environment, game elements and gameplay.

The proposed approach integrates the innovative concepts and technologies of virtual and augmented reality and gamification with the data science. This technology integration is based on user activities that can generate knowledge-oriented business intelligence which can lead into strategy development for sponsors engagement.

The way VAR, cognitive science and data science are integrated in this approach results into a multidisciplinary combination of systemic thinking that can effectively create blue ocean strategy out of a very conservative market and industry, reducing on the other hand, blue ocean risks via the multidimensionality of operations and multiple revenue streams.

2   The Virtual Reality Phenomenon

Virtual reality cannot actually be considered a new idea as it has its roots many decades in the past. It is a concept imagined years ago, with many artists trying to envision what such a concept could be like through sci-fi movies, books and other types of media. However, in recent years, the advancements of technology together with the the accessibility and affordability of VR reality headsets, makes it no longer thoughts of fiction. Today there is a tremendous increase of the people fully interested in the future of Virtual Reality either by observing the progress of this new technology eagerly or by contributing in its development by investing and designing projects in virtual worlds.

According to the World Economic Forum with data from the Goldman Sachs, virtual reality is a market of nearly 25 billion$ in 2018, expected to reach 100 billion$ in the next seven years (2025), driven by the creative sectors with games to be dominant among them, at 27% of the creative sectors [1]. Figure 1 presents the growth of VAR until 2025 indicating the share of the creative sector, while Fig. 2 indicates the share of games in it [1].

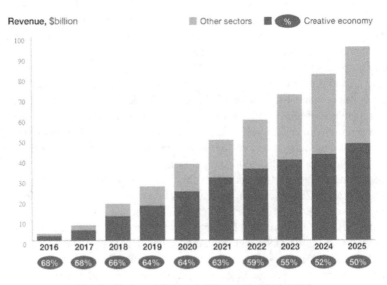

Fig. 1. Projected VA and AR growth 2016–2025

Virtual reality starts to gain blue ocean characteristics as more people and organizations try to adapt their work and function in a virtual world, hoping to be the first ones to do this successfully, efficiently and effectively.

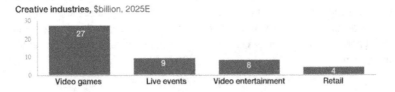

Fig. 2. Breakdown of creative industries.

3   The Virtual and Augmented Reality

Virtual reality, as known today, is the computer-generated simulation of a 3D environment that can be accessed via the use of a headset. Within this environment commonly, the player can perform physical interactions using extra hardware such as body tracking sensors or controllers. Due to the VR interactivity, the market that popularised Virtual Reality is the gaming industry since video games are the most interactive form of media. However, this does not limit interactive VR experiences to just games as there are quite successful interactive VR movies in the making such as the Beat Saber and the Moss.

Augmented reality, unlike Virtual Reality, takes place in the real world and simulates the addition of 3D computer-generated elements within reality. For that reason, it is also referred to as Mixed Reality since it combines both worlds, digital and physical.

As of today, augmented reality applications are a much less interactive than Virtual Reality since they cannot use any advanced sensory hardware. Due to that, most Augmented Reality experiences are portable, affordable and very easily accessible by using smartphones as their preferred platform. Two very successful AR experiences are Pokemon Go and Ingress with significant contribution to the AR popularity.

Even that most of the success of Virtual and Augmented reality has been in the entertainment industry, the application of these technologies can extend to any field. In the medical field for example, Hermes Pardini Laboratories and Vaccination Centers are using Virtual Reality to make the process of vaccination a pleasant experience for the children partaking it. By immersing the children into a fictional environment, the nurse can mask the needle injection as something much more fun and interesting such as a fire fruit used for an energy shield [2]. In the field of education, Discovery VR is trying to bring a new form of documentaries in the classroom. By needing only smartphones and cardboard VR headsets, it aims for affordability and accessibility while retaining the quality expected from an educational documentary experience [3].

4   VR a Blue Ocean Strategy

The concept of Blue Ocean as defined by Kim & Mauborgne is premised on the quote 'help my ocean is turning red" where businesses find themselves in a highly competitive market space, needing a new strategy to survive and grow [4]. This challenge of high competitiveness is addressed by the blue ocean strategy which practically reflects to the name itself as Blue Oceans are new markets while Red Oceans are the saturated ones [5].

The ideology behind blue ocean strategy is that industry player can restructure the industry and the market in order to create new untapped markets. This can create new endless opportunities that business can enjoy in form of profit maximization and risk minimization. These restructures can also occur with the introduction or adaptation of disruptive technologies like the VR which can open room for untapped markets and give businesses competitive edges. Most of the blue ocean strategy emerges from within red oceans by expanding existing industries [4]. The key characteristics of the blue ocean can be summarized in Table 1.

Table 1. Blue and Red Ocean strategy characteristics.

Blue ocean strategy	Red ocean strategy
Compete in existing market space	Create uncontested market space
Beat the competition	Make the competition irrelevant
Exploit existing demand	Create and capture new demand
Make the value-cost trade-off	Brake the value-cost trade-off
Align the whole system of a firm's activities with its strategic choice of differentiation or low cost	Align the whole system of a firm's activities in pursuit of differentiation and low cost

The threats and opportunities on going blue can be considered quite similar in any industry even in the most innovative and fast-growing ones such as the gaming.

The innovation of the VR technology and its areas of application can create experiences with significant success, audiences and impact in the economy and the society and form blue oceans generated. Today's VR early stage, in corporate and serious games, can impact existing markets and generate new ones. In existing markets, VR can be a disruptive technology that puts existing companies in the famous Disruption Innovation Model, also known as the Innovator's Dilemma [6]. The model indicates (Fig. 3) when the enhancement of a product performance trajectory leaves room at the lower spectrum, for smaller companies and disruptive technologies to enter the market.

Overtime, VR being a disruptive technology through incremental improvements can compete with existing (also known as incumbent) technologies available in the market. Therefore, the incumbents would be forced to cater to a higher end of the market and move up the product performance threshold.

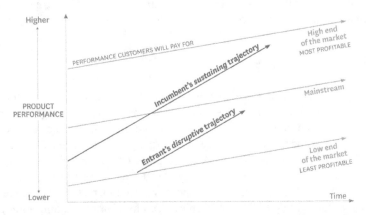

Fig. 3. Product performance varies as the time period increases and the incumbents are forced to move to a higher end of the market.

With VR being such a recent phenomenon and its adaptability to any field of work or entertainment, it does generate blue oceans. Today more companies are racing to be the first ones to set the pace for VR and take advantage of the minimum competition due to the limitation of the created VR content. Taking also into consideration the staggering growth of VR hardware, ideas that are impossible this year can be possible the next. However, going blue on VR needs serious thought and strategy. It is wise to take careful steps in a rapidly evolving field. The way a VR experience can survive the test of time is to be made easy, be updatable and designed with the future as its target.

Overly ambitious projects can turn against the designers and be replicated better and cheaper within the next years due to the staggering evolutionary pace of VR and AR technologies. The goal for a company/project on going blue in VR is to set the VR pace of its field, but if the pace can't be kept, others will take the lead.

5 The F1 Economics

Formula One (F1) is probably the most impressive automotive sport today. It has its roots in the early 1930 in Europe but not realized due to the second world war. In 1946 discussions for the idea started again and the first races began towards the establishment of a drivers' championship and on May 1950 the first world championship race was held at Silverstone. The 'formula' in the name refers to a set of rules all participants and cars must comply. Prior Formula 1, the sport was originally known as Formula A. Since then Formula One demonstrated amazing development, success and growth.

Today the global sport of F1 is valued at $10 billion. It is the most watched annual sporting series in the world with more that 425 million TV viewers. The revenue of F1 in the past 18 years reach $18 billion, even outstepping its closest rival, the FIFA World Cup. On 2014 the revenue of F1 was $1.765 billion and the net profit $520 million. However, this is not the case for the F1 teams where most of them don't make any profit. A mid-range F1 team is expected to spend around $243 million per season from which $55 million in operations, $63 in salaries, $61 in research and development, and $60 in manufacturing. In 2014 Mercedes spend $375 million, McLaren $334, Red Bull $317, Williams $318, Williams $208, Lotus $186, Toro Rosso $170, Force India $133 and Manor $130 million. Ferrari and Sauber did not file public accounts [7]. Figure 4 indicates the 2014 average cost of a formula one car which is 9.765 million with the engine to costs cost $5.3million on average but can go up to $7 million.

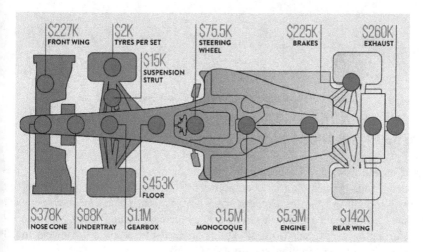

Fig. 4. Component cost and total cost of a Formula One car

The main revenue stream of an F1 team, around 40% comes from sponsorship. The total revenue that can be generated from sponsorships on a car can be $74 million on an average (shown in Fig. 5), with the cost for a sponsor on the rear wing and the sidepod rear to be at £25 million on each. This is 7.5 times the cost of the car, but it is no enough to cover the overall operations of an F1 team [7].

In 2014 more than 300 brands sponsored F1, spending close to $1.5 billior annually. The biggest sponsor was Marlboro ($2.023 million), Vodafone (692 million) Petronas (690 million), Shell (564 million) and Mobil1 (496 million) [7].

The revenue of F1 teams is related to the wins they have each season but also to the wins they had at the last four years. Ferrari finished 4th, in the championship of 201 with net profit of $71.8 million ($12.4 million less than the 2014 champion, McLaren) However, due to the premium (historic) payment of $92,2 million, the total profi for Ferrari was 164 million, when the champions McLaren netted only 100.7 (4th place in profit). F1 is a winner takes it all industry and sponsorships are heavily with the performance or the car, the team and the driver.

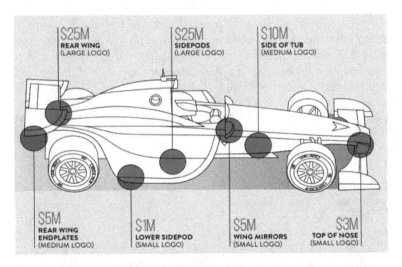

Fig. 5. Sponsorship costs per area on a Formula One car

6   The F1 VR Blue Ocean

Based on the Formula One economics, the impressive Formula One industry nee significant support to sustain its operations and continue to offer its innovation to t automobile industry and the excitement to its fans. An increase on the sponsorsh revenue can contribute but this has to be with an increase of the F1 followers who see to be dropping every year (Fig. 5). The sport lost 40 million TV viewers in 20 dropping down to 352 million viewers [8] (Fig. 6).

It is an imperative need for F1 to seek blue oceans in order to sustain and increa the sponsors for all if its teams and not only for its winners, despite the decline of t industry. Games through immersive technologies like VR can help on attracting ne age target groups to the F1 world creating new generations of fans.

F1 can find in the VR gaming business a Blue Ocean if consumers can receive advanced services to enrich the F1 experience. VR as an immersive technology offers special tools to increase the F1 end user experience. While players are into the game, they can be able to feel the power and the speed of the car, they can measure their driving skills against the driver of the physical car, and also measure their on-game driving capabilities and senses such as perception, attention and strategy. The game Lab of the Turku University of Applied Sciences in Finland developed the prototype of the NeuroCar VR evaluation toolkit for driving inspection which is a step towards this direction (Fig. 7).

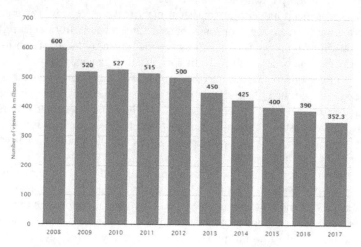

Fig. 6. Number of TV viewers of F1 racing worldwide from 2008 to 2017 (in millions)

NeuroCar can be used to detect the right-side perceptual bias in aging in a laboratory setting during a virtual driving task [9]. This toolkit combines cognitive neuroscience and game technology for efficient and objective screening of driving performance and spatial perception, supporting medical professional's estimations on driving ability.

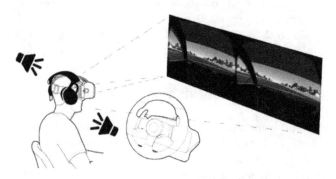

Fig. 7. NeuroCar system description of portable evaluation tool [10]

7 A VR Gamification Approach for Reinventing the F1 Sponsorship Model

A solution to the F1 sponsorship challenge can be the creation of an online VR multiplayer F1 game which can and expose the sport to younger age groups. Creating a popular game is the most efficient way to reach millennials and generate a strong brand connection with this new target group. The game designed for this research enables any Formula One team to create a large portfolio of virtual sponsorships. It includes various areas allocated for sponsorships which increase the overall revenue streams and the awareness of a given F1 team to the younger generations. The proposed approach supports a sponsorship model in a consistent manner the same way it is done in the actual sport. Furthermore the user gains benefits and rewards from entering tournaments while enjoying a competitive experience monitored by the F1 Team itself.

The selected approach is expected to spur existing gamers to use this advanced technology with sponsors tapping into the technology's marketing potential. Sponsors can also be small start-up organizations or corporate organizations aiming to use F1 VR for product/service placement or brand awareness. The game has virtual adverts, colors and logos that can customize the appearance of the player's virtual F1 car, offering this way more opportunities for the sponsors to swift from the traditional advertisement methods to new and modern ones [11].

The proposed approach also provides benefits for the F1 VR users by offering the ability to race with the actual, but virtual, F1 cars, on the actual tracks, and monitored by the actual F1 teams. The racing competitions are local, regional and international and can be set by either by the gamers themselves or by the F1 team. The link between gamers and the actual F1 world merges the virtual reality (gameplay) with the actual reality (viewing), offering a unique feeling and challenge for achievement and success for everyone and anywhere regardless their age, race, sex, financial or social status.

The overall approach works under the concept of a global F1 'try-out' for the digital racing platforms for projects that already exist in various car manufacturers such as McLaren and Ginnetta [12, 13]. The significant difference in this case is the gamer are ordinary people instead of professional ones. Furthermore, the international dimension of the game can lead F1 VR to enter the e-sports word with international teams composed from amateur gamers using the minimum VR equipment. This affordable VR F1 esport can be considered as a shared value global social innovation that provides opportunities to existing and new F1 followers for personal development and recognition, but also to the sponsors who can experience the invention an exponential growth of non-existed target groups within F1.

8 Managing F1 VR Gamification Sustainability

The proposed F1 VR game contains two main features which are the player mode and the spectator mode. Each mode has the ability to provide marketing and advertisement opportunities to sponsors and additional revenue to the F1 teams. The game also provides revenues from tournament registrations, and from elements that can be purchased during the game, such as car skins, boosters, extra time on the track, etc., that can be of high demand due to the desire gamers have for victories. Other monetization practices that can increase the revenue of the F1 teams are the access modes and the microrotations that can be applied before or during the preparation of the car, before the competition or during the gameplay at the pit-stops for car improvements or driver tips. Another revenue stream can be the pricing mode of the game which can change from fermium to premium.

Dynamic and localized advertisement is also another source of income that can be applied during the virtual game. Under this concept and according to the location of the competition or the nationality of the gamer, the brands presented to the gamer's eyes can be either from the location of the competition or from the driver's country. In the first case sponsors from the tournament location can be advertised in the game promoting their brand or their products to international gamers (drivers) and spectators. On the other hand, the gamer (driver) can be viewing brands and products from his/her country which are easy for him/her to get associated.

This dynamic advertisement can increase the number of sponsors not only in volumes but in also in types. F1 advertisement have been a privilege for the very established organizations due to the massive costs required for a spot on the physical car. Therefore, companies operating at national or regional level without being multinational organizations, had no chance to enter the F1 world. The proposed approach provides the opportunity for any organization to be promoted by an F1 team regardless their size, financial or international activity.

The success and sustainability of the proposed approach relies also on the data collection during the game from the gamers and the spectators. Such data can be the demographics related to the gamers from all over the world, but also the preferences gamers have in the customization of their cars, selection of tracks and other activities they do before or during the game. Colors specifically selected during a customization activity can indicate the likeness of the gamers to pay attention to brands or products with the colors they like. Clothing options and dress styles can provide similar information. Such data, non-related with the game but with the psychology of the gamer, is analyzed in the ranking of the sponsored brands or products to be viewed more.

The sustainability of the proposed approach, and not only, is based on understanding the gamer in the first place, and not the industry needs. In order to achieve such a key principle, a number of science disciples need to be integrated with the VR technology and the F1 industry. Access modes, dynamic localization, and data management can certainly support a successful monetization strategy, but more than that they can support a successful sustainability strategy in order for the game to last for long.

Risks and Considerations. SWOT and BCG Matrix Analysis

Innovation can be considered as a gambling practices as no secure and safe path can be predicted with certainly. However, it is a required gamble on the search for blue oceans where development and prosperity can take place. A SWOT analysis has been performed on the proposed approach in order to identify its sustainability characteristics.

Two key strengths the proposed approach possess are related with the popularity of the industry and the technology. Formula 1 VR, as presented, provides physical and social engagement to all type of gamers from any social status while at the same time can be adjusted to various gaming platforms such as the X-box, PlayStation, and PCs. Due to this flexibility, more game users can be attracted, hence, lucrative to sponsors.

Despite the technology being associated with these strengths, there are weaknesses as well with the most important one to be the domination of the VR industry by models that do not support head-mounted displays. Another weakness is the high cost of the more advanced VR tools and equipment which impacts the game development and use.

The opportunities, on the other hand, provided in the proposed approach can be summarised in four points. (1). Users can experience racing with Formula 1 tracks and cars and enjoy the customisation features offered. (2) There is an increasing market share for head-mounted displays (HMD) such as the Oculus Rift and the Sony PlayStation VR which are more favourable for the Formula 1 VR at affordable prices. (3). The data collection and analysis provided by the game can benefit sponsors is various ways, with each way to be a new source of income for the F1 team. (4) The social impact of the game can increase the number of gamers globally.

One of the key threats of the proposed approach is related with the perception of the people on the VR technology as it still sounds expensive, remains a new concept for many applications, and can be considered an investment risk.

Placing the characteristics presented in the SWOT analysis into a Boston Consulting Group growth-share matrix, the proposed approach would be considered as a Question mark. With a high market growth in the fast-growing VR market, but low market share in the interest of F1 in VR, the proposed approach is capable for the best which can be the fast adaptation, and the worse which is low adaptation. However, the market does exist and the opportunities flow.

10   Areas of Further Research. Wearable Computing

Virtual reality games allow gamers to immerse into the gaming world and control their characters by moving their bodies in the real world. Scientific research indicates that virtual reality can provide motivational physical activities [14] which can increase enjoyment while decreasing tiredness [15]. When it comes to gaming in VR, the VR glasses are not enough to fully immerse the player in the virtual reality, as they also need to feel and touch the objects the see. [16] This can be achieved with wearable computing which becomes a trend in gaming and of course in VR gaming further more

Wearable VR technology provides the opportunity to measure brain waves, heart rate, eye movements etc. This paper will extend on the integration of wearable technology in F1 VR games. Even that this technology is not directly related with sponsorship, it is capable to bring the whole IT industry in F1 by offering a unique experience on a target group that can utilized by the F1 sponsors. We believe that that the IT companies behind F1 data analytics doesn't reach visibility as much as they could. VR games and wearable computing could open new possibilities to visualize car performance, driver performance, and achieve significant engagement with a whole new world of F1 fans.

11   Conclusions

The magnificent world of F1 can be by itself an attraction and source of inspiration for a number of products and services from which F1 can benefit. However, despite the many products and services already being developed under this thinking, F1 still struggles to assure descent profitability to the teams that composed it. Event that the F1 brand name is capable to sell by itself, the reality indicates that this does not seem to the case. The same thinking can also be applied on the VR technology. The impressive experience that can be achieved with VR requires a normalization of the equipment standards and costs for the technology to blossom.

Combining F1 and VR for the generation of innovative and immersive games that can help F1 teams increase sponsorships is a challenge that goes beyond understanding the technology or identifying the potential target groups. The challenge relies on the way new target groups can be identified, continuously engage and grow exponentially.

In order to achieve such goals a deeper understanding of the games is required in a wholistic scientific approach, not restricted to the technology and marketing sciences.

The proposed approach integrated data science, cognitive science, management science and behavioural science in an attempt to better understand the needs and drives of the massive user groups that can race within the VR F1 game around the clock globally. This work attempted to create a chain of relations from the problem to the solution. The increase of the F1 industry revenue is related with the increase of the sponsors, which is related with the increase of the gamers, which is related with the understanding the gamers phycology, needs and drives in order to generate incentives to follow a sport not in their prime interest. F1 and VR, if applied within the context of understanding the human first, can then create the momentum needed to identify blue oceans and achieve the desired results. The paper presented the sponsorship challenge of F1, the opportunity of VR, and their integration though a social shared-value oriented game.

References

. Augmented and virtual reality: The promise and peril of immersive technologies. https://www.mckinsey.com/industries/media-and-entertainment/our-insights/augmented-and-virtual-reality-the-promise-and-peril-of-immersive-technologies
. Ogilvy Brazil, VR Vaccine. http://lobo.cx/vaccine/
. Discover Education. Virtual Reality in the classroom. https://www.discoveryeducation.co.uk/discoveryvr
. Kim, W.C., Mauborgne, R.: Blue Ocean Strategy, Expanded Edition How to Create Uncontested Market Space and Make the Competition Irrelevant. Harvard Business Review Press, Boston (2014)
. Kim, W.C., Mauborgne, R.: Blue Ocean Shift: Beyond Competing – Proven Steps to Inspire Confidence and Seize New Growth. New York (2017)
. Harvard Business Review. What Is Disruptive Innovation? https://hbr.org/2015/12/what-is-disruptive-innovation
. Raconteur. The Business of Formula 1. https://www.raconteur.net/infographics/the-business-of-formula-1

8. Forbes, F1. https://www.forbes.com/sites/csylt/2018/01/06/f1-tv-audience-reverses-by-40 million-under-revised-measurement-system/#26a66c2d3a52
9. Hämäläinen, H., Izullah, F.R., Koivisto, M., Takio, F., Luimula, M.: The Right-side Perceptual Bias in Aging Determined in a Laboratory and during a Virtual Driving Task Scand. J. Psychol. **59**, 32–40 (2018)
10. Luimula, M., Besz, A., Pitkäkangas, P., Suominen, T., Smed, J., Izullah, F.R., Hämäläinen H.: Virtual evaluation tool in driving inspection and training. In: Proceedings of the 5th IEEE Conference on Cognitive Infocommunications, Gyor, Hungary, pp. 57–60 (2015)
11. Nichols, G., Savage, M.: A social analysis of an elite constellation: The case of Formula 1 Theory Culture Soc. **34**(5–6), 201–225 (2017)
12. Project CARS X Ginetta. https://www.projectcarsesports.com/ginetta.html
13. Project CARS X McLaren. https://www.mclaren.com/formula1/2018/mclaren-shadow project/
14. Harris, K., Reid, D.: The influence of virtual reality play on children's motivation Can J. Occup. Ther. **72**(1), 21–29 (2005)
15. Plante, T., Aldridge, A., Bogden, R., Hanelin, C.: Might virtual reality promote the moo benefits of exercise? Comput. Hum. Behav. **19**, 495–509 (2003)
16. Wearable VR Gaming Devices To Increase Physical Activity. https://www.wearable technologies.com/2017/09/wearable-vr-gaming-devices-to-increase-physical-activity/

Enhancing Multimodal Interaction for Virtual Reality Using Haptic Mediation Technology

Ahmed Farooq[⊠], Grigori Evreinov, and Roope Raisamo

Tampere Unit of Human Computer Interaction (TAUCHI),
Faculty of Information Technology and Communication,
Tampere University, Tampere, Finland
{Ahmed.Farooq,Grigori.Evreinov,Roope.Raisamo}@TUNI.fi

Abstract. As our interaction in virtual space expands from 2D to 3D, the absence of meaningful touch output restricts our ability to explore new virtual frontiers. The core limitation of not being able to reach out and feel or interpolate an object or sense its texture and form, within a virtual environment, hinders the intuitiveness of the interaction experience. Although, in recent years, tactile feedback has been introduced, as a necessary component of multimodal interaction, the resolution and type of output is still very primitive compared to visual and auditory modalities. For this reason we have developed a radical new approach called 'Haptic Mediation' through which it is not only possible to actively monitor signal integrity and skin sensitivity but also the applied actuation signal, dynamically adjusting the actuation to ensure reliable perception of intended information. In this research, we have extended this technique to develop a self-sensing and actuation haptic glove prototype.

Keywords: Human-systems integration · Multimodal interaction · Haptics · Wearables

Introduction

Human beings interact with external objects or systems using a combination of their core senses. Utilizing these senses, it is possible to interpolate the components of a given system and develop the most efficient method of communicating with it. This ability to learn and adapt, creates the basis of interaction that can be extended to similar environments and systems. When an external system utilizes commonly used interaction techniques (i.e. door knob being rotated clockwise or anticlockwise to open a door), the user of the system is easily able to transition into the particular interaction paradigm, even if the environment or its surroundings vary considerably. This interaction metaphor was first put to the test in the designing and development of 2D virtual environments. In fact, most 2D virtual environments (operating systems such as Windows, MAC or Linux etc.) and the objects within them (i.e. files, folders, menus, etc.) were created to extend the user's understanding of similar interaction, in the physical world. This is still the case today and it essentially means that, simple input tools, such as a pointer (mouse) and a text entry device (keyboard), can be sufficient for interacting with complex virtual environments. It was not until the introduction of

Springer Nature Switzerland AG 2020
Ahram (Ed.): AHFE 2019, AISC 973, pp. 377–388, 2020.
https://doi.org/10.1007/978-3-030-20476-1_38

direct touch based input, that more sophisticated and multichannel interaction tools were needed for effective system interaction.

Furthermore, before the advent of touchscreen based devices, most computing systems, and their virtual environments, primarily engaged the users' visual and auditory senses [1]. This meant that in any environment system usability heavily depended on the user's ability to extrapolate methods of interaction, using only these two modalities. These days, direct touch input complemented with tactile output have become necessary components to support interaction that is more natural. In the last decades, multimodal interaction was extended to include touch output (haptic feedback), however, the resolution and type of output is still very rudimentary. In fact, haptic interaction is used as a form of "confirmation feedback" in most multimodal interaction systems, rather than an active medium of interaction. This may be due to the power and computational requirements of providing a comprehensive haptic-information-channel. For this reason, most touchscreen devices use basic vibrational signals to create rudimentary low-resolution event based feedback, which can only be useful in the presence of other modalities (visual and auditory).

With virtual and augmented reality becoming a more popular approach in system interaction, as well as social interaction, there is a need to redevelop the role of haptics in virtual environments [2]. It may no longer be possible to focus solely on visual and auditory interaction, ignoring the role of haptics. Complex virtual environments such as Virtual Reality (VR) and Augmented Reality (AR) require more comprehensive tactile input and output. For this reason, primitive tactile signals, used in previous systems need to give way to more precisely calibrated actuation, which is specifically designed for various parts of the body. Furthermore, haptic feedback needs to be redeveloped to create natural tactile actuation as compared to encoded haptic signals, for complex multimodal VR environments. And most importantly, to ensure immersive interaction in VR and AR system, there is a need to develop new interaction devices that can continuously input and output real-time haptic information along with visual and auditory information between the user and the VR system, with similar efficiency, as was done by the mouse and keyboard for 2D virtual environments.

2   Using the Haptic Mediation Concept in VR/AR Environments

In our previous work, we have tried to overcome some of the limitations in creating and providing effective haptic feedback, by introducing a new concept known as "Haptic Mediation". Haptic mediation is a process of effectively relaying the actuation signal from the source (actuator) to the necessary point of contact (receptive fields of the skin), mitigating environmental noise and other internal and external inefficiencies within the system [2]. This is done by actively monitoring signal integrity and skin sensitivity to the applied actuation signal, dynamically adjusting the actuation source to ensure reliable perception of intended signal. Using this approach, we have illustrated [3, 4] that it is possible to actively monitor user interaction, and relay sensible haptic information, to the point of contact in the most efficient manner. Therefore, it is possible to deliver precisely calibrated actuation signals (i.e. vibration, electrical pneumatic or temperature based) with the help of (an active or passive) medium, which is calibrated for such an information exchange. In our research we have implemented this concept into a number of 2D interaction devices to gauge the effectiveness of the concept and the results have been quite positive [5, 6]. It is also possible and desirably to incorporate this new technology of "haptic mediation" into VR/AR systems, enhancing user interaction into a more natural and fluid experience.

Conventional techniques of tactile simulation provide direct or indirect actuation signals to the skin, which may not always generate the immersive experience in VR/AR environments. Contrary to how visual and auditory modalities are presented, haptic feedback needs to be precisely calibrated and channeled to provide specific actuation for specific points on the skin (points of contact with virtual objects/environments), to induce the necessary sensory illusions. Haptic Mediation (HM) may be an ideal mechanism to facilitate the exchange of actuation signals from an electro-mechanical source to the bio-mechanical receptors in the skin. In fact, an active haptic medium can calibrate the haptic signal with reference to environmental noise and skin sensitivity requirements by creating a real-time feedback loop between the generated signal and the signal received at the point of contact.

In our previous research [2], we demonstrated that not only can this approach be used to generate tactile illusions [7], but also through haptic mediation it may be possible to stimulate mechanical proprioceptive [8] feedback to the skin. By applying actuation to the muscle tendons (e.g. elbow and shoulder) through haptic mediation, we can induce illusory sensations of complex movements far more accurately than previously possible [9]. As demonstrated by Thyrion and Rolls, strategically applied actuation to various joints and tendons can induce spatial and kinematic perception. Furthermore, research by Rolls et al. [10], shows that it may even be possible to induce virtual two-dimensional movement simulating the sense of straight lines, letters, numbers, and geometrical figures in the horizontal plane. Moreover the authors using direct vibrotactile actuation, through user testing, also achieved 3D figures resembling spirals. We think that with precise and more calibrated actuation signals using haptic mediation, these techniques and their application can be dramatically improved.

Similarly, other actuation techniques, such as funneling and saltation (or hopping effect) [11] can also be improved considerable. By enhancing the simultaneous tactile stimuli presented in two locations on the touch-surface or directly onto the human skin, through haptic mediation, the funneling effect can be made more pronounced. Moreover, because the virtual location of the illusory tactile sensation (phantom object), is somewhere between the pairs of actuators, it can easily be controlled by changing the parameters (such as intensity, phase and duration) of actuation. Thus, dynamically altering the virtually simulated object, which can be ideal for VR and AR interaction. We postulate that by utilizing haptic mediation and existing techniques of generating tactile illusions, it is possible to develop tools that can augment a wide range of haptic illusion for VR/AR environments. In this research, we specifically focus on enhancing funneling and saltation effects through a pair of customized prototype haptic gloves.

3   Advance Haptic Glove: System Design

To enhance funneling and saltation effects we developed a pair of custom haptic gloves with embedded actuation components. The gloves were used to provide localized actuation as well simulated phase shifts and funneling effects between the fingers and the thumb. One of the gloved was layered with a liquid medium to mediate the signals provided by the actuation components while the other provided direct action to the skin. This section elaborates the design and interaction with each of the gloves (Fig. 1).

Fig. 1. Various implementations of the Haptic Mediation Glove (HMG) and the Haptic Glove (HG).

3.1   Glove Design

Both gloves had five Tectonic voice coil actuators (TEAX09C005-8) attached to each fingertip and thumb as well as one actuator on the inside of the palm (between the index finger and the thumb). The actuation components were fixed at the tip of each finger of the glove in such a way that when worn the user would have some contact directly with the actuators or indirectly through the liquid medium (as seen from Fig. 1). The glove with liquid medium was a custom sown multi-layered glove that had a thin latex glove embedded within it. The latex glove was sealed and filled with inert low viscosity transparent oil (as seen from Fig. 1 bottom right). The actuators were attached on the top of the latex glove and the entire assembly was sown into a stretchable cloth in the shape of a glove.

The second glove was similarly designed to the first one. Small pockets were sown inside the glove at the fingertips and the palm area for each actuator. Each actuator was kept in position using these tightly sown thread pockets to ensure the placement and actuation remained constant. As with the first glove, the wiring of the actuators was internally covered using a cloth lining, to ensure the user did not accidentally pull them and sever the connections (Fig. 2). All the actuator wiring was channeled down towards the wrist band of the glove and extended outward. A single extra-flexible shielded twisted pair LAN cable was used to connect each actuator separately to a four channel D-class amplified and signal generator This meant that each actuator could be used independently from the others, with a maximum four actuators running at the same time.

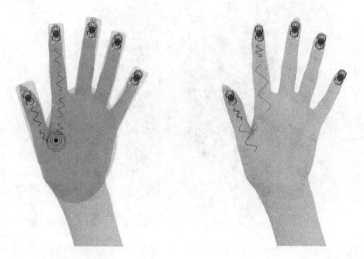

Fig. 2. Internal design of the Liquid Mediation layer, Actuator placement and signal attenuation for the two gloves.

2   Glove Based Interaction

To maximize the quad-channel configuration both gloves were connected to the signal generator and amplified one at a time. Each channel corresponded to one of the actuators positioned on one of the fingers. We then applied actuation pulses with duration of 1, 2 and 3 s each within the frequency range of 10–250 Hz. The intention was to define specific signals that can generate funneling or saltation effects. We also adjusted to amplitude of the signal between 3–12 V at fixed 1A. During the internal testing, we observed that actuator attachment and proximity to the fingertips yielded clear and sharp signals to the skin at even lower voltages. However, frequency, delay and placement (traveling path of the signal) yielded the maximum perceptual variations. Therefore, we decided to use these three parameters as the key variable in the user study.

To test the effectiveness of 'Haptic Mediation' technique for virtual reality interaction, we developed two gloves with identical actuation components (Fig. 3). One of the gloves utilized an embedded liquid mediation layer while the other provided actuation directly to the skin contact (fingertip). Actuation components operated at the same signals to generate similar actuation effects across the different experiment conditions, the only difference was how the feedback signals were delivered to the skin contact.

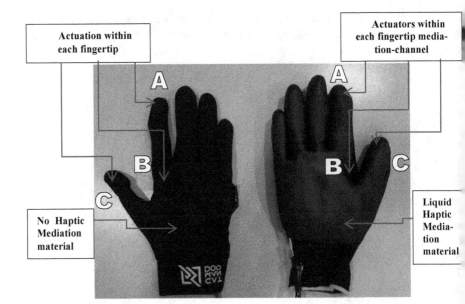

Fig. 3. (*Left*) Glove without Liquid Mediation (HM-Glove), (*Right*) Glove with Liqui
Mediation (H-Glove).

4   User Study

4.1   Testing Methodology

In our internal pilot testing, we observed that some signal parameters affected the tacti
illusion of funneling and saltation more than others. These included frequency, dela
and the placement (traveling path of the signal) of actuation components. We narrowe
down out actuation path to the index finger, the palm and the thumb. This was require
due to two key reasons. Firstly, the limitation of having only a quad-channel amplifi
and signal generator made it quite challenging to accommodate all five-fingers and the
palm accurately at the same time. Secondly, if all five fingers were used in the stud
alongside the three conditions of each of the three signal parameters (Freq, delay
placement), the user test would become extremely long and difficult to comple
Therefore, for the sake of collecting focused data we limited the user testing to th
index finger, the webbing between the thumb and the index finger (inside of the pal
and the thumb itself (as seem in Fig. 3). This approach ensured we had three segmer
or discrete points of the signal between the tip of the index finger and the tip of t
thumb (A, B & C).

For each of the signal parameters, we selected three subsequent values. With fr
quency, we choose to have 50, 150 and 250 Hz signals as it represented low, mid a
high tactile signal thresholds most commonly applied to the fingertips. To vary t
delay between the different actuator signals we first ensured that the duration of o
segment (either A, B or C) of the signal remained close to 250 ms, irrespective of t
frequency of the signal (50, 150, 250 Hz). This meant that each actuator was switche

on for $\sim$ 250–260 ms. We then added delays of -50, 50 and 150 ms, where -50 ms meant that the second segment (second actuator) was triggered 50 ms before the end of the first actuation segment. Delays of 50 ms and 150 ms were when the second segment (second actuator) was turned-on after the 250 ms actuation created by the first actuator. Additionally, the last parameter that was varied was the placement or sequence of the segments being turned on. This meant that participants felt the signal sequence as A-B-C (Index-Palm-Thumb), C-B-A (Thumb-Palm-Index) or B-C-A (Palm-Thumb-Index).

The user study was conducted with 20 participants (17 male and 3 female) and they were provided each feedback signal three (redundant) times after which they were asked to rate the signal and its type in a questionnaire. The participants rated nature (type) of the signal (discrete/continuous) between the three segments (A, B & C). They also evaluated the perceived signal strength as well as the pleasantness of the feedback signal. In total 27 different (unique) feedback signals were provided to each participant for each glove. This was done in such a way that every parameter (Freq., Delay, Sequence of combination) and its sub-value was presented to the user for each glove. This meant that the participants were provided with different frequencies as a sub-parameter (i.e. 50 Hz) within which they were provided the three delay sub-parameters i.e. -50, 50 & 150) and the further three sequence combinations (i.e. A-B-C, C-B-A, B-C-A), with a total of 9 sub-conditions for each parameter value of Frequency. For all three parameter values of Frequency, the total conditions were 27 (each repeated 3 times for redundancy making 81 feedback signals per glove). This approach was used as it provided the most comprehensive method of delivering actuation signals within related signal parameters (Freq., Delay, and Sequence).

4.2 Results of the User Study

Participants rated the provided signal in three categories: nature, sequence or direction and pleasantness. Data for the nature of the signal provided information on how the users perceived the signal (discrete vs continuous). The sequence of the signal was collected by having the users identify which segment the signal traveled through and in what order. While the pleasantness rating was collected to analyze how users perceived the signal variation over the frequency spectrum. The nature of the signal and pleasantness data was collected using 5-point likert scale, whereas sequence was entered in start-Finish order (i.e. A->B->C). The likert scale data from both gloves was analyzed using a Pearson Chi-Square test, while the sequence data was first broken down into number of errors (1, 2 or 3) using each glove and then analyzed using Fisher's Exact test. Finally, the experimenter also conducted a freeform interview to review the participants reasoning for each section, focusing on sub-values of the three parameters.

If we look at user rating of nature of signal we saw that the although there were variance between the HM-Glove (HMG) and the H-Glove (HG) there were no statistical differences at 50 Hz either for -50 ms, 50 ms, or 100 ms. This meant that the participants were able to identify each signal as a unique (discrete) feedback signal irrespective of the glove technology (Liquid Haptic Medium). The results were similar for 100 Hz feedback signal as well, but there was a trend for -50 and 50 ms of delay signal that approached statistically significant difference.

A larger sample size would be able to validate this further. Essentially, at a frequency of 100 Hz, the difference between the three signals became less distinct for participants using the HM-Glove (HMG), as they started to perceive the discrete feedback signals as one large signal, which was especially visible from the results for signal applied at 100 Hz, at −50 ms (Fig. 4).

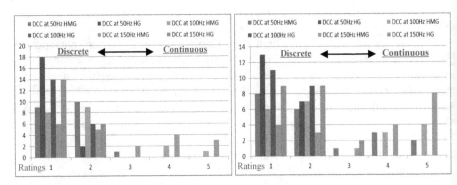

Fig. 4. Subjective ratings of the applied signals averaged over 20 participants (Discrete VS Continuous) over the three frequencies (50 Hz, 100 Hz, 150 Hz) with 100 ms (left) and 50 ms (right) delay between each signal for both gloves.

However, once the signal was provided at 150 Hz the results show (Fig. 5), participants using the HMG perceived the three signals as one continuous signal. However, this was not the case for the HG, where participants could identify the three signals independently with reasonable accuracy. Using Pearson Chi-Square test, we identified that there was a statistically significant difference between the results for the two gloves (HMG and HG) for both −50 ms and 50 ms of delay at 150 Hz. Results were not very clear for the 100 ms delay; however, we think this is because of the smaller sample size (Fig. 5). Interestingly enough, during the interview, the participant pointed out that the HMG felt more sensible throughout the entire hand as compared to the HG, which created a perception of higher intensity and a non-locality of the signal source.

Looking at the number of errors in identifying the correct sequence of feedback signals on either glove, we see a similar trend (Fig. 6). When the signal was provided at 100 ms delay, the users were able to identify the sequence quite accurately for both HMG and HG. However, the error rate rose when the delay was reduced to 50 ms. Although there was no statistically significant difference between the two gloves for 100 ms or 50 ms for all three frequencies, we did see a specific trend that HMG increased the error rate enforcing the hypothesis that participants observed the applied signals as one large distributed signals, compared to individual localized signals. The result was more prominent at −50 ms of delay for all frequencies (Fig. 7), especially at 150 Hz, where there was a statistically significant difference between HMG and HG. When questioning the participants regarding their selection, the overwhelming response was that due to actuation signal in the HMG covering most of the hand; it was difficult to isolate one signal from the other. Participant also mentioned that the

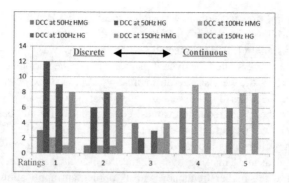

Fig. 5. Subjective ratings of the applied signals averaged over 20 participants (Discrete VS Continuous) over the three frequencies (50 Hz, 100 Hz, 150 Hz) with −50 ms delay between each signal for both gloves.

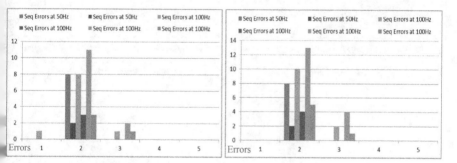

Fig. 6. Error rates for the perceived Sequence of applied signals (number of errors) averaged over 20 participants over the three frequencies (50 Hz, 100 Hz, 150 Hz) with 100 ms (left) and 50 ms (right) delay between each signal for both gloves.

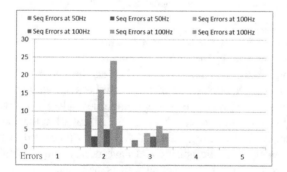

Fig. 7. Error rates of the perceived Sequence of the applied signals (number of errors) averaged over 20 participants over the three frequencies (50 Hz, 100 Hz, 150 Hz) with −50 ms delay between each signal for both gloves.

perceived the same lower frequency signal (especially 50 Hz) as a higher intensity signal on the HMG as compared to the HG, while this was completely opposite for the higher intensity signal (150 Hz) especially when the delay was −50 ms.

Looking at how users rated the overall pleasantness of the various signals applied to the two gloves (Fig. 8), we also see an interesting trend. As expected, no clear variations were seen between the different delay values; however, there were difference with reference to the three frequency values. For signal 100 Hz and below, there were only minor differences as participants rated the HMG higher than the HG. However, for 150 Hz frequency there was a clear deference between the two gloves, as users rated HG to be less pleasant. This was over all the three-delay parameter (Fig. 8), which meant that due to the higher frequency and its unpleasant application directly to the skin, the users perceived this type of feedback as too strong and somewhat disconcerting. Furthermore, the participant added in the interview that the acoustic noise of the actuators, which was higher in the HG added to the unpleasantness during interaction.

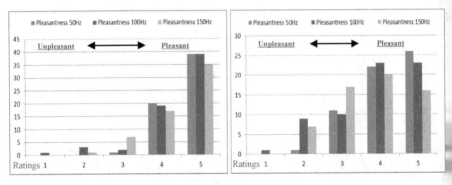

Fig. 8. Subjective ratings of pleasantness for the applied signals averaged over 20 participant over the three frequencies and delay parameters for HM-Glove (left) and Haptic Glove (right)

However, in the HMG where the actuators were not connected to the skin directly the medium acted as a filter to the 150 Hz applied signal, integrating the signal to remove the unpleasantness of the feedback. During the interview, most participant were asked about the overall strength of the signal between the two gloves, and their perception variations between them were quite low. Retrospectively, perceptual strength of the signal could also have been measured (rated) in the main questionnaire as higher frequency feedback signals could have been perceived at higher intensit signals. Measuring the acoustic noise could have been another useful metric to distinguish the two gloves and their actuation signals, as in the post-test interview all the participants mentioned the lack of acoustic noise on the HMG was one of the key reasons for the higher pleasantness rating.

5 Discussion and Conclusions

In this paper, we presented a comparison of two types of haptic gloves for VR interaction: A conventional glove (H-Glove with embedded actuators that directly contact the skin) and a glove layered with tactile mediation mechanism (HM-Glove with a liquid mediation layer between the embedded actuators and the skin). Both the gloves were embedded with identical actuation components and were provided similar electrical signals to general compatible tactile feedback. As with previous results [2–6], the use of haptic mediation enhanced the actuation signal over a larger area while dampening the acoustic noise of the actuation components. This research also showed that an efficient mediation mechanism can improve the overall haptic feedback experience.

The HM-Glove used in the research, provided increase area of actuation as compared to H-Glove using identical actuation components. The glove also integrated multiple actuation sources, creating complex tactile signals, which were perceived as far more pleasant. Moreover, as the application of a haptic medium is able to integrate specific actuation signals and frequencies, it is possible to create a far richer and more natural feedback mechanism as compared to conventional direct skin actuation methods. This opens up tactile simulation in a way that is not possible conventional techniques. This is because an ideal medium cannot only amplify and localize certain actuation (lower frequency) signals but also integrate other (higher frequency) signals to simulate complex textures and object tactility without introducing unpleasant actuation frequencies.

Although further evaluation may be needed to ascertain an improved configuration and technology (i.e. structure and chemical properties of the mediation material or gel-structure), the results clearly, indicate the effectiveness of the proposed approach utilizing haptic mediation in glove-based VR interaction. Furthermore, it may also be possible to combine a number of actuation signals to provide a complex tactile feedback without limiting individual signal parameters due to inefficiency of transfer or intensity variations within similar frequency bandwidths. In our future research, we plan to explore how it may be possible to combine two or more signals within a complex haptic stimuli yet keeping the signal perceptual integrity, thereby achieving some form of complex haptification

Acknowledgement. The research was supported by a Post-Doctoral mobility grant awarded to the PI by the Finnish Culture Foundation and was carried out in collaboration with the Virtual and Augmented Reality Production and Use (VARPU), a Business Finland project.

References

1. Nukarinen, T., Kangas, K.J., Rantala, J., Pakkanen, T., Raisamo, R.: Hands-free vibrotactile feedback for object selection tasks in virtual reality. In: Presented at the 24th ACM Symposium on Virtual Reality Software and Technology (2018). https://doi.org/10.1145/3281505.3283375
2. Farooq, A.: Developing technologies to provide haptic feedback for surface based interaction in mobile devices, Ph.D. Thesis, University of Tampere, Faculty of Communication Sciences (2017). http://tampub.uta.fi/handle/10024/102318

3. Farooq, A., Evreinov, G., Raisamo, R., Takahata, D.: Evaluating transparent liquid screen overlay as a haptic conductor. In: Proceedings of IEEE SENSORS Conference (2015). https://doi.org/10.1109/icsens.2015.7370186

4. Farooq, A., Evreinov, G., Raisamo, R.: Evaluating different types of actuators for liquid screen overlays (LSO). In: Proceeding of IEEE Symposium on Design, Test, Integration and Packaging of MEMS/MOEMS, DTIP'16, pp. 97–102 (2016). https://doi.org/10.1109/dtip. 2016.7514847

5. Evreinov, G., Farooq, A., Raisamo, R., Hippula, A., Takahata, D.: Tactile imaging system. United States Patent 9672701 B2 Filed: 07/08/2015 Published: 06/06/2017. Assignee TAMPEREEN YLIOPISTO (Tampereen yliopisto, FI) and FUKOKU CO., LTD. (Ageo-shi Saitama, JP) (2017)

6. Evreinov, G., Farooq, A., Raisamo, R., Hippula, A., Takahata, D.: Haptic device, United States Patent 9789896 B2 Filed: 07/08/2015 Published: 10/17/2017. Assignee TAMPEREEN YLIOPISTO (Tampereen yliopisto, FI) and FUKOKU CO., LTD. (Ageo shi, Saitama, JP) (2017)

7. Lederman, S.J., Jones, L.A.: Tactile and Haptic Illusions. IEEE Trans. Haptics 4(2), 273–294 (2011)

8. Albert, F., Bergenheim, M., Ribot-Ciscar, E., Roll, J.-P.: The Ia afferent feedback of a given movement evokes the illusion of the same movement when returned to the subject via muscle tendon vibration. Exp. Brain Res. 172, 163–174 (2006)

9. Roll, J.-P., Albert, F., Thyrion, C., et al.: Inducing any virtual two-dimensional movement in humans by applying muscle tendon vibration. J. Neurophysiol. 101(2), 816–823 (2009)

10. Thyrion, C., Roll, J.-P.: Predicting any arm movement feedback to induce three-dimension illusory movements in humans. J. Neurophysiol. 104, 949–959 (2010). https://doi.org/10. 1152/jn.00025.2010

11. Leonardis, D., Frisoli, A., et al.: Illusory perception of arm movement induced by visuo-proprioceptive sensory stimulation and controlled by motor imagery. IEEE Haptic Symp. 421–424 (2012)

Mapping the Monetization Challenge of Gaming in Various Domains

Evangelos Markopoulos[1]([envelope]), Panagiotis Markopoulos[2],
Mika Liumila[3], Younus Almufti[1], and Vasu Aggarwal[1]

[1] HULT International Business School, 35 Commercial Road,
Whitechapel, London E1 1LD, UK
evangelos.markopoulos@faculty.hult.edu,
{yalmufti2015,vaggarwal2016}@student.hult.edu
[2] University of the Arts London, Elephant and Castle, London SE1 6SB, UK
p.markopoulosl@arts.ac.uk
[3] Turku University of Applied Science, Joukahaisenkatu 3,
20520 Turku, Finland
mika.luimula@turkuamk.fi

Abstract. The cost of developing successful games for either entertainment or business purposes is a high-risk investment but mandatory due to the nature of the sector. However, there are discrete and innovative ways that minimize the investments risk and assure profitability without losing the player's engagement. Gaming monetization can be approached from direct or indirect financial charges based on the scope of the game and its target group. As of today, no monetization practice can be considered as a silver bullet as they are all affected by geographical, cultural, social, economic and other factors. This paper attempts to define the major monetization elements in the gaming industry. It also attempts to define the major gaming categories and subcategories and associate on them the monetization elements and techniques. Furthermore, it creates a map for the development of gamification monetization approaches per case which can contribute towards effective gaming investments management.

Keywords: Gaming · Gamification · Serious games ·
Investment management · Entertainment games · Monetization · Ethics ·
Management

Introduction

The growth of the computer and digital games industry has turned out to be non-predictable in market size and number of applications. At the same time the discipline of games design and games development entered unmapped technoeconomic areas where challenges and threats are can be as surprising as the opportunities. The wide range of games (serious entertainment, simulations, etc.) and the limited lifespan they have in terms of player engagement turns the development of computer games into high-risk but also high-value investments.

Games development turned out to be complex technological projects and initiatives that create masses of users, require serious investments, utilize tremendous computing

Springer Nature Switzerland AG 2020
Ahram (Ed.): AHFE 2019, AISC 973, pp. 389–400, 2020.
https://doi.org/10.1007/978-3-030-20476-1_39

resources, and demand continuous design and artistic perfection in order to remain engaging. The cost of developing successful games for either entertainment or business applications is a high-risk investment but mandatory as well due to the nature of the sector. However, there are discrete and innovative techniques and practices to minimize such investment risk and increase profitability.

Game monetization practices is a field in gaming economics that grows faster than the games themselves. In entertainment games, monetization can be approached with direct or indirect financial charges while on serious games monetization is measured by the contribution of the game to effectiveness of the organization, the performance of the employers, their training effectiveness and their overall positive contribution to the operations of the organization.

Applying monetization techniques in any type of game involves socioeconomic and psycho-economic elements as well. The psychology of the gamer plays a significant role on the adaptation of a monetization technique. It must also be noted that monetization the outside entertainment games environment must take into consideration legal dimensions as well. When designing monetization models for serious games parameters such as labour rights, data protection rights, confidentiality performances scores, etc., must be considered as well.

This paper attempts to define the major monetization elements in the gaming industry. It also attempts to define the major gaming categories and subcategories and associate on them the monetization elements and techniques and create a map for the development of gamification monetization approaches per case that can contribute towards effective gaming investments management.

2   The Gaming Industry Economics, and the Rise of the UK

The last 40 years games have tremendously changed due to the massive technological advancements on the information technology sector and the rapid commercialization of computers. Computer games play today a significant role in the life of people regardless their age, sex or social or financial status. This phenomenon has been analyzed over extensive research indicating the impact games have in modern society [1]. Nearly 77% of the families in the United States own videogames, and nearly 46.6% of the employees in Germany play games during working hours.

The computer games industry grows with exponential rates. The industry in the United States had a value of 16.9 billion USD in 2016 and 17.6 billion in 2017 with 4. billion to be from consumers spending. The industry is expected to reach the value of 20.3 billion by 2020 in the US with the action shooting games to be the most popular.

An impressive increase is also recorded on the gamers as well. In 2016 there were 200 million gamers in North America and 912 million in Asia Pacific. Gender does not seem to differentiate games. Women are quite active in games with significant presence in some (69% play match 3 genre) and less presence in others (6% in racing games).

The is tremendous growth of the markets created new business success stories. On 2017, the Chinese entertainment company Tencent generated revenues of 18.1 billion U.S. dollars. [2]

Impressive development is also indicated in the UK gaming industry as well. The gamers in the UK in 2017 were 32.4 million making the 5th largest video game market, (China, USA, Japan and Germany were the top four). Consumer spending increase in the UK as well, reaching £4.33 in 2016 and £5.11 in 2017. This growth is reflected in the growth of the computer game companies in the country. On 2018, 2,261 active games companies were recorded in the UK, developing all types of games from mobile, PC and console, to VR/AR, esports and Artificial Intelligence. [3]

3 Game Categories and Design Elements

Various disciplines in the gamification industry are differentiated in terms of their applications and the type of users they have. However, an overlap exists in the combination of competitive and design elements with the motive of increasing user engagement and support learning [4, 5].

Primarily games can be categorized into entertainment and serious/business games. The entertainment games include, Massively Multiplayer Online games (MMO), simulations, role playing (RPG), first & third person shooters (FPS/TPS) which have been spread across various genres like puzzle, strategy, action and racing. This gamified approach includes various design elements created for pleasure and entertainment with minimalistic support and existence of predetermined learning outcomes.

Serious Games, on the other hand, is an upcoming trend with significant potential to revolutionize business management and operations. Serious games, are games carrying game-inspired design elements, designed not for entertainment but gamified structurally to either gamify a corporate's content (Structural Gamification) or turn the content into a game (Content Gamification) with a predetermined objective [5].

A serious game, comprised of several design elements (points, adaptation to difficulty and a narrative) can be similar across other serious game s [6].

Under a structural gamification approach, in serious games, the corporate content is transformed into a template – based approach with plug-and-play functionality within the game and minimalistic structural changes to the native content. This can be achieved by using design elements to a design structure of a game that applies to different corporate contents. Content Gamification on the other hand involves a one-time unique structure created for specific corporate content, incorporating various design elements, unique to the content type. For example, business simulations (or gamesims') [7] for inventory management training the Purchase Department would constitute gamifying the inventory management content.

In between the entertainment and cooperate games a new category exists which emphasizes on the learning experience of any game. The educational games are based on gaming mechanics which increase the ability to learn new skills [8]. Today gamification has applications in the educational sector, which can potentially disrupt the dynamics of education due to lack of active engagement and motivation amongst students of any age and type, in cognitive learning processes [5]. Educational games can incorporate competitive design elements with motives for fulfilling the market gap of lack of engagement. Research indicates that gamification in education including science, mathematics, cultural heritage, foreign languages, health, computer science, software engineering, business and logistics [9].

Improving upon the antecedents of the user engagement such as the player's cognitive abilities on task management, sense of control would lead to the user incrementally improving the in-game performance on the outcomes of education [10] and the predetermined objectives [11].

4 Monetization Practices

The gaming industry evolution can be characterized by innovation which in most cases require high budget with uncertain success. Game monetization refers to the strategy and practices that assure return on investment in game development projects and initiatives.

In monetization, there are fundamental revenue-generation approaches that can be interlinked. One of them is the utilization of the advertising space. This tackles the way a firm can utilize web spaces to sponsor advertisements that can generate revenue. This kind of classic approach has been adopted by many industries with online presence. Subscription is considered to be another base of revenue generation and it can manifest itself in seven different ways based on the involvement of the user, the duration and other options. The decision on which subscription a user should select is highly depended on the needs of the user. Various types of subscriptions can be used together this is widely used in mobile applications such as Headspace.

Monetization practices and strategies can also be developed around the utilization of the game's data. Data driven monetization refers to the use, selling or access of important data to generate revenue. This is one of the main monetization models used in websites such as Facebook that allows data access to other firms [12].

While the data-oriented monetization utilizes the information of users the agency oriented monetization focuses on access to the platform and increase its popularity in various marketplaces. Ebay is such an example which allows access only through commission.

More monetization practices and approaches are the agent oriented and lifestyle oriented ones. The Agency-based monetization is usually interlinked with subscription and has been used effectively in human resources management and recruiting firm [10]. The Lifestyle based monetization is not commonly used in gamified products as it usually refers to the selling of merchandise or direct to consumer-based models.

However indirect marketing can be achieved in gamification through which life style products can be promoted as part of the game scenario. Table 1 presents these five monetization orientations that can be utilized by a game from many dimensions such as before, during and after game-play.

Table 1. Categorization of monetization practices.

Monetization base	Example
Ad space	Banner Ads
	Affiliate Marketing
	Native Advertising
	Branded Content
	Ad Free
Data base	DaaS (data as a service)
	Packed Data
	DIY DaaS
	Optimization
Subscription base	Virtual Goods (microtransactions)
	Free DIY or Paid DFY
	Pay What You Want
	Pay Per Device, License, User
	Freemium
Agency base	Commission
	Subscription Fee
	Pay to: Offer, List, Get Noticed, Cashback
Lifestyle base	Direct to Consumer
	Branded Merchandise

Mapping the Monetization Practices to Game Domains

The monetization practices in the context of the different game categories have also different applications that can be either used as standalone practices or in various combinations towards achieving unique pathways for revenue generation and profitability.

In serious games two subscription-based monetization approaches can be used effectively. The first one is the freemium business model that can be used primarily on the structural gamification applications. The term is composed from the words "free" and "premium" indicating that a service is offered for free on the basic operations, but for premium add-on services, functionalities or related services, a fee is charged [13].

Through the application of a freemium price model, serious games can monetize various design elements that can provide basic operations such as employee training and productivity enhancement. However, combinations of other operations and functionality can be sold in form of a premium 'add on' allowing serious games to be agile in use and cost, and therefore more applicable in different corporate departments. For companies like 'Tencent' who invested in the freemium business model and generated revenues over $18.1 billion in 2017 [14], monetization through the freemium model turned out to be very attractive.

The second subscription base that can be used is 'Pay Per User', which works in team-oriented products and services. Depending on the organisational size, the potential for monetization exists where large number of employees need to be involved.

Therefore, implementation and adaptation of a 'Pay Per User' coupled with a freemium subscription base would allow sustainable and affordable monetization opportunities.

These monetization practices can also be adapted in the entertainment games domain which currently benefits from virtual goods with various microtransactions or free games. A successful example is Fortnite, which generated revenues of $318 million in May 2018 [15]. The specific approach allows users to get premium add-ons such as various game-skins and Experience Point (XP) boosts. These features, however, don't give a competitive advantage but support a lower user churn rate and product loyalty.

Education games on the other hand could benefit significantly from the 'Pay Per Device' subscription approach which allows education institutions to be charged a fee for each student who participates in the game. This subscription base could work well on a tier-based system, according to the size of the education institution. Another potential subscription base would be 'Pay Per License' where institutions would pay to be receiving each gamified solution being offered by the provider.

6   Niche Markets and Monetization Approaches

The evolution of gamification and the monetization practices are aligned with the evolution of the technology and the areas of applications games can contribute. The rapid development of the gaming industry does not allow monetization models to be settled down in the PC, console and mobile gaming regardless the type of game (serious, entertainment, educational). This has caused challenges for startups who enter in gaming business with new ideas. Reaching the visibility needed in heavily compete market places is very demanding, thus the probability for success can be significant lower.

New market places are being born and dying every now and then even in the gaming industry. A new promising market is the Virtual Reality games where monetization faces various challenges primarily due to the cost of VR game operations. Early adopters in VR have not yet been as successful as technology providers have been expected. Even the new technologies have the same type of challenges Microsoft Kinect motion detection sensor have had earlier. Both VR glasses and Kinect sensor have been designed for consumer markets.

However, monetization models in non-consumer markets are totally different making traditional game companies face difficulties on apply other monetization models. The prime reason for these difficulties is because they try to utilize game technologies by being focused on projects rather than service innovations. Figure presents an example of monetization model for exergaming [16].

Digital rehabilitation markets are fragmented with thousands of single solution This led to the established of the DigiRehab consortium which unites the Finnish digital rehabilitation industry into open business ecosystems. This monetization practice relies on joint digital physical therapy and gamified solutions for exercise, aware IoT, user profile, and analytics management provided by various companies in business ecosystem. Currently, research is been executed in prevention and rehabilitation.

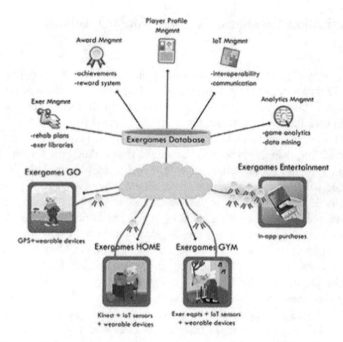

Fig. 1. Monetization model in digital physical therapy and gamified solutions for exercise.

This monetization model can be scaled from the healthcare sector to the technology industry. It has been experienced that safety training solutions for example in VR, require various expertise areas such as exercise (including health, safety, environment, and quality), pedagogy, player profile, natural language understanding, and analytics management.

To manage such challenges, strategic multi-disciplinary partnerships need to be created like the Finnish National Safety Training Consortium which has been established by the Turku University of Applied Sciences and participated in various applications (Fig. 2).

Fig. 2. VR Safety Training solutions from Turku University of Applied Sciences.

7 Monetization Geo-Socio-Techno-Economic Influence Factors

Besides the type of games and their area of application there are many other external factors that impact monetization strategies and approaches. Socio-geographic and socio-economic influences on how gamers, interact with each other and with the games, influence the game monetization practices selected per case.

Aspects such as the financial status of the player's income can affect the way gamers spend. Surprisingly people who earn more than £90,000 per year are less likely to spend on game microtransactions despite their financial ability [17]. On the other hand, people with associate degrees or high school diplomas tend to spend most in-game purchases. This difference in income and education highlights a very specific demographic point and the psychological aspect of superiority. It is hypothesized that people with medium income buy more in-game content as it provides them with a feeling of superiority. This need can be correlated with the Maslow's hierarchy of needs as the main aspect people with medium income fail into [18].

Riot Games, the multibillion-dollar company behind "League of Legends" is known to adjust the prices for its digital currency, known as Riot Points, on its online store based on the average wealth of each geographic region and/or the GDP per country [19]. In Europe for example, the smallest amount of money someone can spend is 2.50€ (2.2£) for 400 Riot Points while the maximum is 50€ (43.75£) for 9.300 Riot Points. In the USA however the company has different pricing limits where the minimum is 5$ (3.82£) for 650 Riot Points and the maximum is 100$ (76.4£) for 15.000 Riot Points. Similar to the European pricing limits is also Mexico despite the fact that is next to the USA. In Mexico, the minimum is amount a gamer can spend is 55 Mex (2.2£) for 555 Riot Points while the maximum is 1.320 Mex$ (52.8£) for 15.780 Riot Points. Table 2 indicates the pricing differences under one currency, the Great Britain Sterling, and also presents the cost per Riot Point in the minimum and the maximum purchase limits.

Table 2. Riot Points buying value in various geographic regions.

County	Minimum purchase	Minimum riot points	Cost/Riot point in £	Maximum purchase	Maximum riot points	Cost/Riot point in £
EUROPE	2.50 € EUR	400	0,0055	50 € EUR	9.300	0.0047
USA	5.00 $ USD	650	0,0058	100 $ USD	15.000	0.0050
MEXICO	55.00 $ MXN	555	0,0039	1,320 $ MXN	17.780	0.0029

Geo-economic factors can also be combined with geo-cultural factors not only direct game monetization of the game access costs but also on the game devices. Asia mobile gaming is very common. Asian players either through mobile games handheld consoles, enjoy playing video games while being outside, during public transportation or in between work or school breaks [20]. The same cannot be said for the western market however. While mobile and handheld gaming still exists in the west, it is not considered a legitimate method of gaming by many people. Game developers must be very careful where and when they release their new projects in order to predict its success more precisely.

A recent and interesting example of game release which did not go as expected is the way Blizzard entertainment presented their newest mobile game, Diablo Immortal, to the western audience.

Diablo has been a successful series that slowly gained a loyal PC fanbase. Blizzard is a company which has all of its greatest games on PC, and recently on consoles. However, the decision to release the newest Diablo game as a mobile exclusive had a severe impact on Blizzard's profits with its share-value to drop significantly, more than 30%, as indicated in Fig. 3 [21].

Fig. 3. Blizzard Activision stock after revealing on Nov. 2, 2018 the release day of Diablo Immortal on mobile.

3 Monetization Pre and Post Conditions, Addiction and Ethics

Selecting the best modernization approach in games of any kind is highly related to the type of game itself and its players target group.

Microtransactions based game monetization is one of the most common method used today by most game developers and the one with the most critical pre and post conditions to apply. Microtransactions are highly related to the player's culture, financial ability to pay the game effectively, geographic locations, social status and much more. They are applied to large target groups in an indirect and discrete way trying not to disrupt the gameplay, the gaming ethics and the gamer's communities. However, microtransactions are not always quite innocent and there can be serious ethical issues on the way they are used as game pre- and post-conditions.

A key requirement to identify the proper monetization strategy is the device games are played on (gaming console, computer, mobile phone, etc.). Paywalls for mobile games, subscriptions for PC games and loot-boxes for console games are the most popular monetization practices with different pre-conditions and post conditions but all under the microtransaction strategy.

The most common paywall is the patience paywall were the player is being stopped from progressing further into the game by making the resources or time demanded to progress extremely high or difficult. Players will either spend weeks trying to make it past the absurd hurdles set before them or they can simply pay to get over them instantly. Mobile developers deploy multiple paywalls within each game, each spaced apart from the next, in order to profit from impatient players. What makes this method ethical is that technically the game can certainly be finished without paying, but what is "unethical" is that without paying the game can be time consuming and frustrating. For the paywalls to work, the game must "shower" the player with rewards and then slowly give them less until they finally face the wall, creating this way a form of addiction. Once the player gets past the paywall this loop starts again providing the player with an addicting feeling accomplishment.

Subscription is another method of monetization used in many games operating on all devices, but it is mostly used by PC game developers. It is very similar to a paywall but instead of having to pay once to get over it, the players need to pay monthly or yearly in order to maintain their right to progress or even play. Some games offer optional subscriptions that only provide benefits while others make them mandatory. The pre-conditions for a successful subscription system are either stopping the player from progressing until a subscription is bought or letting the player progress slowly but keeping many of the useful and beautiful in-game features away, while constantly promoting them. After the subscription is bought the player feels the need to maintain what they obtained hence being pressured to keep their subscriptions active.

Outside of paywalls and subscriptions, there is another method commonly used in the console gaming industry that rose to popularity in recent years. That method is the loot-boxes and it was popularized recently as a way to make the purchase of game cosmetics more fun and addicting for the players, while profitable for the developers by integrating the sense on gambling. Loot-boxes contain numerous in-game cosmetic that usually dont affect the gameplay. A loot-box always costs the same but the contents within it are random. This is a very common trend when it comes to monetization methods as it creates a form of addiction. One of the oldest forms of addiction is gambling, and that is what loot boxes capitalize on. Gamer's hopes that the next loot box they get will possibly have what they want is what keeps them purchasing more. A precondition for loot-boxes to sell is to offer items worth the gamble. A simple way of making somethings value inflate is by keeping them rare. Rare items are desired so much that in some cases these items are sold in community markets between players for incredible amounts of money [21].

Gaming monetization is a very serious part of any game development project. Game ethics on the other hand are closely related to the monetization strategy and practice adopted by the game designers. What is legal might not always be ethical, but since it is legal it can be considered ethical and this is where monetization pre- and post-conditions and shall be further developed with more attention to the game and the gamers.

9   Areas of Further Research

Identifying monetization techniques and aligning them with various gamification domains is the first step towards understanding gamification economics. This paper presented an overview of the monetization concept and introduced specific methods, practices, precondition and post conditions towards monetizing a gamified application. However, the concept of gamification monetization extends furthermore in a micro and macro analysis, decision, strategy and ethics.

The implementation cost of a gamified application impacts significantly the monetization approach that will follow. The return on investment on gamification projects is vital to the developers due to the high cost of the technology involved, and the short life span of operations, since players/clients demand continuous updates and alignment with the gaming trends. On the other hand, the high cost of gaming projects shall not be passed only on the players/ clients and monetization practices shall involve more game sponsorships, and this is an area this research will extend.

Solving this equation of how cost distribution is spread on various revenue streams is more of a strategy than a financial challenge. This research will also extend towards a further and more in-depth analysis of the monetization strategy gamified applications and projects could adopt. Future work will target the development of a monetization strategy generator which will compose the right monetization approach per activity type within the game specifications and operations.

10   Conclusions

Gamification and monetization are two closely related concepts with a distant background. The creative dimension of gamification has to be aligned with the practical dimension of monetization.

Creative arts and economics cannot be aligned effectivity all the times as the cost of creativity might not be covered effectively with the monetization practices of the gamified project. After all, art can be expensive. However, the growth of the games industry and the high volumes of funds being involved, together with the tremendous revenues generated, allows many monetization practices to be considered valid, effective and ethical. On the other hand, the increase of the competition in the gaming industry and the increase of the type of gamified applications will request more specific and more ethical monetization strategies and methods in the very near future. This is something this research has indicated and approached with the geographical, social, technical, economical and other influence game factors that can determine the pre-conditions and the post-conditions of the monetization practices to be selected.

The paper approached this mapping process and resented the base for understanding the monetization challenge in gamification per type of game and game device. Further research will be conducted to extend this work on more advanced areas where gamification monetization can be modelled under a structured methodology and framework.

References

1. ESA: Essential Facts About the Computer and Video Game Industry. http://www.theesa. com/wp-content/uploads/2014/10/ESA_EF_2014.pdf
2. Statista: Video Game Industry – Statistics & Facts. https://www.statista.com/topics/868 video-games/
3. The Association of UK interactive entertainment: The games industry in numbers. https: ukie.org.uk/research
4. Salen, K., Zimmerman, E.: Rules of Play: Game Design Fundamentals. MIT Press Cambridge, MA (2004)
5. Kiryakova, G., Angelova, A., Yordanova, L.: Gamification in Education. https://www.sur ac.za/english/learning-teaching/ctl/Documents/Gamification%20in%20education.pdf
6. Lundgren, S., Björk, S.: Game mechanics: describing computer-augmented games in term of interaction. In: Proceedings of TIDSE (2003)
7. Hays, R.T.: The effectiveness of instructional games: a literature review and discussion Technical Report. Naval Air Warfare Center Training Systems Division. Florida (2005)
8. Business Insider. "Gamification" Techniques Increase Your Employees' Ability To Lear By 40%. http://whttp://www.businessinsider.com/gamification-techniques-increase-you employees-ability-to-learn-by-40-2013-9
9. Caponetto I., Jeffrey Earp J., Ott M.: Gamification and Education: A Literature Review http://www.itd.cnr.it/download/gamificationECGBL2014.pdf
10. Dicheva, D., Dichev, C., Agre, G., Angelova, G.: Gamification in education: a systemat mapping study. Int. Forum of Educ. Technol. Soc. **18**, 75–88 (2015)
11. Lameras, P., Arnab, S., Dunwell, I, et al.: Essential features of serious games design higher education: linking learning attributes to game mechanics. Br. J. Educ. Technc (2017)
12. Inverse. This is How Facebook Actually Makes All Its Money. https://www.inverse.con article/44566-how-does-facebook-make-money-mark-zuckerberg
13. Liu, C.Z., et al.: Effects of Freemium strategy in the mobile app market: an empirical stud of google play. J. Manag. Inf. Syst. **31**(3), 326–354 (2014)
14. Newszoo. Top Public Video Game Companies | By Revenue . https://newzoo.com/insigh rankings/top-25-companies-game-revenues/
15. Recode. Fortnite is generating more revenue than any other free game ever. https://ww recode.net/2018/6/26/17502072/fortnite-revenue-game-growth-318-million
16. Luimula, M., Ailio, P., Ravyse, C., Katajapuu, N., Korpelainen, R., Heinonen, A., Jamsa, 1 Gaming for health across various areas of life. In: 9th IEEE International Conference Cognitive Info-communications CogInfoCom 2018, pp. 247–252 (2018)
17. Priceonomics: Gender, Income & Education: Who Plays Video Games? Priceonomics Da Studio. https://priceonomics.com/gender-income-and-education-who-plays-video-games/
18. Taormina, R.J., Gao, J.H.: Maslow and the motivation hierarchy: measuring satisfaction the needs. Am. J. Psychol. **126**(155). University of Illinois Press (2013)
19. The league of legends on line store: https://na.leagueoflegends.com/en/ Mashable. A lit over half of the world's mobile game revenue comes from Asia. Nov. 17, 2015. https mashable.com/2015/11/17/mobile-game-revenue-asia/?europe=true#snOPqkWS6uq7
20. NASDAQ: ATVI. https://www.google.com/search?q=NASDAQ:ATVI&tbm=fin#scso WxDhW5LmHoKz9QOT14fwCQ2:0,_CYNUXKu6AviU1fAP_6Ws4AI2:0
21. Polygon. https://www.polygon.com/2018/1/30/16952248/counter-strike-global-offensive agon-lore-skadoodle-skin-sale-opskins

Research on Location of Emergency Sign Based on Virtual Reality and Eye Tracking Technology

Sun Guilei[(⊠)]

Department of Safety Engineering, China University of Labor Relations,
Beijing 100048, China
sunguilei@126.om

Abstract. In order to analyze the attention to exit sign during the emergency state, virtual reality technology (VR) was used to simulate escape scene and 3D Max was used to design the scene. To obtain the number of gaze points and the gaze duration of subjects, eye tracker was utilized to get data and spss 21.0 was taken advantage of to analyze the data. Results show that the position of emergency exit sign has significant influence on the recognition. And the exit sign of the height of 1.0 m on the front of the observer's line of sight is the most beneficial to discovery and identification. Moreover, the height and the position of exit sign have no significant influence on reading time.

Keywords: Eye tracking · Exit sign · Height · Fixation time · Identification

1 Introduction

In emergency escape time, the correct choice of escape route is an important guarantee to improve the chance of escape. Emergency exit sign is used as an important mark for escape. The recognition of the exit sign is susceptible to environmental factors such as smoke, high temperature and noise, which makes it impossible to quickly extract the indication information.

At present, researches on safety signs mostly focus on the design, individual characteristics and situational factors. Quantitative research focused on the identification of traffic signs, such as Lai [1] studied on the effect of color schemes and message lines on driver attention distribution based on virtual driving environment. Most scholars studied through qualitative methods such as subjective evaluation and few scholars researched on the identification of signs for escape. For example, Wang [2] pointed out that sadness was not easy to identify safety signs, and fear can increase exit signs; Yin [3] pointed out that the stimulus clues with auxiliary words have a regulating effect on return inhibition; Tian [4] found that the psychological inertia, the location of safety signs and auxiliary texts have significant effects on their effectiveness. Jiang [5] pointed out that the safety signs should be placed on the same side of the mine entrance and exit. In addition, Zhang [6] studied on visual attention characteristics of safety logo same shape and color. Cheng [7] applied the smart chip and wireless sensor network

Springer Nature Switzerland AG 2020
Ahram (Ed.): AHFE 2019, AISC 973, pp. 401–408, 2020.
https://doi.org/10.1007/978-3-030-20476-1_40

technology to the research and development of the new intelligent fire emergency light indicating system.

In China's standard, "Fire Emergency Lighting Knowledge System" GB17945-2010 [8], although the corresponding specifications are proposed for the emergency sign, it does not indicate the setting height and the place in the fixed area clearly. In order to accurately obtain the attention of the escape personnel to the different positions and height in the emergency state and the response time of the extracted information, virtual reality technology was used in this experiment to put the subjects into a realistic. In the escape situation, the simulation of the emergency escape scene is realized, and the escape sign is set. The eye movement data of the subject can be obtained by the eye tracking equipment.

2   Experiment

2.1   Experimental Equipment

EyeLink II eye tracker was used to monitor eye movement data. The highest sampling frequency of the device was 500 Hz, the noise limit was < 0.01° RMS, and the average gaze position error was <0.5°. Eight active optical motion capture cameras were set up to capture body position information. For positioning, each motion capture device captures a sector-shaped area with an angle of 120 degrees and an effective capture distance of 20 ms. The specific placement position is shown in Fig. 1.

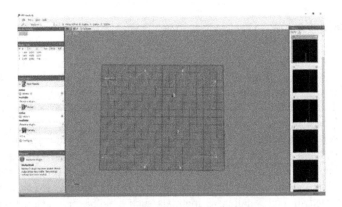

Fig. 1. Capture instrument placement

2.2   Subjects

In order to study the influence of height and position to signs through the eye movement parameters of the subjects, the target stimulation experiment was carried on 35 subjects, 18–25 years old undergraduate university students, normal visual acuity, no color blindness, participated the experiment. 5 failed to escape. Therefore 30 subjects, 15 males and 15 females, with an average height of 1.70 ms, were effective. Subjects had not had any relevant eye movement experiments before this experiment

2.3   Stimulus Materials

Taking "Safety signs and guideline for the use" (GB2894-2008) [9] as a reference, the size of the exit sign is fixed at $359 \times 149 \times 23$ mm, which is used to analyze the position and height effects on the subject in reading efficiency.

2.4   Scene Settings

A space (20 meters long, 15 meters wide, 5 meters high) in a room was selected as the experiment place, and 8 active optical motion capture cameras were set up in the place to capture body position information for positioning. The scene of the virtual reality simulation was a corridor with two return-shaped layouts. The designed scene was 20 meters long and 11 meters wide, and the corridor is 2 meters wide. 12 exit signs were set as shown in Table 1 and Fig. 2. A smoke scene was set at the upper corner as shown in Fig. 2(b).

Table 1.   Scene logo design

Position	Height from ground/m	Remarks
A	0.5, 1, 1.5, 2	Exit sign
B	0.5, 1.5	Exit sign
C	1, 2	Exit sign
D	0.5, 1.5	Exit sign
E	1, 2	Exit sign
F	2.3	Smoke scene

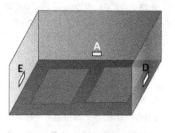 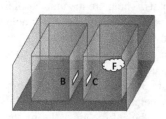

(a)   at A, E and D   (b) at B, C and F

(c) at A, height is 0.5 m   (d) at B and C, height is1.5 m   (e) at D and E, height is 0.5 m

Fig. 2.   Exit Sign location diagrams

2.5   Procedure

Participants entered the experimental scene to identify the escape signs of different orientations and heights and could be disturbed by smoke and noisy sound. For the subjects, the right directions should be chosen to escape. A circular scene was used in the experiment to identify three markers of the same height but at different positions. After correct identification and action, the subject would find an emergency door finally and escape from the scene. The specific experimental task steps are shown in Table 2.

Table 2. Escape program

No.	Position of subject	Height of sign	Sign information
1	A	0.5, 1.5	To the right
		1.0, 2.0	To the left
2	D	0.5, 1.5	To the right
3	F	Smoke stimulation	
4	B	0.5, 1.5	To the right
5	E	1.0, 1.5	To the left
6	C	1.0, 2.0	To the left
7		Success! End of escape!	

3   Data Analysis

3.1   Number of Gaze Points at Different Heights and Positions

During the experiment, the subject watched the stimulating material and formed fixation points. The more the number of fixation points, the more attractive the sign to the subject or the more difficult the stimulating material to read [10]. In this experiment, the exit sign was single-form and easy to recognize and all the subjects were familiar with it. Therefore, the more gaze points, the more gaze content that attracts the attention of the subject.

Figure 3 shows the average number of fixation points when reading exit signs at different heights and positions. Since the exit signs used are consistent in the complexity of the information, the reason for this difference should be the interest and attention of the subjects to the different position signs. It can be seen from the three positions, in front, lateral anterior and corner, the average number of gaze points of the sign directly in front of the subject is 6.57, followed by 4.82 at the corner, and finally is 4.60 at the lateral anterior sign. It indicates that the sign in front receives the highest attention of the subject because the front sign can be exposed in a short time. In addition, the focus of eyes is mostly in front of the escape road, which is easy to read. As shown in Fig. 3, the gaze point formed at the height of 1.0 m is the most, totaling 17.5. Therefore, the attention to the height of 1.0 m is also the most.

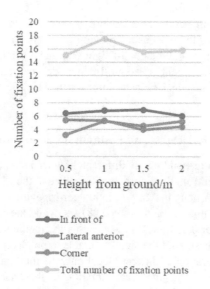

Fig. 3. Fixation points for signs in different positions and height

The data was processed with spss 21.0. The results are shown in Table 3. The F alue of the position factor is equal to 7.60, which is greater than the preset Fa value (a 0.05, Fa = 3.88), and the P value is equal to 0.02, which is less than the preset a alue. It indicates that there are significant differences between the average fixation ɔints and the different locations of emergency sign. The position of the sign has a gnificant influence on the recognition efficiency of the subject to identify the escape gn. Therefore, the escape sign placed the front of the intersection, which means at the ɪd of the corridor, can improve the chance of escape and enhance the possibility to ɪd the escape information.

Table 3. Multivariate analysis of variance results

Source of difference	SS	df	MS	F	P-value	F crit
Position	9.33	2.00	–	7.60	0.02	5.14
Height	1.11	3.00	0.37	0.60	0.64	4.76
Error	3.68	6.00	0.61	–	–	–
Total	14.12	11.00	–	–	–	–

The F value of the height factor is equal to 0.60, which is less than the preset Fa lue, and the P value is equal to 0.64, which is greater than the preset a value. It licates that the height of escape signal has no significant effect on the recognition iciency in a state of emergency. Although the data analysis does not support the height has a significant impact on the number of gaze points, the number of gaze points at a height of 1.0 meter is the most as the previous analysis. In the process of escaping people's vision will shrink, and the line of sight will move habitually. Therefore, when the sign is placed in the height of about 1.0 meter, it can attract the attention and interest of the participants, which helps to improve the identification to the information of escape sign.

3.2 Effects of Different Heights and Positions on the Duration of Fixation

The gaze durations corresponding to different heights and positions are shown in Fig. 4. As it can be seen from the figure, the difference in the gaze duration at different positions is very little. Among the gaze duration corresponding to different heights, the 1.5 m high sign has total gaze duration of 0.2814 s, while the 1.0 m high sign has the lowest total gaze time of 0.2240 s. It indicates that the reading efficiency at the height of 1.0 meters is the highest. And the sign is suitable to be placed near the height of 1.0 meters. For the gaze duration at different locations, the total gaze duration in front of the subject was 0.3086 s, the total gaze duration in the lateral front direction was 0.3115 s, and the total gaze duration at the corner was 0.3537 s. It indicates that the sign at the corner is not conducive to the identification of escape information because the reading efficiency is relatively low. Therefore, the sign should be placed directly in front of the line of sight of the subject meanwhile avoiding being placed in the lateral anterior.

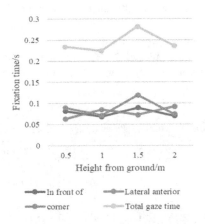

Fig. 4. Fixation time for signs in different positions and heights

In the test of covariance equivalence, the significance level of the data test is greater than 0.05, which means there is no significant difference. After two-way variance test as shown in Table 4, it is found that the F values affected by the position and height factors are equal to 0.26 and 0.53, respectively. Both are smaller than the Fa value (a = 0.05, Fa = 3.88), and the P values are equal to 0.76 and 0.67, respectively, which are greater than the preset a value. It indicates that the positions and heights have significant difference to the gaze time.

Table 4. Multivariate analysis of variance results

Source of difference	SS	df	MS	F	P-value	F crit
Position	–	2.00	–	0.26	0.76	5.14
Height	–	3.00	–	0.53	0.67	4.76
Error	–	6.00	–	–	–	–
Total	–	11.00	–	–	–	–

In this experiment, the size of the sign is small, and the information is easy to be recognized, therefore the gaze time is extremely short, and the deviation of the data has a great influence on the result. It may be the reason why the height and position have no significant influence on gaze duration. In the data analysis of Fig. 4, it is found that the subject has the shortest gaze duration and the corresponding reading efficiency at 1.0 m, which is consistent with the conclusion that the placement position and height are studied by the number of gaze points.

4 Conclusions and Prospects

Eye tracking technology and VR technology were used in studying the attention to the exit sign. Two stimulus variables, the position and the height of the exit mark, were analyzed. The conclusions are as follows:

(1) The position of the exit sign has a significant influence on the number of gaze points, but the height has no significant influence on the number of gaze points.

(2) The position of the exit sign and the height of the setting have no significant influence on the gaze time of the subject.

(3) The subjects have a shorter gaze duration which means the highest reading efficiency at the height of 1.0 m in the front, so the exit signs should be placed at a position 1.0 m high at the end of aisle.

The shortcomings of this paper are that the effect of size and brightness of the exit sign on reading efficiency has not studied and the subjects are all undergraduate students. It will be further developed in the follow-up work.

Acknowledgements. The presented work has been supported by General project of China University of Labor Relations (19YYJS003) and Teaching Reform Project (zyjs201804).

References

1. Lai, C.J.: Effects of color scheme and message lines of variable message signs on driver performance. Accid. Anal. Prev. **42**, 1003–1008 (2010)
2. Wang, C.X., Lv, S.R: Research on effect of emotions on recognition of safety sign. China Saf. Sci. J. **26**, 15–20 (2016)
3. Yin, Z.Y., Shi, F.Q., Li, N.W., Niu, L.X.: Effects of safety cues on inhibition of attention return of miners different in degree of burnout. China Saf. Sci. J. **09**, 37–42 (2017)

4. Tian, S.C., Chen, Y., Zou, Y., Li, G.L.: Analysis of factors influencing effectiveness of mine safety signs. China Saf. Sci. J. **26**, 146–150 (2016)
5. Jiang, W., Shi, S.H., Liu, Y.N.: Studies on exit sign design of the mine entrances. Appl. Mech. Mater. **543–547**, 4626–4629 (2014)
6. Zhang, K.: Research on Visual Attention Characteristics of Safety Logo Frame Shape and Color. Henan Polytechnic University, Jiaozuo (2014)
7. Cheng, K.W.: Research on Key Technologies of Intelligent Wireless Fire Emergency Light Indicator System. Hangzhou University of Electronic Science and Technology, Hangzhou (2013)
8. GB 17945-2010. Fire emergency and evacuate indicating system
9. GB 2894-2008. Safety signs and guideline for the use
10. Sun, G.L., Li, Q., Fu, P.W., Ran, L.H.: Analysis and optimization of information coding for automobile dashboard based on human factors engineering. China Saf. Sci. J. **28**, 68–74 (2018)

Healthcare Design Based on Virtual Reality Technology: Enhance the User's Engagement in the Rehabilitation Process

Yu Qiao[1(⊠)] and Siyi Yang[2]

[1] Department of Industry Design, Academy of Arts and Architecture,
Central South University, Changsha, China
77252501@qq.com
[2] School of Public Administration, Central South University, Changsha, China
17655460@qq.com

Abstract. This study on VR technology in healthcare design aims to enhance the engagement and provide an interactive way for the whole rehabilitation process. By presenting the real-time data to the users and collecting their feedback, users' participation and achievements increase. From the study and analysis on the existing postoperative knee rehabilitation training, we establish typical user profiles, analyze their pain points and specific needs, test it by employing the users' data we collected, build the data modeling and then propose corresponding solutions. A working prototype of the device is used to test and illustrate the device's availability in the generation and calculation of motion data. Based on this, in the research, the further application of VR in the field of medical rehabilitation is promising, and a variety of possibilities concerning the combination of VR as a technology and a kind of future life style from the essence of things are put forward.

Keywords: Rehabilitation · Training · Knee joint · Healthcare ·
Data visualization

Introduction

Virtual reality technology has become a tool for assisting the balance of exercise learning and gait rehabilitation. The conventional rehabilitation training methods are limited by the technical level of medical personnel. Single training methods, the boring training process, lack of personalization and other problems need to be solved urgently. According to human brain plasticity theory [1], mirror neuron theory [2] and continuous passive training theory [3], a rich training environment can promote the process of rehabilitation [4]. Studies have shown that the rehabilitation effect will be greatly improved if we can provide various forms of information feedback from the course of training, give patients' subjective initiative into full play, and give hints or suggestions in line with the patient's state, etc. [5] In recent years, virtual reality technology has been developed and perfected, so that traditional rehabilitation training has broken the limitation of environment and space, which has made the process and the mode of training more diversified and made it an important treatment tool to assist sports

Springer Nature Switzerland AG 2020
Ahram (Ed.): AHFE 2019, AISC 973, pp. 409–415, 2020.
https://doi.org/10.1007/978-3-030-20476-1_41

learning balance and gait rehabilitation. How to further verify and strengthen the virtual reality technology in the field of medical rehabilitation is a major research hotspot at present.

Based on the analysis of the existing problems and simulation solutions to the current methods of joint rehabilitation, this paper expounds the development, the trend, the concept and new characteristics of virtual reality technology, and clarifies the design methods of joint rehabilitation treatment equipment based on virtual reality technology. In this paper, the possibility of making an equipment working prototype of Arduino MCU and the processing programming language as the basic tools is put forward.

In the comparative studies of MIT and Rutgers University, it was shown that the virtual reality rehabilitation cases had better effect than the basic therapy when the rehabilitation therapy of the game was added [6]. The patient's main inner drive allowed them to get very good feedback on the patient's performance when repeating the test.

2 Methods

The output equipment will be entertaining and visualized in the process of rehabilitation treatment, which gives users a positive, perceptible, quantifiable and interactive experience in the process of rehabilitation treatment in an innovative way. At the same time, the doctor can improve the patient's rehabilitation treatment experience by designing the remote data monitoring in the system, quantifying the rehabilitation effect and the process, and providing relevant treatment suggestions tailored to the patient's need. In turn, the cost of outpatient rehabilitation treatment in terms of the personnel equipment and venues should be optimized; the rehabilitation medical experience needs to be enhanced; scientific guidance of the rehabilitation process should be provided. Thus, the rehabilitation process could be accelerated to ease the tension between doctors and patients.

2.1 Analysis

Based on the previous theory of research and analysis, in this part as the discovery line to the design, the author summed up and organized its cocoon and analyzed 10 key words on the needs of patients with joint diseases. From the angle of doctor-patient relationship, the author adopted user interviews, brainstorming, user portraits and other design methods. In terms of the priority, we divided them into 4 levels (Table 1).

Analysis on the shape and scale design of knee rehabilitation equipment: by collecting information of 69 patients, according to the human engineering scale measurement based on the knee joint and a series of shape design, we defined the shape scale of the rehabilitation equipment as follows (Fig. 1).

Table 1. Analysis of the needs from patients with knee joint disease

Priority	Keywords
1	Safe, comfortable Treatment quantification Complete medical records
2	Professional guidance Instrument autonomy Remote data monitoring
3	Feedback Remote data monitoring
4	Personalization

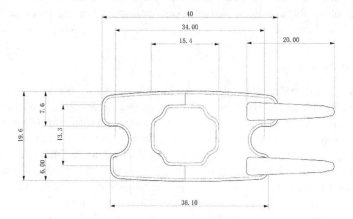

Fig. 1. Diagram of the size of knee rehabilitation equipment

2   Prototype Design

Motion data collection based on the MPU-6050 sensor: Gait, as a biological characteristic, is an individual behavior characteristic formed by the acquired living habits of
human beings. It has the unique advantages of not being affected by distance, non-
aggression, difficult camouflage and small environmental impact[1], and has attracted
much attention in the field of biometric identification. Patients with knee problems will
be different from disease-free persons in the matter of gait, and the collection and
analysis of gait data can be compared with its walking characteristics.

The MPU-6050 six-axis motion sensor is small in size, low in cost, easy to
experiment, and has strong anti-jamming. Besides, it can accurately test the acceleration data and tilt angle data of the knee joint during the movement. After connecting the
MPU-6050 sensor with the Arduino microcontroller, the program is written to define its

Gulihua, Choi Chang, Gao Song, Xu. Gait signal acquisition system based on MPU-6050 [J]. Journal
of Shenyang University of Technology, 2015, 37 (02): 176–182.

six-axis data and set it in a serial plotter to present it in the form of a line diagram (Fig. 2).

Based on the working principle of the Arduino single chip microcomputer, an MPU-6050 six-axis sensor, two HC06 Bluetooth sensors and a serial communication module are needed. The four pins on the MPU-6050 sensor, GND, SCL and SDA are connected to the 5 V, GND, A5 and A4 of the motherboard, respectively. And the yellow indicator light of the MPU-6050 sensor is on after power on, that is, the module can be used normally. At the same time, it is necessary to define two HC06 Bluetooth modules, one of which is sent as Bluetooth through the "at" instruction. Four pins of Rxd, TXD, VCC, and GND are connected with 11, 10, ICSP and GND of the motherboard, respectively. And the blue indicator of the sensor remains lit after power up, indicating that the module is running normally. Another piece is used to receive Bluetooth, and the serial module is connected to the computer USB interface. Two Bluetooth blue indicator lights are often lit, indicating that data can be sent and received (Fig. 3).

2.3 Physical Design

The design of knee rehabilitation treatment equipment shown in this paper is inspire by the application of the wearable lower extremity exoskeleton rehabilitation robot. According to our experience, wearable exoskeleton can effectively help patient complete exercise training, but there are also high costs, large volumes and other disadvantages. Through the research and analysis based on the theory of knee joint rehabilitation and the basic theory of human engineering, the products of knee joint rehabilitation equipment with small size, light weight and low costs are developed (Fig. 4). The surface material of the equipment is hydrophilic and breathable. It has good elongation shrinkage; can adapt to the buckling of the knee joint during the movement; plays the role of the protection buffer through the varicose material. Since its working prototype is based on the Arduino open source platform, it has good flexibility and development prospects.

3 Prototype

Calculation and Treatment of Data Generation of Knee Rehabilitation Equipment: base on the processing of motion data as well as the connection of the Arduino serial port and the processing serial port, the motion data of the hardware part can be transferred to the processing window. The time of motion is represented by the horizontal axis, and the vertical axis denotes the motion value. The three points of X, Y and Z correspond to the three-axis data of the MPU-6050 sensor module, respectively, which is shown by the motion trajectory of the point (Fig. 5).

To save data storage space, this article sets the motion time period to 90 s. When you click the start key in the lower-left corner, you can start testing; when the horizontal data range is exceeded, the display of the data stops.

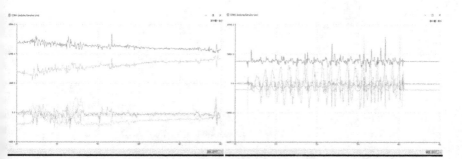

Fig. 2. Serial plotter data diagram

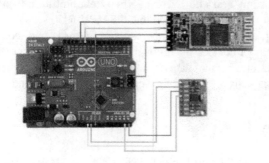

Fig. 3. Arduino circuit design diagram

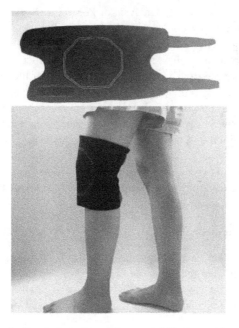

Fig. 4. Physical product of knee rehabilitation equipment

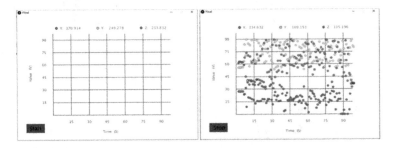

Fig. 5. The generated data graph from the processing window

The Data Generation Test and Error of Knee Rehabilitation Equipment: Through a number of experimental tests, we find that the MPU-6050 sensor has good durability and anti-jamming, and the current known cause of error is poor contact.

When there is poor contact, the following conditions may occur (Fig. 6) because of the inability to perform normal data detection and the graphics generated by the static movement of single-axis data along the vertical axis of the maximum data within 90 s

4 Conclusion

By visually processing the effect of data generation, immersive visual feedback based on data generation results will be displayed in the virtual reality system when the device is worn, and the rehabilitation exercise training is carried out. Combining with the theoretical research and design analysis of knee rehabilitation training, knee joint rehabilitation equipment in accordance with virtual reality technology is devised. The device can gather the user's knee gait data and generate more intuitive data graphic through the cross processing of Arduino MCU and the processing programming language, which is of great significance for the visualization of the rehabilitation process

(a) Based on the theory of knee joint rehabilitation and the research and analysis of patients with knee joint disease, auxiliary knee joint rehabilitation training equipment was designed, and a working prototype was made to realize the data collection and analysis of the knee joint gait when combined with the basic theories of rehabilitation medicine and ergonomics.

(b) Data visualization processing is based on the process of programming software. The gait data collected based on the working prototype is sent to the device via the Bluetooth module for processing and data visualization.

Due to the limited time, there is still much to be done in depth and refinement:

(a) For the movement data of the knee joint, we should not only collect gait data but also take the knee joint rehabilitation training database as reference. Moreover, we should take into account the reception and stiffness of rehabilitation equipment and knee contact and introduce the relevant algorithm to carry out more in-depth data analysis.

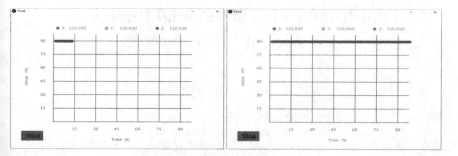

Fig. 6. Processing window data error diagram

(b) The product can be extended to a lightweight, modular, wearable exoskeleton device which integrates devices and virtual reality technologies more organically into one.

Acknowledgement. Thanks to the support from National Social Science Fund of Art Project, Approval number 17CG214, 2017, the project I'm in charge of Research on medical product design based on virtual reality technology

References

1. Xiaohui, Z., Fei, C., Linsheng, W.: Effect of continuous passive exercise on four head muscle tension and lower limb motor function in hemiplegia patients with stroke. Contemp. Chin. Med. **21**(31), 34–36 (2014)
2. Xian, L., Bing, X.: Enriching the environment and rehabilitation of stroke. Theory Pract. Rehabil. China **18**(1), 47–52 (2012)
3. Wang, F., Liu, H.: Application of Nursing rehabilitation guidance training combined with continuous passive exercise in postoperative knee fracture. Lab. Med. Clin. **12**(13), 1866–1870 (2015)
4. Zhang, G.: Effects of rich environments on the plasticity of the central nervous system. Chin. J. Rehabil. Med. **21**(3), 280–283 (2006)
5. Xu, J.X.: Development of a progressive buckling trainer for knee rehabilitation [J]. Med. Health Equip. **37**(3), 45–47 (2016)
6. Ching, S.J., Hehanwu, et al.: Evolution, development and prospect of virtual reality technology. J. Syst. Simul. **16**(9), 1905–1909 (2004)
7. Cail, C.J.S., Yan, J.H., et al.: Brain plasticity and motor practice in cognitive aging. Front. Aging Neurosci. **6**(2), 167–189 (2014)
8. Antonino, C.: Mirror neurons(and beyond)in the macaque brain: an overview of 20 years of research. Neurosci. Lett. **540**(6), 3–14 (2013)

An Investigation of Material Perception in Virtual Environments

Mutian Niu[✉] and Cheng-Hung Lo

Xi'an Jiaotong-Liverpool University, 111 Ren'ai Road,
Suzhou Dushu Lake Science and Education Innovation District,
Suzhou Industrial Park, Suzhou 215123, People's Republic of China
{Mutian.Niu, CH.Lo}@xjtlu.edu.cn

Abstract. Material representation has always been an important part of visual effects in industrial design. And the judgment and recognition of product material often remain on the rendering effect drawings of the 2D display. However, it cannot fully intuitive performed, even sometimes cannot identify the specific material composition. As a device to simulate the real environment, VR strengthens people's immersive experience by its 3D sense of space. The purpose of this study is to explore whether the material perception in VR is different from that in traditional 2D mode, and to determine whether VR can be used as a tool for users' material perception in the future. The study found that VR provides the users with stereoscopic visual effects not seen on a 2D display. This feature seems to deepen the perception of material, which may facilitate the design of industrial products, furniture design, automotive interior and so on.

Keywords: Material perception · Virtual reality · CAD modelling · Rendering process

1 Introduction

Industrial design mainly solves the problems of product shape, color and material, and the ultimate goal of the design is to meet the needs of consumers. When people evaluate the design visualiser of industrial design, the evaluator can only perceive the material on the surface of product design sketch through the eyes [1]. Because of the manufacturing process and other reasons, there are great differences between the surface material and the material displayed in the effect drawing of the actual product [2]. The process of product design has already changed from product-centered to user-centered. Designers begin to attach importance to the users' experience and feelings of product design. Target users as the center of product design evaluation have been adopted by more and more design industries. User experience has become an important way to help product improvement. Vision is an important way for human beings to understand the world. The objective material world acts on people's eyes and forms information through the brain, thus making people feel and understand. In our daily life, people will face or contact various materials, such as plastic, metal, wood, marble, water, jam, cotton and so on. These materials have different physical and optical properties, which determine how we interact with them (or avoid interacting with them,

T. Ahram (Ed.): AHFE 2019, AISC 973, pp. 416–426, 2020.
https://doi.org/10.1007/978-3-030-20476-1_42

3]. Material perception helps us to decide which interaction method is more appro-priate before we contacting the objects. Ashby [4] stated that we recognize certain materials to form cognitive perception by touching them, such as the temperature and texture of the surface through contact [4]. Material perception is a kind of feeling and impression of material based on the visual perception of material surface characteris-tics, such as texture and smoothness. At the same time, material perception is also produced by the brain's comprehensively processing of the surface characteristics of materials perceived through vision, such as texture and smoothness. In fact, realistic image synthesis depends not only on illumination, but also on accurate simulation of the virtual scene, and the challenging task is to assign appropriate material descriptions to each object in the scene [5].

Vision is a channel for human beings to acquire information for cognition and thinking. Every object in life is made of some kind of materials, and we usually know what it is by observation. Nature provides us with a neural architecture that recognizes the basic elements of certain materials in images even without training [6]. Early work on visual perception focused on the physiological and neurological characteristics of the human visual system, such as contrast and color. For example, in the study of different aspects of color constancy, Brainard [7] found that the color and brightness of objects remained distinctly unchanged under substantial changes in light [5, 7]. Obein et al. [8] proved the similar invariance of perceived glossiness under varying illumi-nation, that is, glossiness invariance [8]. Xiao and Brainard (2006) showed that the appearance of color is indeed slightly affected by gloss [9]. Ngan et al. [10] proposed that L2 metric on spherical images rendered with different BRDF under the light as an alternative model for image gloss and sharpness perception. It can be proved from the perceptual point of view that people perceive materials by observing objects made of the materials, rather than consciously considering the psychological model of abstract reflectance function [10]. Fleming et al. [11], for example, has found that humans can recognize irregular transparent objects, such as ice, even though it is difficult to measure with machines and formulas because of the varying refractive index. He chose thick transparent objects as research subjects to explore how humans evaluate trans-parent materials. He thought that human beings were evaluated by observing distorted images of refracted materials. [11] Previous studies have been carried out in many literatures. They believe that human vision depends on a series of images related to material properties [12–16].

People's visual system is very sensitive to the appearance of materials, and the imperfect approximation cannot meet the user's requirements [17]. Virtual reality technology is a new cognitive tool which emphasizes perception ability on the basis of human feeling. Recently, virtual reality technology has been applied more and more widely, such as in psychological research. Because VR can create more ecologically stimulus programs and reflection scheme than traditional experimental devices, and can help experimenters better control the environment [18]. At the same time, the immersion and visual fidelity provided by virtual reality can stimulate some psychological reactions the testers. Immersion is considered as an important performance measure of virtual systems, and the reason for immersion is that users have a sense of existence or hal-lucination similar to real objects in the virtual environment. Virtual objects are not only similar to the real objects, but also more realistic than the real objects in order to achieve immersive effect [19]. Designers demand immersion because the higher fidelity of the virtual world causes wider responses from participants. In fact, immersive environments seem to be easier for participants to remember the surroundings [20].

One of the main features of VR system is the introduction of stereo depth, which gives users an illusion that they can see objects in virtual space [21]. Virtual reality provides near-real visual effects, which may be a good tool for material perception and the experience of designing effect maps. Vangorp et al. [5] and Bonneel et al. [22] have taken advantage of the changes in user's perception of vision and auditory under different conditions of rendering quality in virtual environments. The real and high quality rendering effect of these materials is the core element of the overall reality and immersion provided by the virtual environments [5, 22]. Wilson and Alessandro [18] have discussed the ability of human visual recognition and perception of space and motion in an immersive virtual environment provided by virtual reality. And they applied VR as a visual medium in psychological research, detecting the possibility of complex participants' reactions and behaviors in the virtual environment. It is found that VR can present visual stimuli along 3D planes, which is more conducive to stimulate participants' behavior than traditional experimental schemes [17]. Although there have been some explorations in the field of visual perception of VR, most of them focus on the user's virtual experience, and few studies focus on material perception in the virtual environment.

In this paper, we present a comparative experimental study conducted to analyse the visual effect of virtual reality compared to the feeling of 2D traditional display for the material perception process of geometrical 3D models. It is hypothesized that the two display modes would bring different viewing experiences to the users, henceforth resulting different perceptions of the same materials. This paper is organized as follow Sect. 2 presents the object of our comparative experiment and the process of rendering materials using computer-aided design and game engine. We illustrate the comparison process and specific steps between VR and traditional 2D display mode. We also record the experimental results of the two display modes based on participants' feedback. The resulting data set is analyzed using a wide range of statistical techniques in Sect. 3. We summarize and propose future works in Sect. 4.

2 Methods

The purpose of this study is to test whether users feel differently when they use different devices (computer and VR) to observe the object with the same material. Therefore, the experimental process is mainly to use software to render different materials, recording the user's feelings after using the devices to observe the material separately.

2.1 Selection of Experimental Objects and Materials

As far as the choice of experimental target is concerned, the material which is greatly changed by environment or illumination should be avoided as much as possible. Identification of a material belongs to a certain category can be based on the properties and parameters of the material contained in the relevant information [23]. Berzhanskaya et al. [24] have experimentally proved that the spatial distribution of surface gloss perception is inconsistent, and it is affected by specular reflections [24]. For translucent materials like ceramics, the information such as specular highlights, rendering and background environment has a great influence on the estimation of glossiness. Although BRDF tries to separate reflectivity and material-related information, this technique does not take into account the influence factors of texture and geometric shape. The same reflectivity properties can be observed on the surfaces of different materials [25]. In addition, transparent materials cannot be generalized by a single reflection function [26, 27]. Due to the complex optical properties of translucent or transparent materials, we include only opaque materials in this study, and the tested object in the experimental stage is a kettle that consists of stainless steel and plastic materials. In order to ensure the objectivity and scientific rigor of the follow-up rendering task, we carried out the following rendering and testing process according to a picture of a kettle (as the Fig. 1 shown), so as to avoid the increase of rendering difficulty and visual viewing error caused by the influence of illumination and environment on the real objects.

Fig. 1. The physical picture of the kettle as the test object.

2   Material Rendering Process of Computer and Virtual Reality

The focus of this study is on the difference of material perception under different devices, so the rendering process is mainly to create the geometry to represent the object and give the corresponding material for the subsequent contrast stage. In the field of industrial design, spheres are often used as the basic models to visualize material parameters. The sphere presents all possible surface directions to the observer. Its convexity eliminates the need for its own shadow and mutual reflection. But Fangorp et al. [5] indicated that the geometric shape of the object would affect the material perception, and that the sphere was not necessarily the simplest shape to distinguish, which depends mainly on the type of material or the shape similar to the target [5]. Therefore, in order to avoid the influence of object shape on material perception, we choose to create two geometries, cubes and spheres, to be the models to test the effect of plane and surface on material perception.

3Ds max is used as the software for 2D display that builds basic geometric models according to requirements, and then using light and camera to adjust the visual effects of each perspective. For rendering images, we used the VRay plug-in to make the materials of our 3D model objects become more real. Because VRay allows us to adjust the lighting and materials properties on the models [28]. According to the picture of kettle by the previous statement and was used to adjust the parameters of material attributes in VRay in order to simulate the material appearances. Afterwards, the adjusted materials were assigned to the cube and sphere, and 4 pictures were rendered for participants as a test image of the computer (as shown in Fig. 2).

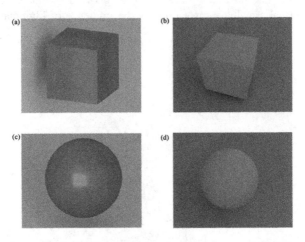

Fig. 2. The test image of 2D display: (a) cube made of stainless steel, (b) cube made of plastic, (c) sphere made of stainless steel, (d) sphere made of plastic.

For the test image of the VR device, we use HTC VIVE as the VR display. Relying on Unreal Engine4 (UE4) to create geometry and rendering tasks. UE4 contains rendering code and design tools that can be used to build 3D models. The Unreal Engine source code can help designers to simulate whole new scenes, whether indoors or outdoors, and visualize 3D scenes by the perspective or stereo view of Unreal Engine [29]. Like 3D max, we used UE4 to build 3D models of cubes and spheres, and use its own renderer to adjust stainless steel and plastic materials depends on the picture of kettle. In order to minimize the error of material presentation of two different renderers in the contrast process, the material attributes and lighting positions in UE4 are the same as the details of the VRay adjustment, and then four projects are generated (shown in Fig. 3).

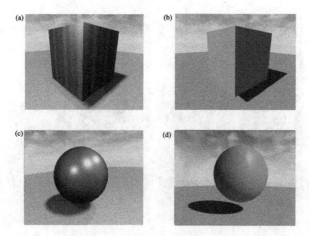

Fig. 3. The test image of Virtual Reality: (a) cube made of stainless steel, (b) cube made of plastic, (c) sphere made of stainless steel, (d) sphere made of plastic.

2.3 Experimental Procedures

The participants were selected primarily because of their knowledge in product design and the relevant field. We recruited 30 participants aged 19–34 (18 males and 12 females) for the test, which took place in the Xi'an Jiaotong-Liverpool University. They were students and researchers from the industrial design and architecture education backgrounds and only 9 of them had VR experience.

Introduced the experiment process to the participants and show them the picture of the kettle (Fig. 1). Ask them to observe and remember the visual effects of the material in the picture. Because the study wants to distinguish the influence of geometric shape on material quality, this experiment was divided into two parts: one was to test the effect of the cube material performance displayed in VR and 2D, the other is about the sphere material perception test. In the test of cube model, we first asked participants to compare VR and 2D display drawings of stainless-steel material, and they were asked to record how close they thought they were to the actual stainless steel material (on the kettle's picture). As measured on the 10-point scale, where 1 represented little similar, and 10 represented much similar. After that, the participants also had to compare the perception of plastic cubes in VR and computer viewing, and still assess the similarity between the material content presented by the two devices and the picture of the kettle. The testing process is shown in the following Fig. 4. After the first test was completed, we conducted the second part of the sphere test. The steps of the two experiments are the same, only the differences from the models. We still requested participants to evaluate the similarity of plastic materials based on the 10-point scale.

After the experiment, the researcher addressed participants with questions about the experience of observing materials on different devices. These questions aim to deepen our understanding of the participants' perception, determine and their experience during viewing.

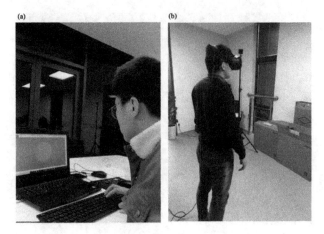

Fig. 4. The testing process of the participant: (a) observe the material on the 2d display image, (b) observe the material on the VR device.

3 Results

As we have pointed out, the main objective of this study is to explore whether users will have different experiences in material perception by VR and 2D display. Therefore, it is necessary to determine whether and to what extent the images presented by the two devices are close to the actual material, and the analysis of the test objectives also focuses on this. Factorial Analysis of Variance (ANOVA) is used to analyze the test data of two kinds of equipment and materials. The factorial ANOVA determine whether the individual and interaction of each factor have a statistically significant impact on changing user perception of material [30]. The experimental data do not take into account the gender impact.

3.1 Cube

Table 1 shows the results of participants' material perception on the cube model. The table heading of the rightmost column is the statistical significance, we used the P value to represent it. If $p < 0.05$, significantly changes in the corresponding factor affects the participants' perception of material. So, as shown in Table 1, the P value of the device is 0.000, which means that it had a significant impact on the participants' material perception.

Figure 5 displays the mean plots of the material perception on the cube model. Device 1 represent VR display and Device 2 represents 2D display. And the Material 1 and 2 are stainless steel and plastic. It can be emphasized that participants feel the content presented by VR is closer to the real material on the picture than that presented by the traditional 2D display.

Table 1. Factorial ANOVA results of the cube.

Source	Type III sum of squares	Df	Mean square	F	Sig.
Corrected model	12.617[a]	3	4.206	4.828	0.003
Intercept	4563.333	1	4563.333	5238.463	0.000
Material	1.200	1	1.200	1.378	0.243
Device	11.408	1	11.408	13.096	0.000
Material* device	0.008	1	0.008	0.010	0.922
Error	101.050	116	0.871		
Total	4677.000	120			
Corrected total	113.667	119			

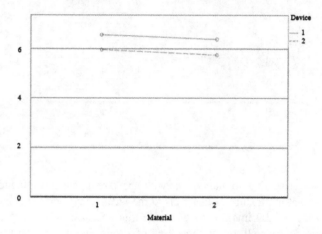

Fig. 5. Mean plots of the cube testing.

Sphere

Table 2, it shows that the variations of the device significantly affect participants' ling of the material for sphere model, because the P value is 0.001.

Table 2. Factorial ANOVA results of the sphere.

Source	Type III sum of squares	Df	Mean square	F	Sig.
Corrected model	11.106[a]	3	3.702	3.738	0/113
Intercept	5286.769	1	5286.769	5338.543	0.000
Material	0.252	1	0.252	0.255	0.615
Device	10.502	1	10.502	10.605	0.001
Material* device	0.352	1	0.352	0.356	0.552
Error	114.875	116	0.990		
Total	5412.750	120			
Corrected total	125.981	119			

The mean plots of the effect of material perception on sphere model is shown Fig. 6. It can be seen that the participants chose VR more than computers according content similarity. And according to the mean plots of the data, we can find that t material performance of the sphere is more helpful than that of the cube in the proce of user identification.

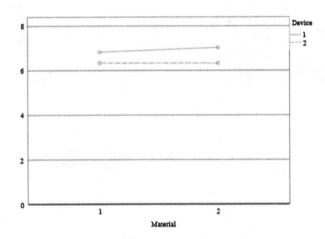

Fig. 6. Mean plots of the sphere testing.

In addition, according to interviews with the participants, we found that mc people believe that the texture of materials in the virtual environment is more obvio and real than that of 2D images. At the same time, because of the stereoscopic a spiritual effects, the overall visual perception of VR is also better than the tradition model. But some participants stated that even though the visual effect VR was mc realistic, it was not as delicate as the 2D image.

4 Conclusion

Visual material perception is one of the main methods for people to understand t objects, including physical and optical properties of the objects. Influenced by il mination and shape, people also face some challenges in the process of identifyi materials, especially in the design of today's 3D models, which is very common. Ma studies have confirmed that VR can enhance user experience and interaction in virt environments. As a tool of the immersive experience, VR has been widely used in t field of design, making use of its virtual visual effect to create a realistic environme We studied the difference between material perception effect in the virtual environme and the traditional 2D view. This paper attempts to explore whether VR can be used a tool for designers and users to perceive material by its immersion and stereo sen Based on the material of testing object picture, we used 3D Max and Unreal Engine manipulate and render the contents for computer and VR display respectively. In orc to ensure the rendering effect, the material parameters are treated the same. Participa were invited to observe the material with two devices, compared with the previous picture, and scored according to the similarity. The experiment found that the perception of plastics and stainless steel in the material test of cube or sphere was more obvious and direct than that in the 2D display. Some participants said that the material texture observed in VR was clearer, which helped them to identify material types more quickly. In future research, we will increase the number of participants and improve the experimental data and methods. At the same time, the experiment will involve more material comparison, such as rubber, paper, wood, leather, or the same type of material comparison, such as copper, iron, alloy and so on, to explore whether VR has a better visual effect on a certain type of material.

References

1. Papagiannidis, S., See-To, E., Bourlakis, M.: Virtual test-driving: the impact of simulated products on purchase intention. J. Eta. Cons Serv. **21**(5), 877–887 (2014)
2. Bangbei, T., Gang, G., Jinjun, X.: Method for industry design material test and evaluation based on user visual and tactile experience. J. Mech. Eng. (2017)
3. Fleming, R.W.: Material perception. Ann. Rev. Vis. Sci. **3**(1), 365–388 (2017)
4. Ashby, M.F., Johnson, K.: Materials and Design: the Art and Science of Material Selection in Product Design, 2nd edn. Butterworth-Heinem, Oxford (2010)
5. Vangorp, P., Laurijssen, J., Dutre, P.: The influence of shape on the perception of material reflectance. ACM Trans. Gr. **26**(3), 77 (2007)
6. Cichy, R.M., Khosla, A., Pantazis, D., Torralba, A., Oliva, A.: Comparison of deep neural networks to spatiotemporal cortical dynamics of human visual object recognition reveals hierarchical correspondence. Sci. Rep. **6**, 27755 (2016)
7. Brainard, D.H.: Color constancy. In: Chalupa, L.M., Werner, J.S., (Eds.) The Visual Neurosciences, pp. 948–961. MIT Press (2004)
8. Obein, G., Knoblauch, K., Vienot, F.: Difference scaling of gloss: nonlinearity, binocularity, and constancy. J. Vis. **4**(9), 711–720 (2004)
9. Xiao, B., Brainard, D.H.: Color perception of 3D objects: constancy with respect to variation of surface gloss. In: Symposium on Applied Perception in Graphics and Visualisation, pp. 63–68. ACM (2006)
10. Ngan, A., Durand, F., Matudik, W.: Image-driven navigation of analytical BRDF models. In: Rendering Techniques, pp. 399–407 (2006)
11. Fleming, R.W., Jäkel, F., Maloney, L.T.: Visual perception of thick transparent materials. Psychol. Sci. **22**(6), 812–820 (2011)
12. Fleming, R.W., Bülthoff, H.H.: Low-level image cues in the perception of translucent materials. ACM Trans. Appl. Percept. **2**(3), 346–382 (2005)
13. Fleming, R.W., Dror, R.O., Adelson, E.H.: Real-world illumination and the perception of surface reflectance properties. J. Vis. **3**(5), Article 3 (2003)
14. Ho, Y.X., Landy, M.S., Maloney, L.T.: Conjoint measurement of gloss and surface texture. Psychol. Sci. **19**(2), 196–204 (2008)
15. Motoyoshi, I., Nishida, S., Sharan, L., Adelson, E.H.: Image statistics and the perception of surface qualities. Nature **447**(7141), 206–209 (2007)
16. Nishida, S., Shinya, M.: Use of image-based information in judgments of surface-reflectance properties. J. Opt. Soc. Am. **15**(12), 2951–2965 (1999)

17. Adelson, E.H.: On seeing stuff: the perception of materials by humans and machines. Human Vision & Electronic Imaging VI. International Society for Optics and Photonics (2001)
18. Wilson, C.J., Alessandro, S.: The use of virtual reality in psychology: a case study in visual perception. Comput. Math. Methods Med. **2015**, 1–7 (2015)
19. Xiang, F.U., Jian-Guo, J., Bao-Long, G.: A mechanical cad system based on virtual reality. Comput. Simul. **22**(9), 161–166 (2005)
20. Sutcliffe, A., Gault, B., Shin, J.E.: Presence, memory and interaction in virtual environments. Int. J. Hum Comput Stud. **62**(3), 307–327 (2005)
21. Wann, J.P., Rushton, S., Mon-Williams, M.: Natural problems for stereoscopic depth perception in virtual environments. Vis. Res. **35**(19), 2731–2736 (1995)
22. Bonneel, N., Suied, C., Viaud-Delmon, I., Drettakis, G.: Bimodal perception of audio-visual material properties for virtual environments. ACM Trans. Appl. Percept. **7**(1), 1–16 (2010)
23. Fleming, R.W., Wiebel, C., Gegenfurtner, K.: Perceptual qualities and material classes J. Vis. **13**(8), 9 (2013)
24. Berzhanskaya, J., Swaminathan, G., Beck, J., et al.: Remote effects of highlights on gloss perception. J. Percept. **34**(5), 565–575 (2005)
25. Sharan, L., Liu, C., Rosenholtz, R., et al.: Recognizing materials using perceptually inspired features. J. Int. J. Comput. Vis. **103**(3), 348–371 (2013)
26. Ikeuchi, K.: Numerical shape from shading and occluding contours in a single view (1979)
27. Horn, B.K.P., Brooks, M.J.: The variational approach to shape from shading. J. Comput. Vis. Gr. Image Process. **33**(2), 174–208 (1985)
28. Davis, B., Oken, E., Kennedy, S., Peterson, M.T., Su, S.W., King, D., et al.: 3d Studio Max 3 Magic (2000)
29. Qiu, W., Yuille, A.: Unrealcv: connecting computer vision to unreal engine. In: European Conference on Computer Vision, pp. 909–916. Springer, Cham (2016, October)
30. Collins, L.M., Dziak, J.J., Kugler, K.C., Trail, J.B.: Factorial experiments: efficient tools for evaluation of intervention components. Am. J. Prev. Med. **47**(4), 498–504 (2014)

Author Index

Springer Nature Switzerland AG 2020
Ahram (Ed.): AHFE 2019, AISC 973, pp. 427–429, 2020.
https://doi.org/10.1007/978-3-030-20476-1

CPSIA information can be obtained
at www.ICGtesting.com
Printed in the USA
LVHW080551180619
621490LV00002B/11/P